Materials Horizons: From Nature to Nanomaterials

Series Editor

Vijay Kumar Thakur, School of Aerospace, Transport and Manufacturing,
Cranfield University, Cranfield, UK

Materials are an indispensable part of human civilization since the inception of life on earth. With the passage of time, innumerable new materials have been explored as well as developed and the search for new innovative materials continues briskly. Keeping in mind the immense perspectives of various classes of materials, this series aims at providing a comprehensive collection of works across the breadth of materials research at cutting-edge interface of materials science with physics, chemistry, biology and engineering.

This series covers a galaxy of materials ranging from natural materials to nanomaterials. Some of the topics include but not limited to: biological materials, biomimetic materials, ceramics, composites, coatings, functional materials, glasses, inorganic materials, inorganic-organic hybrids, metals, membranes, magnetic materials, manufacturing of materials, nanomaterials, organic materials and pigments to name a few. The series provides most timely and comprehensive information on advanced synthesis, processing, characterization, manufacturing and applications in a broad range of interdisciplinary fields in science, engineering and technology.

This series accepts both authored and edited works, including textbooks, monographs, reference works, and professional books. The books in this series will provide a deep insight into the state-of-art of Materials Horizons and serve students, academic, government and industrial scientists involved in all aspects of materials research.

Review Process

The proposal for each volume is reviewed by the following:

1. Responsible (in-house) editor
2. One external subject expert
3. One of the editorial board members.

The chapters in each volume are individually reviewed single blind by expert reviewers and the volume editor.

Manish Kumar
Editor

Sustainable Perovskite Multiferroic Materials

A Plethora of Applications

Springer

Editor
Manish Kumar
Experimental Research Laboratory
Department of Physics
Atma Ram Sanatan Dharma College
University of Delhi
New Delhi, Delhi, India

ISSN 2524-5384 ISSN 2524-5392 (electronic)
Materials Horizons: From Nature to Nanomaterials
ISBN 978-981-96-7418-3 ISBN 978-981-96-7419-0 (eBook)
https://doi.org/10.1007/978-981-96-7419-0

This Springer imprint is published by the registered company Springer Nature Singapore Pte Ltd.
The registered company address is: 152 Beach Road, #21-01/04 Gateway East, Singapore 189721, Singapore

Contents

About the Editor

Dr. Manish Kumar obtained his M.Tech. degree in Ceramic Engineering from Indian Institute of Technology, Banaras Hindu University, Varanasi, and Ph.D. degree in Physics from Institute of Science, Department of Physics, Banaras Hindu University, Varanasi, India. At present, he is serving as **Assistant Professor** at Department of Physics, ARSD College, University of Delhi, New Delhi, India, since last more than 12 years. His research area focuses on environmental-friendly multifunctional materials for energy and memory devices, materials for green energy solutions, multiferroics, magnetoelectrics, composites, perovskite materials for photovoltaic applications, supercapacitors, white light emitting materials, solar simulation via different softwares, and DFT calculation. **Recently, he was listed among top 2% scientists worldwide by Stanford University Ranking-2021** and published more than **125 research publications** in international journals of high repute, **1 authored book on White Light Emitting Materials, 5 edited books (in press)** in Springer Nature, **30 international book chapters**, and **1 Indian Patent** with **h-index of 28, i10-index of 68** including more than **2550 citations** as per Google Scholar. He has delivered the talks in a number of national and international conferences. He is also Member of various research committees of different universities. In addition, he has completed two research projects as Co-investigator funded by University of Delhi and one is under process as Principal Investigator. He is guiding number of Ph.D., M.Sc. and under graduate students as supervisor and/cosupervisor. Dr. Kumar is serving and/served as an editor in number of highly reputed journals such as Scientific Reports (Nature), Materials Today Communications (Elsevier), High Performance Polymers (SAGE), Materials Today: Proceedings (Elsevier) and Springer Nature Books. He has organized one national conference (RAFM-2020) as a co-convener and three international conferences (RAFM-2022, RAFM-2024 and APGC-2025) as a convener.

Chapter 1
Sustainable Perovskite Multiferroic Materials: An Introduction

G. D. Dwivedi

1 Overview of Multiferroics

In 1820, Hans Christian Oersted was working on an experiment which involved a battery and a magnetic compass. He recorded a deflection in the magnetic needle whenever the electric current was switched on/off. This seemingly small experimental finding opened a new horizon of research, which still continues to draw the attention of scientists and engineers alike. This was the first report of interaction between electric and magnetic properties. At that point in time, these two fields were considered independent to one another. In 1865, Maxwell developed some equations, named as Maxwell's equations, to combine electricity and magnetism. The electron spins are responsible for observed magnetic properties whereas charges of electrons and protons are responsible for electric properties of matter. The contrasting origins of electricity and magnetism present a huge challenge to unify these two fields. Landau and Lifshitz, in 1959 [1], envisioned two possible phenomena in solids: (i) a linear correlation between magnetism and deformation, piezomagnetism, and (ii) a linear correlation between magnetism and electricity. Antiferromagnetic Cr_2O_3 became the first system, which was predicted to exhibit coupling between magnetic and electric properties [2]. Later, it was demonstrated that antiferromagnetic Cr_2O_3 indeed exhibits the coupling between magnetic and electric properties [3, 4]. This coupling was named linear magnetoelectric coupling. Later on, various compounds such as $DyPO_4$ [5], $FeSb_2O_4$ [6], and many others [7–15] were also found possessing linear magnetoelectric coupling. The materials exhibiting two or more primary ferroic ordering—such as ferroelectricity, ferromagnetism, and ferroelasticity—simultaneously within a single phase were termed as "multiferroics" [16]. But, nowadays materials exhibiting only magnetism and ferroelectricity, simultaneously are called

G. D. Dwivedi (✉)
Department of Physics, Faculty of Science, Kalinga University, Atal Nagar-Nava Raipur, Chhattisgarh, India
e-mail: gopeshwar.dwivedi@kalingauniversity.ac.in

M. Kumar (ed.), *Sustainable Perovskite Multiferroic Materials*, Materials Horizons: From Nature to Nanomaterials, https://doi.org/10.1007/978-981-96-7419-0_1

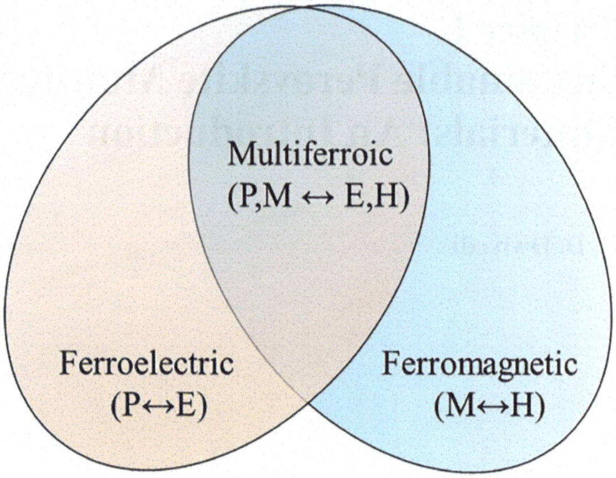

Fig. 1 A schematic diagram of multiferroic system combining ferroelectric and ferromagnetic order. Courtesy Ref. [17]. Credit: APS/Alan Stonebraker

multiferroics (Fig. 1) [17]. Boracites were the first system that exhibited multiferroic properties [8].

For the next couple of decades, scientists around the world tried to find multiferroics with rare success. This shows that the multiferroic materials are rare in nature. Hill [18] and Khomskii [19] addressed this problem and found that *d*-orbital electrons of transition metals (TM) suppress inclination towards the off-center distortion. The bulk $BiFeO_3$ was already a well-known multiferroics, but it was observed that thin-film $BiFeO_3$ exhibits multi-fold increase in multiferroic properties of $BiFeO_3$ [20]. Furthermore, $TbMnO_3$ and $TbMn_2O_5$ were introduced by Tokura et al. [21] and Hur et al. [22], respectively, as a new class of multiferroic systems. These materials do not just show coexistence of magnetism and ferroelectricity simultaneously in single phase, but ferroelectricity in systems arises due to their own internal magnetic field. These three materials are at the forefront of multiferroic systems and are often used for the demonstration of multiferroicity. Multiferroic materials with magnetoelectric coupling provide a better application potential in memory devices, four-state logic, sensors, and many more [23–27].

1.1 Symmetry Consideration

Every ordering requires some kind of symmetry breaking. For a material to exhibit ferroelectricity, inversion symmetry must be broken, allowing for the presence of a spontaneous electric polarization. This is because an electric dipole requires an asymmetric charge distribution, which cannot occur in a system with inversion symmetry. In ferroelectrics, space inversion causes reversal of polarization (P) while it does not affect magnetization (M), i.e., magnetization remains invariant (Fig. 2a). In ferromagnets, time reversal causes the switching of magnetization (M) direction whereas polarization (P) remains invariant, i.e., in ferromagnets, the time reversal

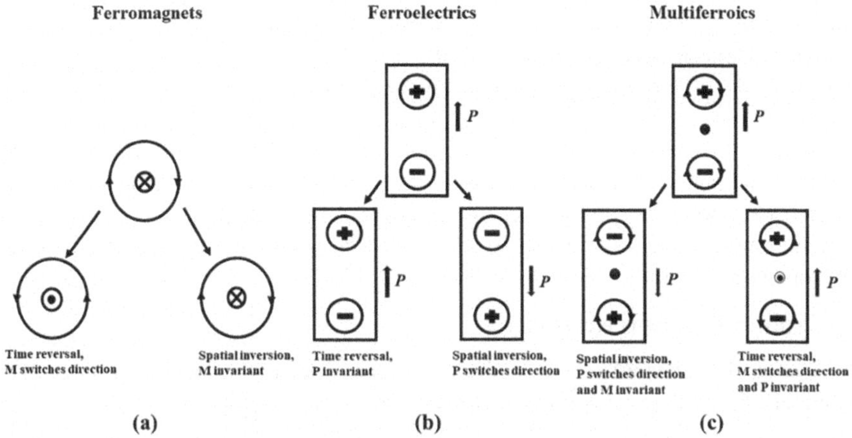

Fig. 2 Different symmetry breaking in ferromagnets, ferroelectrics, and multiferroics. Reproduced with permission Ref. [28]

symmetry is broken (Fig. 2b). Hence, it is obvious that multiferroic materials lack both symmetries, space as well as time inversion symmetry (Fig. 2c) [28].

1.2 Thermodynamics of Multiferroics

The thermodynamic study of multiferroics helps in understanding phase transitions, coupling mechanisms, and the stability of different ferroic phases.

(i) **Thermodynamic Framework for Multiferroics**

The free energy of a multiferroic system can be written in terms of order parameters associated with different ferroic properties, typically the electric polarization P and magnetization M:

$$F = F_0 + \alpha P^2 + \beta P^4 + \gamma M^2 + \delta M^4 + \varepsilon P^2 M^2 \tag{1}$$

where F is the total free energy, and the coupling term $\varepsilon P^2 M^2$ accounts for magnetoelectric interaction [29]. The coefficients α, β, γ, δ, and ε depend on temperature and external fields.

(ii) **Phase Transitions in Multiferroics**

Thermodynamic models describe phase transitions in multiferroics, such as:

(a) **Second-order transitions**

Second-order transition occurs when the first derivative of a thermodynamic potential is continuous, but a second derivative (such as heat capacity, magnetic susceptibility) exhibits a divergence or discontinuity. These follow Landau theory, where a soft phonon mode triggers ferroelectricity [30].

(b) **First-order transitions**

First-order transition is characterized by a discontinuous change in the first derivative of a thermodynamic potential (such as entropy, magnetization) with respect to an external parameter like temperature or pressure. This behaviour can be observed in certain multiferroics like $BiFeO_3$, where electric polarization and magnetization develop abruptly [31].

(iii) **Spin-driven ferroelectricity**

In materials like $TbMnO_3$, ferroelectricity is induced due to the spiral magnetic order through the inverse Dzyaloshinskii-Moriya (DM) interaction [21].

(iv) **Entropy and Specific Heat Behaviour**

Entropy (S) and specific heat (C_P) provide insights into multiferroic phase transitions. The entropy change, ΔS, across a transition can be estimated using:

$$\Delta S = \int \frac{C_P}{T} dT \tag{2}$$

For example, in $BiFeO_3$, a significant entropy change is observed at the Néel temperature (~640 K), indicating strong spin–lattice coupling [28].

(v) **Magnetoelectric Coupling and Free Energy**

The thermodynamic stability of a multiferroic phase is governed by magnetoelectric coupling, which is represented by the term:

$$F_{ME} = -\alpha P \cdot M \tag{3}$$

This term determines how polarization changes under an applied magnetic field and vice versa [32]. Large magnetoelectric coefficients are desirable for practical applications in non-volatile memory.

1.3 Magnetoelectric Coupling

Magnetoelectric (ME) coupling in multiferroic materials refers to the interaction between the material's electric and magnetic properties. This coupling enables the control of magnetization via an external electric field and control of polarization via an external magnetic field. The general formulation of the ME effect is given by the expression of the free energy $F(E, H)$:

$$F(E, H) = F_0 - P_i^S E_i - M_i^S H_i - \frac{1}{2}\varepsilon_0 \varepsilon_{ij} E_i E_j - \frac{1}{2}\mu_0 \mu_{ij} H_i H_j - \alpha_{ij} E_i H_j$$

$$-\frac{1}{2}\beta_{ijk}E_iH_jH_k - \frac{1}{2}\gamma_{ijk}H_iE_jE_k \quad (4)$$

where, P_i^S and M_i^S are spontaneous polarization and magnetization, ε_{ij} and μ_{ij} are the dielectric and magnetic susceptibilities, α_{ij} represents the linear ME effect, coupling the electric and magnetic fields, and higher-order ME effects involve β_{ijk} and γ_{ijk}, which account for quadratic and nonlinear effects.

Magnetoelectric coupling can be classified into different types based on its origin and interaction mechanism. The main types include:

(i) **Linear Magnetoelectric Coupling**

The linear magnetoelectric effect is described by the equation:

$$P_i = \alpha_{ij}H_j \quad (5)$$

$$M_i = \alpha_{ij}E_j \quad (6)$$

where P_i is the electric polarization, M_i is the magnetization, E_j is the applied electric field, H_j is the applied magnetic field, and α_{ij} is the magnetoelectric coupling tensor. The coupling coefficient α_{ij} determines the efficiency of electric-field control over magnetization or vice versa.

(ii) **Quadratic and Higher-Order ME Coupling**

Higher-order contributions arise from terms like:

$$P_i = \alpha_{ij}H_j + \beta_{ijk}H_jH_k + \gamma_{ijk}E_jH_k + \ldots \quad (7)$$

$$M_i = \alpha_{ij}E_j + \beta_{ijk}E_jH_k + \gamma_{ijk}E_jE_k + \ldots \quad (8)$$

where β_{ijk} represents quadratic ME effects, and γ_{ijk} describes effects involving both electric and magnetic fields.

(iii) **Composite Magnetoelectric Coupling**

In multiphase composite materials, the ME effect results from the mechanical interaction between a magnetostrictive and a piezoelectric phase. The effective ME coupling follows:

$$S = s \cdot T + d^T \cdot E + q^T \cdot H \quad (9)$$

$$D = d \cdot T + \varepsilon \cdot E + \alpha \cdot H \quad (10)$$

$$B = q \cdot T + \alpha^T \cdot E + \mu \cdot H \quad (11)$$

where s, d, and q are material-dependent tensors describing strain, electric displacement, and magnetic induction.

(iv) **Spin–Orbit Coupling and Inverse Dzyaloshinskii-Moriya Interaction (DMI)**

The Dzyaloshinskii-Moriya interaction (DMI) arises in materials lacking inversion symmetry, leading to a noncollinear spin structure. This mechanism can induce an electric polarization P due to spin–orbit coupling:

$$P \propto r_{ij} \times (S_i \times S_j) \tag{12}$$

where r_{ij} is the unit vector connecting spins S_i and S_j. This mechanism is prominent in cycloidal spin structures observed in $BiFeO_3$ and $TbMnO_3$.

(v) **Exchange Striction**

Exchange striction is a mechanism where the electric polarization arises due to spin-dependent exchange interactions. It can be described by the Hamiltonian:

$$H_{ex} = -J_{ij}(S_i \cdot S_j) \tag{13}$$

where J_{ij} is the exchange coupling. If the exchange interaction depends on lattice distortions, a coupling between electric and magnetic properties is established.

For a chain of spins with nearest-neighbour interaction, the induced polarization can be estimated as:

$$P \propto \sum_{\langle i,j \rangle} J_{ij}(S_i \cdot S_j) \tag{14}$$

where the sum is taken over neighbouring spin pairs. This mechanism is responsible for multiferroicity in systems like RMn_2O_5 (R = Tb, Dy).

(vi) **Spin–Lattice Coupling and Lattice Modulation**

In some materials, lattice distortions due to magnetic order modify the electric polarization. The free energy F of such a system can be expressed as:

$$F = F_0 + aP^2 + bM^2 + cP^2M^2 \tag{15}$$

where the term cP^2M^2 accounts for magnetoelectric coupling.

Calculation of the minimization free energy leads to:

$$P = -c\frac{M^2}{2a}, M = -c\frac{P^2}{2b} \tag{16}$$

which shows that the presence of magnetic order (nonzero M) induces an electric polarization P, and vice versa.

(vii) **Magnetostriction and Piezoelectric Coupling**

Magnetostriction is the phenomenon where a material changes shape in response to a magnetic field. In composite multiferroics, strain-mediated coupling between a magnetostrictive phase and a piezoelectric phase results in a strong ME effect.

The induced polarization is given by:

$$P = d_{ijk}\sigma_{jk} \tag{17}$$

where d_{ijk} is the piezoelectric coefficient and σ_{jk} is the strain due to magnetostriction. Such mechanisms are crucial in artificial heterostructures like $CoFe_2O_4/BaTiO_3$.

2 Classifications of Multiferroics

Magnetic properties of systems are governed by the electron spins. The localized spins of partially filled *d* or *f* orbit electrons of transition-metal ions or rare earth ions produce magnetic moment. These magnetic moments form various magnetic ordering depending upon the exchange interactions between neighbouring electrons. Hence, it is obvious that electron spins are the origin of magnetism in all magnetic systems. Whereas, the origin of ferroelectricity is complex and it varies from system to system. Therefore, multiferroics are classified on the basis of the origin of ferroelectricity in the different systems. In general, multiferroics are of two types, type-I multiferroics and type-II multiferroics.

2.1 *Type-I Multiferroics*

Type-I multiferroics are those systems where the origin of magnetism and ferroelectricity are independent. In type-I multiferroics, the cross-coupling between their magnetic and electric order parameters is almost zero. Since there is almost no communication between both ferroic order parameters, the magnetic and ferroelectric transition temperatures appear far from one another. On the basis of the origin of ferroelectricity, the type-I multiferroics are further classified into various subclasses.

(i) **Perovskite multiferroics**

In ferroelectric systems, $BaTiO_3$ and $Pb(Zr,Ti)O_3$ are the most famous and most used ferroelectric materials, and both represent perovskite structure. Goodenough [33] and Mitsui [34] formed lists of perovskite materials with magnetic and ferroelectric properties. They found that many perovskites show magnetic behaviour and many of them show ferroelectric behaviour, but there was almost no overlap in both lists. This observation clearly suggests that magnetism and ferroelectricity can rarely coexist in a single system. For magnetism, the presence of partially filled *d*-shells of transition metals is mandatory, whereas perovskite ferroelectric systems possess transition metals with zero *d*-electrons. The absence of *d*-electrons allows off-center displacement of transition-metal ion, whereas the *d*-electrons of transition metal (magnetic) inhibit the off-center displacement [18]. This phenomenon is referred to as the "d^0 versus d^n problem" in multiferroics, where presence of d-electron in magnetic perovskites suppresses ferroelectricity.

(ii) **Ferroelectricity due to lone pair in perovskites**

The Bi^{3+} and Pb^{2+} ions have two unused 6 s electrons in their furthest shell, which are termed as lone pairs. These electron lone pair exhibit strong polarizability, leading to the induction of ferroelectricity in these materials. $BiFeO_3$, $BiMnO_3$, and $PbVO_3$ systems are the most prominent systems where ferroelectricity is produced due to the presence of lone pair of electrons.

(iii) **Ferroelectricity due to charge-ordering**

Ferroelectricity due to charge-ordering usually occurs in those transition-metal systems which contain mixed valence cations and possess either geometric or magnetic frustration. This phenomenon disrupts spatial inversion symmetry, a key requirement for ferroelectric behaviour. By creating inequivalent sites and bonds, charge-ordering can induce improper ferroelectricity without involving ionic displacement [35]. Notable examples include $LuFe_2O_4$ [36], (Pr,Ca)MnO_3 [35, 37–39], and spinel Fe_3O_4 [39–44], which demonstrate ferroelectricity arising from charge-ordering. Among these, the spinel system stands out as particularly significant for charge-order-induced ferroelectricity.

(iv) **Ferroelectricity in hexagonal manganites**

$YMnO_3$ undergoes a ferroelectric transition below 914 K (T_{FE}) and an antiferromagnetic transition below 76 K (T_N). Its ferroelectricity originates from the tilting of MnO_5 blocks, which have magnetic Mn ions at their centers [45]. This tilting of MnO_5 block shifts oxygen ions closer to the smaller Y ion. This action forms Y–O bond which acts as a dipole. The resulting disparity between the "up" and "down" electric dipoles gives rise to ferroelectricity. Consequently, at lower temperatures when Mn spins align, the system turns multiferroic.

2.2 *Type-II Multiferroics*

Type-II multiferroics are known to exhibit multiferroicity under the influence of magnetism. In these systems, the ferroelectricity emerges due to their own magnetic ordering. These materials often exhibit strong interactions between their order parameters and typically display magnetoelectric characteristics. Type-II multiferroics are further divided into two distinct subclasses.

(i) **Spiral magnetic structure induced ferroelectricity**

The $RMnO_3$ (R = Tb, Gd) and RMn_2O_5 (R = Tb, Y, etc.) systems are fascinating materials that exhibit ferroelectricity below their magnetic transition temperatures, highlighting their strong magnetoelectric coupling [21, 22]. This unique property is driven by their spiral or helicoidal magnetic structures, which play a crucial role in inducing ferroelectricity [46, 47]. For example, $TbMnO_3$ transitions from a sinusoidal magnetic phase (40–30 K) with no electric polarization to a helicoidal phase below 30 K, where a finite polarization emerges [48]. However, the mere presence of a spiral magnetic structure isn't enough to generate ferroelectricity. To produce ferroelectricity, spin rotation axis (r) and spiral wave vector (Q) must not have parallel alignment. Intriguingly, the resulting polarization (P) is governed by the cross product of r and Q, following the relationship $P \sim [r \times Q]$.

(ii) **Collinear magnetic structure induced ferroelectricity**

There are not many systems that possess ferroelectricity under collinear magnetic structures, i.e., where magnetic moments are directed along a fixed axis without involving spin–orbit coupling. In these cases, the phenomenon of exchange striction plays a pivotal role in generating ferroelectricity. Ca_3CoMnO_6 is one such system where alternating Co^{2+} and Mn^{4+} ions form a one-dimensional chain [49]. At high temperatures, these ions occupy equivalent positions along the chain, resulting in zero polarization. However, when magnetic ordering sets in, space inversion symmetry is disrupted as the spins adopt a $\uparrow\uparrow\downarrow\downarrow$ configuration. The exchange striction creates inequivalent ferromagnetic ($\uparrow\uparrow$) and antiferromagnetic ($\uparrow\downarrow$) arrangements, which helps the system in developing ferroelectricity.

3 Introduction to Perovskite Structure

The name "perovskite" originates from the mineral $CaTiO_3$, first discovered in 1839 and named after the Russian mineralogist Lev Perovski. But over the period of time, the term has come to represent a broader class of materials that share the same crystal structure as $CaTiO_3$. The widespread interest in perovskites is driven by their exceptional structural flexibility, which allows for extensive compositional engineering. Perovskite systems also play a crucial role in fundamental research, offering insights into phenomena like superconductivity, colossal magnetoresistance, and multiferroicity. These materials provide a rich playground for exploring the interplay between structure, composition, and properties at both macroscopic and microscopic levels. Oxide perovskites like $BaTiO_3$ are widely studied for their ferroelectric and piezoelectric properties.

The perovskite crystal structure is characterized by the general formula ABX_3, where, A is a larger cation, typically an alkali, alkaline earth, or rare earth metal, B is a smaller cation, often a transition metal and X is an anion, usually oxygen (O^{2-}) in oxide perovskites or a halide (Cl^-, Br^-, I^-) in halide perovskites. The ideal perovskite systems should acquire cubic structure with space group $Pm\overline{3}m$. The A cations occupy the corners of the cube, the B cations occupy the center of the cube, and the

X anions occupy the midpoints of the cube edges, forming an octahedral coordination around the B cation. This arrangement creates a network of corner-sharing BX_6 octahedra, with the A cations occupying the interstitial spaces. Nevertheless, in reality, the ideal cubic structured perovskites are not very common in nature. Perovskites are most commonly found in orthorhombic crystal structure with space group *Pnma*, and tetragonal crystal structure with space group *I4/mcm* or *P4mm*. The perovskite structure is named after $CaTiO_3$ system, and the very system itself does not belong to cubic structure. $SrTiO_3$ and $CaRbF_3$ are examples of cubic perovskites.

4 Mechanism of Multiferroicity in Perovskite

There exist many magnetic materials with perovskite structure, and most classic ferroelectrics such as $BaTiO_3$ and (PbZr)TiO_3 belong to perovskite family; hence, perovskites became the obvious target in search of multiferroic materials. As we have already seen from the reported lists from ref [33, 34] that multiferroics are very rare in perovskite system. But, there exist a few exceptional systems which support ferroelectricity and magnetism in single phase. Now, we will discuss some of those systems one by one and will try to understand the mechanisms which allow these systems to afford both ferroic ordering in single phase.

4.1 *Bismuth Ferrite ($BiFeO_3$)*

The most famous system that seems exception to 'the exclusion rule' observed in perovskite systems is $BiFeO_3$. $BiFeO_3$ is most extensively studied single-phase multiferroic material. It is one of the very few room-temperature multiferroic materials, i.e., it exhibits magnetic and ferroelectric ordering above room temperature. $BiFeO_3$ undergoes ferroelectric ordering below 1103 K and a *G*-type antiferromagnetic ordering below 640 K [50–53]. $BiFeO_3$ has Fe^{3+} cations which are responsible for magnetic ordering but there are no transition-metal (TM) cations responsible for ferroelectric ordering. It seems that the "d^0-ness" requirement for the ferroelectric ordering is not being followed in this system but in reality, the origin of ferroelectricity in this system has nothing to do with the presence of "d^0" system. The origin of ferroelectricity in $BiFeO_3$ has different origin as compared to the $BaTiO_3$-type ferroelectrics. The required structural instability for ferroelectricity in $BiFeO_3$ does not come from the B-site transition metal but rather it comes from A-site Bi^{3+} cations. The lone pair electrons of Bi^{3+} cations do not participate in any chemical bonds. This creates a strong polarizability of Bi cations which causes a stereochemically active distortion. This distortion shifts the Bi^{3+} cations away from the center of the oxygen octahedra, generating a spontaneous polarization [54]. The Fe^{3+} cations and oxygen atoms in the lattice also contribute to the polarization through their interactions (Fig. 3). The rotation and tilting of the FeO_6 octahedra modulate the ferroelectric

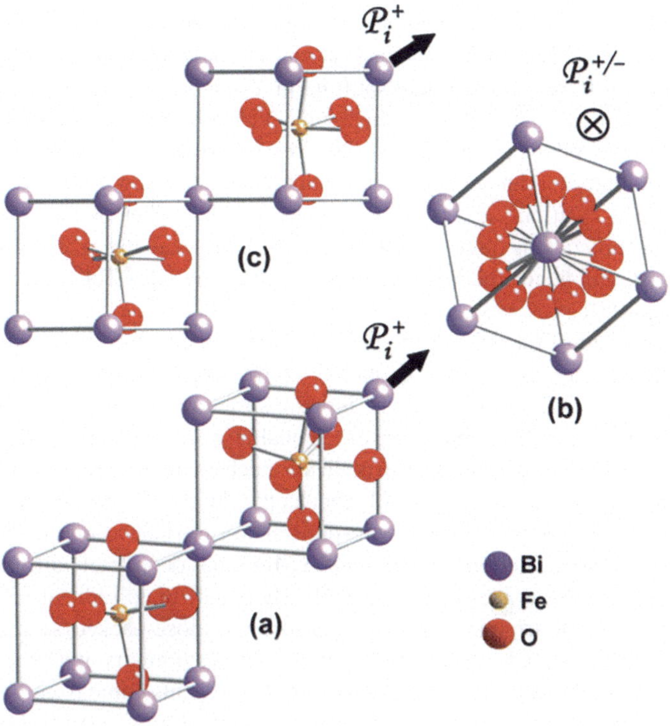

Fig. 3 Schematic illustration of the $BiFeO_3$ (001) unit cell as seen from different perspectives. **a** Three-dimensional representation, **b** unit cell observed along the pseudocubic [111] direction, which aligns with the polarization direction, and **c** a view along the pseudocubic [110] direction, which is perpendicular to the polarization direction. Reproduced with permission; Figure credit: Ref. [54]

properties. The spontaneous polarization in $BiFeO_3$ is ~ 100 μC/cm^2 [54–56], which is very high compared to other ferroelectric materials, making it highly attractive for applications in memory devices and sensors. Similar mechanism is also responsible for ferroelectricity in $BiMnO_3$ system which shows ferroelectric ordering (T_{FE}) below 450 K and magnetic ordering (T_M) below 105 K [57–59].

4.2 *Yttrium Manganite ($YMnO_3$)*

$YMnO_3$ represents another class of material that does not obey "d^0-ness" rule. $YMnO_3$ is considered as hexagonal perovskites, but its crystal structure is distinctly different from the conventional perovskite systems. In $YMnO_3$, the magnetic Mn^{3+} cations are situated in five-fold coordination and not in six-fold coordination. The Mn cations form a trigonal biprism structure where Mn cations are at the center and five oxygen atoms are at the corners of trigonal biprism. Similarly, Y atoms are situated in a seven-fold coordination instead of a 12-fold coordination. Hence, Mn^{3+} cations exhibit different crystal field splitting in $YMnO_3$ from the usual octahedral coordination. In usual octahedral symmetry, the 3d levels are split into triplet t_{2g} and doublet e_g levels. Whereas, in $YMnO_3$, the 3d levels of Mn^{3+} cations are split into two lower doublets and one upper singlet. This special splitting of 3d level, the four electrons of Mn^{3+} cations occupy the two lower doublets whereas the upper singlet remains unoccupied. It is obvious that, in this configuration, the orbital degeneracy does not exist, and hence the Mn^{3+} cations do not act as Jahn–Teller active cations in this system [60].

The mechanism of ferroelectricity in $YMnO_3$ remained a puzzle for a long time, and many ideas were presented to explain the ferroelectricity in this system. It was demonstrated that in the MnO_5 trigonal biprism structure, the Mn^{3+} cations exhibit a small off-center shift from the centre of trigonal biprism. This off-center shift was very small, and definitely it could not explain the mechanism of ferroelectricity. It was apparent that Y–O bonds are responsible for electric dipole moment and not the Mn–O bonds, but Y cations are not directly responsible for ferroelectricity in the system. Hence, it was obvious that ferroelectricity in $YMnO_3$ has completely different origin than other ferroelectrics like BTO and PZT. The small ionic radii of Y^{3+} cations lead titling of MnO_5 trigonal biprism to ensure the close-packed structure. This tilting in $YMnO_3$ system leads to breaking of inversion symmetry and hence leads to ferroelectricity in the system where electric dipole moments come from Y–O bonds (Fig. 4). In $YMnO_3$, the ferroelectric transition temperature (T_{FE}) is ~ 900 K, and antiferromagnetic temperature (T_N) is ~ 70 K [45, 60–62]. Since, the magnetism and ferroelectricity have different origins (which is reflected in their respective ordering temperatures), it is obvious that this system does not show strong coupling between both their order parameters [45].

4.3 *Terbium Manganite ($TbMnO_3$)*

The magnetoelectric materials show that application of external magnetic field could produce electric polarization in systems. We could ask a very genuine question—*if an external magnetic field could produce an electric polarization in the system then is it possible to produce electric polarization with internal field from the magnetic ordering of the system*? The answer is—Yes. In some classes of materials electric polarization is observed due to magnetic ordering of the system. $TbMnO_3$ is one such material which also has ABO_3 structure [21]. $TbMnO_3$ crystallizes in a distorted orthorhombic *Pbnm* space group at high temperatures [63, 64]. It consists of layers of MnO_6 octahedra and Tb ions, where Mn^{3+} cations ($3d^4$ configuration) form a magnetic lattice with antiferromagnetic interactions, and Tb^{3+} cations contribute weakly to magnetic order but play a role in structural distortions. Below 41 K, $TbMnO_3$ exhibits an incommensurate sinusoidal antiferromagnetic phase, with no

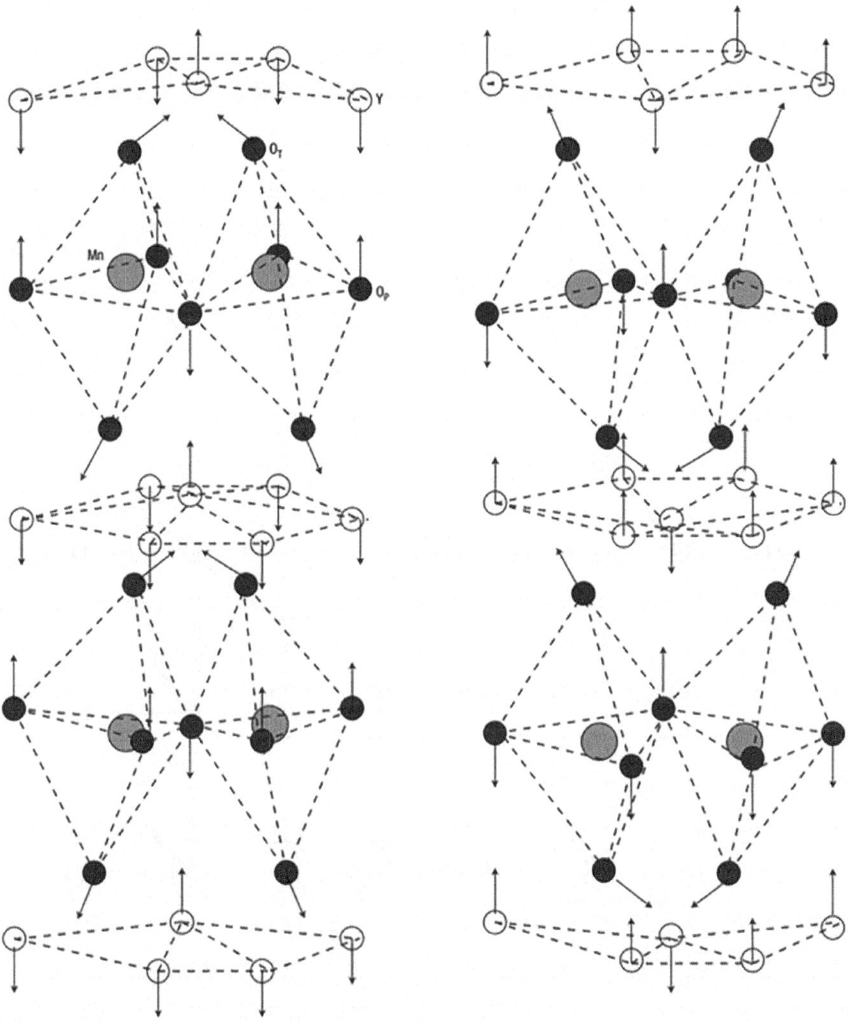

Fig. 4 Three-dimensional schematic representation of $YMnO_3$ in its two enantiomorphic polarized states. The arrows depict the directions of atomic shifts as the structure transitions from the centrosymmetric to the ferroelectric phase. Figure credit: Ref. [45]

net polarization, and below 28 K, the Mn spins develop a cycloidal spiral magnetic order due to competing exchange interactions and spin frustration [65–67].

The cycloidal spin order breaks inversion symmetry due to the noncollinear arrangement of spins. This leads to spin–orbit coupling and Dzyaloshinskii-Moriya (DM) interactions, which are key in generating ferroelectric polarization. The DM interaction between neighbouring Mn spins, $\overrightarrow{S_i}$ and $\overrightarrow{S_j}$, induces an electric polarization, $\vec{P}$, given by:

$$\vec{P} = \vec{r}_{ij} \times (\overrightarrow{S_i} \times \overrightarrow{S_j}) \tag{18}$$

where $\vec{r}_{ij}$ is the vector connecting neighbouring spins. The $\overrightarrow{S_i} \times \overrightarrow{S_j}$ can be written as $\vec{Q}$, where $\vec{Q}$ is spiral wave vector.

The cycloidal nature of the spin arrangement enhances the breaking of inversion symmetry, allowing the polarization to emerge along the *c*-axis (Fig. 5). The Tb^{3+} cations also contribute to ferroelectricity through magnetostrictive effects (where Tb cations distort the MnO_6 octahedra via lattice coupling), and it interact with Mn sublattice to influence the magnetic anisotropy and stability of the cycloidal phase. Similarly, $DyMnO_3$ also exhibits ferroelectricity under internal magnetic field of its own magnetic ordering [69, 70]. The mechanism of ferroelectricity is similar to the one explained for $TbMnO_3$.

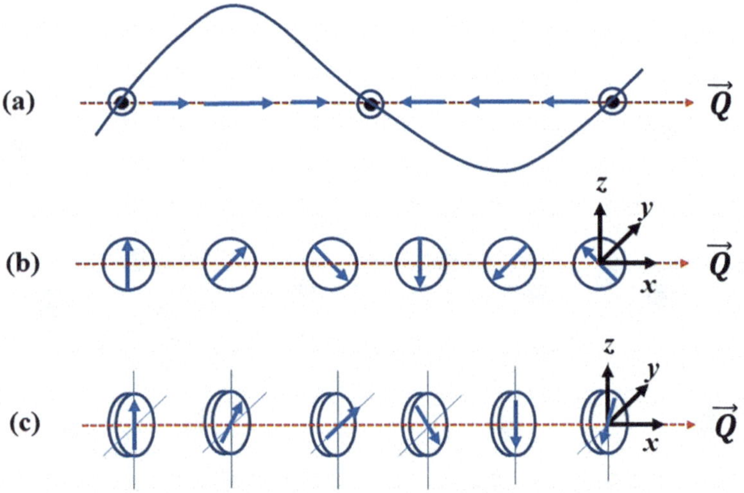

Fig. 5 Schematic illustration of different spin structures relevant to type-II multiferroics. **a** Depicts sinusoidal spin waves, where spins are aligned in the same direction but vary in magnitude. This centrosymmetric structure does not support ferroelectricity. **b** Shows a cycloidal spiral with a wave vector along the *x*-direction ($Q = Q_x$), where spins rotate within the (x, z)-plane, resulting in a nonzero polarization along the *z*-direction ($P_z \neq 0$). **c** Represents a 'proper screw' configuration, where spins rotate in a plane perpendicular to the wave vector **Q**. Despite breaking inversion symmetry, this structure generally lacks polarization, though in some cases, polarization may still arise. [68]. Courtesy Ref. [17]. Credit: APS/Alan Stonebraker

5 Potential Applications of Perovskite Multiferroics

Multiferroic materials have a vast range of application potential in fields of spintronics and data storage systems. $BiFeO_3$ is widely accepted and one of the most used materials in device fabrication. Here, we are naming a few areas where these multiferroic materials have huge application potentials.

5.1 Spintronic Devices

In spintronic applications, the integration of $BiFeO_3$ with ferromagnetic materials has shown promise in controlling magnetic properties via electric fields. For example, in $BiFeO_3/La_{0.7}Sr_{0.3}MnO_3$ heterostructures, electric-field control of exchange bias has been demonstrated [71], which is crucial for the development of spin valves and magnetic tunnel junctions. The BFO/LSMO bilayers exhibited exchange bias effect, enhanced coercivity, and hysteresis loop shifts at low temperatures. The strength of exchange bias depends on the thickness of LSMO, and thin LSMO layers exhibit stronger effects. The phenomenon is attributed to interface interactions between antiferromagnetic BFO and ferromagnetic LSMO. The Au/Ti/BFO/LSMO/STO/GaAs heterostructure showed resistive switching behaviour for Low set/reset voltages (~ 4 V), making it suitable for low-power applications (Fig. 6). The switching mechanism follows space-charge limited conduction (SCLC) and thermionic emission (TE), ensuring non-volatile memory operation without destructive electroforming. A BFO/LSMO heterostructure-based FeFET device demonstrated a high ON/OFF ratio (~ 19), sharp pinch-off characteristics indicating effective gate control, drain current modulation with BFO serving as the gate dielectric and LSMO as the channel. The results confirm efficient field effect modulation, essential for FeFET-based memory and logic applications.

5.2 Non-volatile Memory Devices

With the rapid growth of the Internet of Things (IoT), efficient and low-power non-volatile memory (NVM) is crucial for storing and processing vast amounts of data. Ferroelectric memories (FeRAM) are ideal candidates due to their fast write/erase speeds, high endurance, and low power consumption. $BiFeO_3$'s ferroelectric properties have been extensively studied for non-volatile memory applications. The material's ability to maintain polarization states without power makes it suitable for ferroelectric random-access memory (FeRAM). It has been shown that integrating $BiFeO_3$ with high-k dielectrics like HfO_2 in metal/ferroelectric/insulator/silicon (MFIS) structures enhances memory window and data retention properties,

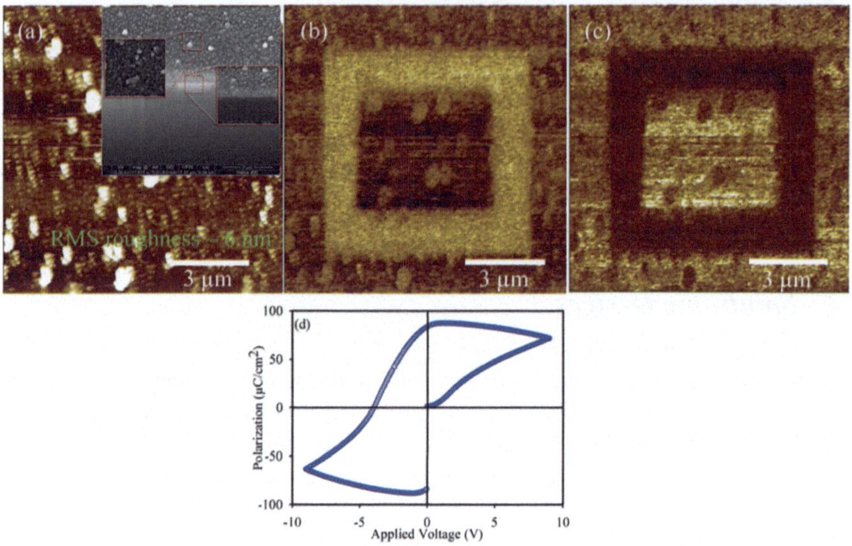

Fig. 6 Observation of ferroelectric domain switching in the BFO/LSMO/STO/GaAs heterostructure. **a** Surface topography, with the inset displaying an SEM image of the uniform sample surface. **b** PFM phase and **c** amplitude measured after poling with 10 V. **d** Room-temperature polarization–voltage hysteresis loops obtained from electrical measurements using the top electrode. The film exhibits a low coercive voltage and high remnant polarization, making it highly suitable for low-power, high-density non-volatile memory applications. Reproduced with the permission of RSC publications [71]

making it a potential candidate for non-volatile memories in Internet of Things (IoT) applications [72].

5.3 Sensors

The multiferroic nature of $BiFeO_3$ allows for the coupling of electric and magnetic orders, enabling the development of advanced sensor technologies. For instance, its magnetoelectric coupling can be harnessed in magnetic field sensors, where an applied magnetic field induces a measurable electric polarization change. Additionally, $BiFeO_3$'s ferroelectric properties have been explored in microactuators and portable electronic devices, leveraging its low electric coercive field and high-temperature stability [73].

5.4 Renewable Energy

$BiFeO_3$ is a promising material for photocatalysis due to its narrow bandgap (~2.2 eV) and stability, but its performance is limited by poor charge carrier transport. It is important to improve charge separation and photocatalytic efficiency of $BiFeO_3$ for hydrogen generation. The $Bi_{0.95}Sm_{0.05}FeO_3$ composition exhibited the highest photocurrent density (0.1061 mA/cm^2), which is 5.6 times higher than undoped $BiFeO_3$ (0.018 mA/cm^2). The Sm^{3+} doping led to a bandgap narrowing from 2.29 eV (undoped $BiFeO_3$) to 2.23 eV ($Bi_{0.95}Sm_{0.05}FeO_3$). In addition, it is also observed that $Bi_{0.95}Sm_{0.05}FeO_3$ has increased light absorption in the visible spectrum due to oxygen vacancies created by Sm^{3+} substitution. Moreover, the doped samples have shown improved photocurrent densities, indicating enhanced photoelectrochemical activity. This makes $BiFeO_3$ suitable for applications like water splitting and solar energy conversion [74].

$BiFeO_3$ has been reported as a direct band gap semiconductor with a gap ranging from 2.1 to 2.7 eV, making it suitable for visible-light absorption. The absorption arises due to charge-transfer transitions between O2p and Fe3d states. The photovoltaic properties of $BiFeO_3$ have been extensively studied, and it has been observed that photoresponse depends on light wavelength, intensity, and film thickness. The devices with thinner $BiFeO_3$ layers exhibited higher open-circuit voltage and short-circuit current density, likely due to fewer charge recombination barriers. Doping $BiFeO_3$ with different metal ions has led to significant improvements in photovoltaic performance; such as Ca-doping has improved charge separation, leading to a power conversion efficiency (PCE) of 0.0072%. Ba-doping has shifted the band gap, increasing absorption in the visible range, with PCE reaching 0.006% and Sm-doping has reduced the band gap to 2.18 eV, significantly enhancing PCE (~0.37%) [75, 76] (Fig. 7).

Thin films of $BiFeO_3$ have demonstrated superior photovoltaic performance compared to bulk materials due to better control over crystal structure and reduced recombination losses. There are various factors which influence their performance. Optimal annealing temperatures lead to smoother surfaces and fewer defects, and enhanced photocurrent. Film thickness of ~ 300 nm provides maximum photocurrent. Introduction of strain via different substrates can reduce the band gap to 2.5 eV, and hence could improve the light absorption. $BiFeO_3/BaTiO_3$ multilayer films have shown enhanced charge separation, achieving 1.43 V open-circuit voltage and a PCE of 0.33%. In $ZnO/BiFeO_3$ heterostructures, films built-in electric fields improved charge carrier dynamics and increased photocurrent to 340 μA/cm^2 [76].

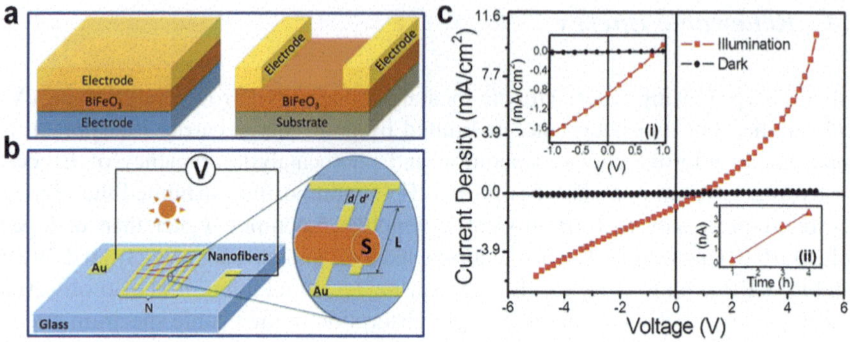

Fig. 7 **a** Schematic representation of thin-film-based photovoltaic devices. The left panel illustrates the device configured as a parallel capacitor, while the right panel shows a design with laterally arranged interdigital electrodes. **b** Diagram of the measurement setup for random $BiFeO_3$ nanofiber-based photovoltaic devices. **c** Photoresponse characteristics of the nanofibers. Inset (i) provides a magnified view of the current density near zero bias, while Inset (ii) presents the averaged photocurrent over multiple measurements for deposition times ranging from 1 to 4 h. Reproduced with the permission of ACS publications Ref. [76]

5.5 Data Storage and Logic Devices

The ferroelectric nature of $BiFeO_3$ is characterized by a significant spontaneous polarization, approximately 90 $\mu C/cm^2$, oriented along the pseudocubic [111] direction. This substantial polarization is pivotal for non-volatile memory applications, where information is stored based on the orientation of electric dipoles.

In Ferroelectric Random-Access Memory (FeRAM), $BiFeO_3$'s polarization states represent binary data, with the 'up' and 'down' orientations corresponding to '1' and '0', respectively. The material's high polarization facilitates distinct and stable data storage, ensuring reliability over extended periods [56].

A defining feature of $BiFeO_3$ is its magnetoelectric coupling the interaction between its ferroelectric and magnetic orders. This coupling allows for the control of magnetic properties via electric fields and vice versa. Magnetoelectric property is instrumental in developing advanced logic devices. In the context of logic gates, $BiFeO_3$ can be utilized to create components where logical operations are executed through the manipulation of its coupled electric and magnetic states. For instance, by applying an electric field, one can modulate the magnetic state of BFO, enabling the design of energy-efficient, non-volatile logic circuits [56].

5.6 Quantum Computing

$BiFeO_3$'s stable and switchable ferroelectric polarization enables the storage of quantum information in non-volatile states, reducing decoherence effects. The ability to manipulate spin currents in $BiFeO_3$ films through electric fields offers a pathway to low-energy spintronic devices, which are integral components of quantum computing architectures [77].

$BiFeO_3$ enables quantum logic operations by exploiting its ability to switch spin states via an electric field, reducing power consumption and enhancing processing speeds. Recent studies indicate that the presence of skyrmions (topologically protected spin structures) in doped $BiFeO_3$ can serve as stable quantum bits [30].

$BiFeO_3$ has been investigated as an interface material in superconductor-ferroelectric hybrid devices, which can enable robust quantum coherence for superconducting qubits. The interaction of $BiFeO_3$ with superconducting materials can lead to new quantum states useful for implementing quantum circuits [78]. $BiFeO_3$-based resistive switching (memristors) can be used in neuromorphic architectures, which are crucial for quantum neural networks. The nonlinear behaviour of $BiFeO_3$ memristors enables complex quantum neural network operations, mimicking synaptic plasticity [79].

5.7 Medical Applications

Recent studies have explored $BiFeO_3$-based nanocomposites for their antimicrobial properties. For instance, a self-assembled $BiFeO_3$@MIL-101 nanocomposite demonstrated significant antimicrobial activity under natural sunlight, suggesting potential for sterilization and infection control [80]. A Cobalt-doped $BiFeO_3$ (Co@BFO) nanocomposite exhibited notable antibacterial effects against pathogens like *Klebsiella pneumoniae* and *Bacillus cereus*, achieving a minimum inhibitory concentration of 10 μg/mL. This indicates its potential as an antimicrobial agent in medical settings [81].

5.8 Hybrid Devices

(i) **Energy Harvesting**: $BiFeO_3$-based composites have been utilized in nanogenerators to convert mechanical energy into electrical energy. For instance, a flexible multi-unit hybrid nanogenerator (M-HNG) combining $BiFeO_3$ with hydrophobic polymers demonstrated enhanced performance, achieving an output power density of approximately 600 μWcm^{-2}. This device was effective in biomechanical energy harvesting and powering small electronics [82].

(ii) **Biosensing**: In biosensor applications, hybrid structures incorporating $BiFeO_3$ have been designed for medical analyte detection. A notable example is an improved surface plasmon resonance (SPR)-based optical biosensor that employs a hybrid structure of Au–Ag–$BiFeO_3$-Graphene on a BK7 prism. This configuration was specifically optimized for the detection of SARS–CoV–2, enhancing the sensor's performance [83].

Table 1 List of major area of application, description of $BiFeO_3$-based devices and related references

Application area	Description	References
Renewable energy (photovoltaics and PEC cells)	Used in photovoltaic cells and photoelectrochemical (PEC) cells	Chen et al. [75], Fei et al. [76], Tokura et al. [77]
Spintronics	$BiFeO_3$-based heterostructures enable spin-dependent transport and memory devices	Tripathi et al. [72]
Non-volatile memory devices	Utilized in ferroelectric RAM (FeRAM) for non-volatile memory storage	Muneeswaran et al. [73]
Sensors	$BiFeO_3$-based materials exhibit high sensitivity applications	Man et al. [74]
Data storage and logic gates	Used in logic gate operations and high-density data storage	Martin et al. [56]
Quantum computing	Used for topological quantum computing due to their quantum phase transitions	Scott [30], de Sousa [77], Pantel et al. [78], Zubko et al. [79]
Medical applications	Used in targeted drug delivery and cancer treatments	Pulvirenti et al. [80], Jalandhara et al. [81]
Hybrid devices	Used in hybrid devices fabrication such as energy harvesting, biosensors and memory devices	Sasmal et al. [82], Mousania et al. [83], Sagheer et al. [84]

(iii) **Memory Storage**: $BiFeO_3$'s distinctive magnetoelectric properties have been explored for next-generation memory storage devices. However, its intrinsically weak magnetoelectric coupling has limited widespread applications. To overcome this, researchers have integrated $BiFeO_3$ with materials possessing enriched ferroelectric and ferro/ferrimagnetic properties, resulting in optimized heterostructures with improved energy and memory storage capabilities [84] (Table 1).

6 Applications of Composite Perovskite Multiferroics

There are many composite systems synthesized using various compositions such as multiferroics and some (anti)ferromagnets/ferroelectrics, (anti)ferromagnets/ ferroelectrics. These kinds of systems become even more interesting if there exists a magnetoelectric coupling. Composite perovskite multiferroics, which integrate both ferroelectric and magnetic properties, have gained significant attention due to their potential applications in various advanced technologies. These materials offer strong magnetoelectric (ME) coupling, allowing electric-field control of magnetism and vice versa, making them highly desirable for next-generation electronic and energy devices. There are many reports which shows application potentials of these composite-based perovskite multiferroics. The following sections outline the key applications of composite perovskite multiferroics.

The combination of photoactive magneto-piezoelectric nanomaterials like $BiFeO_3$ with ferroelectric polymers has led to the development of multi-stimuli-sensitive composite materials. These composites offer new properties that can address environmental and energy challenges. For example, a porous hybrid membrane based on polyvinylidene fluoride (PVDF) and $BiFeO_3$ was synthesized using the doctor blade casting method with the thermally induced phase separation technique, resulting in a material with enhanced functionalities [85].

Layered and Superlattice composites were synthesized by combining non-magnetic ferroelectrics (e.g., $BaTiO_3$) with non-ferroelectric magnets (e.g., $CoFe_2O_4$). This composite system has shown multiferroic properties. $LuFeO_3$/$LuFe_2O_4$ superlattices exhibit magnetoelectric coupling [86]. These results have opened up new possibilities for tunable electronic applications. By precisely controlling the layer thickness and interfacial interactions, one could improve functionality of composite multiferroics. There is one more way to enhance ME coupling is by synthesizing vertically aligned nanocomposites. Their three-dimensional heteroepitaxy could help in improving the interaction between ferroelectric and ferromagnetic components. For example, $BaTiO_3$–$CoFe_2O_4$ and $BiFeO_3$-based VANs are useful in energy-efficient memory devices [87]. Multiferroic nanopatterned hybrid materials can integrate conventional ferromagnetic metals with ferroelectric polymers. These enable magnetic-field-driven switching of ferroelectric polarization, which has potential in low-power logic devices [88, 89].

Composite multiferroics, such as $BaTiO_3$–$CoFe_2O_4$ and PZT–Terfenol-D, exhibit strong magnetoelectric coupling, which makes them highly sensitive to magnetic field variations. These materials are widely used in high-precision magnetic field sensors, accelerometers, and energy harvesters. Composite multiferroic heterostructures enable electric-field-controlled magnetization switching, reducing energy consumption compared to traditional MRAMs. Systems like $BiFeO_3$/CoFe and $Pb(Zr,Ti)O_3$/NiFe demonstrate enhanced magnetoelectric coupling at room temperature. Composite structures with coupled magnetic and ferroelectric phases offer tunable microwave properties, essential for phase shifters, filters, and antennas. $CoFe_2O_4$–$BaTiO_3$ thin-film composites show significant microwave tunability due to strong strain-mediated interactions. Piezoelectric-magnetostrictive composites (e.g., PZT/Terfenol-D) efficiently convert mechanical energy into electrical energy, useful in vibration energy harvesters. These materials enable self-powered IoT devices and wireless sensors. Magnetoelectric heterostructures enable voltage-controlled spintronic devices, reducing power dissipation. Composite systems like $BiFeO_3$–$SrTiO_3$ are explored for quantum computing applications [90].

Magnetoelectric (ME) laminate composites, made of piezoelectric (PZT) and magnetostrictive (Terfenol-D, Metglas) layers to enable magnetoelectric coupling for sensors and memory devices, emphasize their superior performance compared to single-phase multiferroics and particulate composites. ME laminates, particularly $BiFeO_3$/Terfenol-D, enable voltage-driven magnetic switching, offering low-energy alternatives to MRAM (Magnetoresistive RAM). The ability to store information in both polarization and magnetization states creates multi-bit memory elements. Laminates with ferrite and piezoelectric layers allow for voltage-controlled microwave devices, essential for 5G and IoT applications. ME laminates exhibit a 100 × increase in efficiency when driven at their electromechanical resonance frequency (~100 kHz–1 MHz). These composites are integrated into self-powered IoT devices, reducing dependence on external power sources [91].

7 Challenges and Future Aspects

7.1 Challenges

(i) **Weak coupling between ferroic orders**: Achieving strong magnetoelectric (ME) coupling at room temperature is difficult. The mechanisms underlying these couplings are often weak and require optimization.
(ii) **Low transition temperatures**: Many multiferroic perovskites exhibit desired properties at low temperatures, limiting their practical utility in ambient conditions.
(iii) **Material synthesis**: Producing high-quality, defect-free multiferroic films or bulk samples is challenging. Strain, grain boundaries, and doping effects can complicate the material's behaviour. Controlling crystal orientation and achieving phase-pure samples remains problematic.
(iv) **Toxicity and environmental impact**: Some perovskite multiferroics contain lead or other hazardous elements, raising environmental and sustainability concerns.

7.2 Future Aspects

(i) **Room-temperature multiferroics**: Developing materials with strong ferroic and magnetoelectric properties at room temperature is a priority for practical applications.
(ii) **Enhanced magnetoelectric coupling**: Exploring new mechanisms (e.g., strain-mediated or interfacial effects) and materials to enhance ME coupling for efficient energy conversion and storage devices.

(iii) **Multifunctional devices**: Designing devices that leverage multiple ferroic properties simultaneously, such as magnetoelectric random-access memory (MERAM) or multifunctional sensors.
(iv) **Integration with quantum technologies**: Investigating the role of multiferroic materials in quantum computing and quantum sensors, leveraging their ability to couple electric and magnetic states at the nanoscale.
(v) **Spintronic applications**: Exploring spintronic devices, such as spin valves or magnetic tunnel junctions, that utilize multiferroic materials for low-power, high-performance operation.

8 Summary

This chapter briefly introduces multiferroics in general and their various types. The chapter mainly focuses on the various types of perovskite multiferroics and their varying origin of multiferroicity. There exist some other perovskite multiferroics in nature, which have not been discussed here, and most of them also fall in the category discussed in this chapter. The unique ability of multiferroics to manipulate both electric and magnetic fields simultaneously makes them ideal candidates for next-generation technologies like spintronics and multifunctional devices. We have discussed some $BiFeO_3$-based devices for potential applications in various fields such as spintronics, sensors, non-volatile memory devices, renewable energy, data storage, quantum computing, medical applications and few more. However, despite their potential in various field there are some limitations such as weak coupling between order parameters, low transition temperatures, etc. Researchers are working to overcome these limitations by optimizing material compositions, improving structural stability, and discovering new fabrication techniques. As these hurdles are addressed, perovskite multiferroics are expected to play a pivotal role in advancing energy-efficient technologies and functional electronic systems.

References

1. Landau LD, Lifshitz EM (1959) Electrodynamics of continuous media. Fizmatgiz, Moscow. https://www.aldebaran.cz/studium/books/1984_Landau-Electrodynamics.pdf
2. Dzyaloshinskii IE (1960) On the magneto-electrical effect in antiferromagnets. Sov Phys JETP 10:628–629. http://www.jetp.ras.ru/cgi-bin/dn/e_010_03_0628.pdf
3. Astrov DN (1960) The magnetoelectric effect in antiferromagnetics. Sov Phys JETP 11:708–709. http://jetp.ras.ru/cgi-bin/dn/e_011_03_0708.pdf
4. Folen VJ, Rado GT, Stalder EW (1961) Anisotropy of the magnetoelectric effect in Cr_2O_3. Phys Rev Lett 6:607–608. https://doi.org/10.1103/PhysRevLett.6.607
5. Rado GT (1969) Magnetoelectric evidence for the attainability of time-reversed antiferromagnetic configurations by metamagnetic transitions in $DyPO_4$. Phys Rev Lett 23:644–647. https://doi.org/10.1103/PhysRevLett.23.644
6. Gorodetsky G, Sayar M, Shtrikman S (1970) The magnetoelectric effect in the antiferromagnet $FeSb_2O_4$. Mater Res Bull 5:253–255. https://doi.org/10.1016/0025-5408(70)90120-0

7. Rado GT (1964) Observation and possible mechanisms of magnetoelectric effects in a ferromagnet. Phys Rev Lett 13:335–337. https://doi.org/10.1103/PhysRevLett.13.335
8. Asher E, Rieder H, Schmid H, Stossel H (1966) Some properties of ferromagnetoelectric Nickel–Iodine Boracite, $Ni_3B_7O_{13}I$. J Appl Phys 37:1404–1405. https://doi.org/10.1063/1.1708493
9. Santoro RP, Segal DJ, Newnham RE (1966) Magnetic properties of $LiCoPO_4$ and $LiNiPO_4$. J Phys Chem Solids 27:1192–1193. https://doi.org/10.1016/0022-3697(66)90097-7
10. Watanabe T, Kohn K (1989) Magnetoelectric effect and low temperature transition of $PbFe_{0.5}Nb_{0.5}O_3$ single crystal. Phase Trans 15:57–68. https://doi.org/10.1080/01411598908206837
11. Krichevtsov BB, Pavlov VV, Pisarev RV, Selitsky AG (1994) Linear magnetoelectric effect in magnetic garnet thin films. Ferroelectrics 161:65–71. https://doi.org/10.1080/00150199408213354
12. Schmid H (1973) On a magnetoelectric classification of materials. Int J Magn 4:337–361. https://archive-ouverte.unige.ch/unige:32935
13. Rivera JP (1994) The linear magnetoelectric effect in $LiCoPO_4$ revisited. Ferroelectrics 161:147–164. https://doi.org/10.1080/00150199408213364
14. Krichevtsov BB, Pavlov VV, Pisarev RV (1989) Giant linear magnetoelectric effect in garnet ferrite films. JETP Lett 49:535–538. http://jetpletters.ru/ps/1119/article_16956.pdf
15. Rado GT, Ferrari JM, Maisch WG (1984) Magnetoelectric susceptibility and magnetic symmetry of magnetoelectrically annealed $TbPO_4$. Phys Rev B 29:4041–4048. https://doi.org/10.1103/PhysRevB.29.4041
16. Schmid H (1994) Multiferroic magnetoelectrics. Ferroelectrics 162:317–338. https://doi.org/10.1080/00150199408245120
17. Khomskii D (2009) Classifying multiferroics: mechanisms and effects. Physics 2:20. https://doi.org/10.1103/Physics.2.20
18. Hill NA (2000) Why are there so few magnetic ferroelectrics? J Phys Chem B 104:6694–6709. https://doi.org/10.1021/jp000114x
19. Khomskii D (2001) Magnetism and ferroelectricity; why do they so seldom coexist? Bull Am Phys Soc C21:002. https://ui.adsabs.harvard.edu/abs/2001APS..MARC21002K/abstract
20. Wang J, Neaton JB, Zheng H, Nagarajan V, Ogale SB, Liu B, Viehland D, Vaithyanathan V, Schlom DG, Waghmare UV, Spaldin NA, Rabe KM, Wuttig M, Ramesh R (2003) Epitaxial $BiFeO_3$ multiferroic thin film heterostructures. Science 299:1719–1722. https://doi.org/10.1126/science.1080615
21. Kimura T, Goto T, Shintani H, Ishizaka K, Arima T, Tokura Y (2003) Magnetic control of ferroelectric polarization. Nature 426:55–58. https://doi.org/10.1038/nature02018
22. Hur N, Park S, Sharma PA, Ahn JS, Guha S, Cheong SW (2004) Electric polarization reversal and memory in a multiferroic material induced by magnetic fields. Nature 429:392–395. https://doi.org/10.1038/nature02572
23. Gajek M, Bibes M, Fusil S, Bouzehouane K, Fontcuberta J, Barthélémy A, Fert A (2007) Tunnel junctions with multiferroic barriers. Nat Mater 6:296–302. https://doi.org/10.1038/nmat1860
24. Zvezdin AK, Logginov AS, Meshkov GA, Pyatakov AP (2007) Multiferroics: promising materials for microelectronics, spintronics, and sensor technique. Bull Russ Acad Sci: Phys 71:1561–1562. https://doi.org/10.3103/S1062873807110263
25. Vopsaroiu M, Blackburn J, Cain MG (2007) A new magnetic recording read head technology based on the magneto-electric effect. J Phys D: Appl Phys 40:5027–5033. https://doi.org/10.1088/0022-3727/40/17/003
26. Kumar A, Katiyar RS, Scott JF (2010) Novel room temperature multiferroics for random access memory elements. IEEE Trans Ultrason Ferroelectr Freq Control 57:2237–2242. https://doi.org/10.1109/TUFFC.2010.1684
27. D'Souza N, Atulasimha J, Bandyopadhyay S (2011) Four-state nanomagnetic logic using multiferroics. J Phys D: Appl Phys 44:265001. https://doi.org/10.1088/0022-3727/44/26/265001

28. Eerenstein W, Mathur N, Scott J (2006) Multiferroic and magnetoelectric materials. Nature 442:759–765. https://doi.org/10.1038/nature05023
29. Ramesh R, Spaldin NA (2007) Multiferroics: progress and prospects in thin films. Nat Mater 6:21–29. https://doi.org/10.1038/nmat1805
30. Scott JF (2007) Applications of modern ferroelectrics. Science 315:954–959. https://doi.org/10.1126/science.1129564
31. Catalan G, Scott JF (2009) Physics and applications of bismuth ferrite. Adv Mater 21:2463. https://doi.org/10.1002/adma.200802849
32. Fiebig M (2005) Revival of the magnetoelectric effect. J Phys D: Appl Phys 38:R123. https://doi.org/10.1088/0022-3727/38/8/R01
33. Goodenough JB, Longo M (1970) Magnetic and other properties of oxides and related compounds. In: Landolt-Börnstein, Numerical data and functional relations in science and technology, New Series, vol III. 4. Springer, Berlin, p 126. https://doi.org/10.1007/b19968
34. Mitsui T et al (1981) Ferroelectrics and related substances. In: Landolt-Börnstein, Numerical data and functional relations in science and technology, New Series, vol 16. Springer, Berlin, p 1. ISBN: 978-3-540-10484-1
35. Efremov DV, van den Brink J, Khomskii DI (2004) Bond-versus site-centred ordering and possible ferroelectricity in manganites. Nat Mater 3:853–856. https://doi.org/10.1038/nmat1236
36. Ikeda N, Ohsumi H, Ohwada K, Ishii K, Inami T, Kakurai K, Murakami Y, Yoshii K, Mori S, Horibe Y, Kito H (2005) Ferroelectricity from iron valence ordering in the charge-frustrated system $LuFe_2O_4$. Nature 436:1136–1138. https://doi.org/10.1038/nature04039
37. Jardón C, Rivadulla F, Hueso LE, Fondado A, López-Quintela MA, Rivas J, Zysler R, Causa MT, Sánchez RD (1999) Experimental study of charge-ordering transition in $Pr_{0.67}Ca_{0.33}MnO_3$. J Magn Magn Mater 196–197:475–476. https://doi.org/10.1016/S0304-8853(98)00829-4
38. Mercone S, Wahl A, Pautrat A, Pollet M, Simon C (2004) Anomaly in the dielectric response at the charge-orbital-ordering transition of $Pr_{0.67}Ca_{0.33}MnO_3$. Phys Rev B 69:174433. https://doi.org/10.1103/PhysRevB.69.174433
39. Cheong SW, Mostovoy M (2007) Multiferroics: a magnetic twist for ferroelectricity. Nat Mater 6:13–20. https://doi.org/10.1038/nmat1804
40. Rado GT, Ferrari JM (1975) Electric field dependence of the magnetic anisotropy energy in magnetite (Fe_3O_4). Phys Rev B 12:5166–5174. https://doi.org/10.1103/PhysRevB.12.5166
41. Rado GT, Ferrari JM (1977) Linear and bilinear magnetoelectric effects in magnetically biased magnetite (Fe_3O_4). Phys Rev B 15:290–297. https://doi.org/10.1103/PhysRevB.15.290
42. Miyamoto Y, Chikazumi S (1988) Crystal symmetry of magnetite in low temperature phase deduced from magnetoelectric measurements. J Phys Soc Japan 57:2040–2050. https://doi.org/10.1143/JPSJ.57.2040
43. Miyamoto Y, Shindo M (1993) Magnetoelectric measurement of magnetite (Fe_3O_4) at low temperatures and direct evidence for nonexistence of ac mirror plane. J Phys Soc Japan 62:1423–1426. https://doi.org/10.1143/JPSJ.62.1423
44. Miyamoto Y, Ishihara S, Hirano T, Takada M, Suzuki N (1994) Ferroelectricity of magnetite (Fe_3O_4) observed by means of magnetoelectric effect. Solid State Commun 89:51–54. https://doi.org/10.1016/0038-1098(94)90416-2
45. Van Aken BB, Palstra TTM, Filippetti A, Spaldin NA (2004) The origin of ferroelectricity in magnetoelectric $YMnO_3$. Nat Mater 3:164–170. https://doi.org/10.1038/nmat1080
46. Katsura H, Nagaosa N, Balatsky AV (2005) Spin current and magnetoelectric effect in noncollinear magnets. Phys Rev Lett 95:057205. https://doi.org/10.1103/PhysRevLett.95.057205
47. Mostovoy MV (2006) Ferroelectricity in spiral magnets. Phys Rev Lett 96:067601. https://doi.org/10.1103/PhysRevLett.96.067601
48. Kenzelmann M, Harris AB, Jonas S, Broholm C, Schefer J, Kim SB, Zhang CL, Cheong SW, Vajk OP, Lynn JW (2005) Magnetic inversion symmetry breaking and ferroelectricity in $TbMnO_3$. Phys Rev Lett 95:087206. https://doi.org/10.1103/PhysRevLett.95.087206

49. Choi YJ, Yi HT, Lee S, Huang Q, Kiryukhin V, Cheong SW (2008) Ferroelectricity in an Ising chain magnet. Phys Rev Lett 100:047601. https://doi.org/10.1103/PhysRevLett.100.047601
50. Roginska YE, Tomashpo YY, Venevtse YN, Petrov VM, Zhdanov GS (1966) Nature of dielectric and magnetic properties of $BiFeO_3$. Sov Phys JETP 23:47–51. http://jetp.ras.ru/cgi-bin/e/index/e/23/1/p47?a=list
51. Kiselev SV, Ozerov RP, Zhdanov GS (1963) Detection of magnetic order in ferroelectric $BiFeO_3$ by neutron diffraction. Sov Phys Dokl 7:742–744. https://ui.adsabs.harvard.edu/abs/1963SPhD....7..742K/abstract
52. Sosnowska I, Loewenhaupt M, David WIF, Ibberson RM (1992) Investigation of the unusual magnetic spiral arrangement in $BiFeO_3$. Physica B 180–181:117–118. https://doi.org/10.1016/0921-4526(92)90678-L
53. Ko KT, Jung M, He Q et al (2011) Concurrent transition of ferroelectric and magnetic ordering near room temperature. Nat Commun 2:567. https://doi.org/10.1038/ncomms1576
54. Zavaliche F, Yang SY, Zhao T, Chu YH, Cruz MP, Eom CB, Ramesh R (2006) Multiferroic $BiFeO_3$ films: domain structure and polarization dynamics. Phase Transitions 79:991–1017. https://doi.org/10.1080/01411590601067144
55. Bea H, Gajek M, Bibes M, Barthlelmy A (2008) Spintronics with multiferroics. J Phys: Condens Matter 20:434221.https://doi.org/10.1088/0953-8984/20/43/434221
56. Martin LW, Crane SP, Chu YH, Holcomb MB, Gajek M, Huijben M, Yang CH, Balke N, Ramesh R (2008) Multiferroics and magnetoelectrics: thin films and nanostructures. J Phys: Condens Matter 20:434220. https://doi.org/10.1088/0953-8984/20/43/434220
57. Chi ZH, Yang H, Feng SM, Li FY, Yu RC, Jin CQ (2007) Room-temperature ferroelectric polarization in multiferroic $BiMnO_3$. J Magn Magn Mater 310:e358–e360. https://doi.org/10.1016/j.jmmm.2006.10.335
58. Zhai LJ, Wang HY (2017) The magnetic and multiferroic properties in $BiMnO_3$. J Magn Magn Mater 426:188–194. https://doi.org/10.1016/j.jmmm.2016.11.065
59. Khomskii DI (2005) Role of orbitals in the physics of correlated electron systems. Physica Scripta 72:CC8. https://doi.org/10.1238/Physica.Regular.072a00CC8
60. Kumar M, Phase DM, Choudhary RJ (2019) Structural, ferroelectric and dielectric properties of multiferroic $YMnO_3$ synthesized via microwave assisted radiant hybrid sintering. Heliyon 5:e01691. https://doi.org/10.1016/j.heliyon.2019.e01691
61. Kumar M, Choudhary RJ, Phase DM (2015) Structural and multiferroic properties of self-doped yttrium manganites $YMn_{1+x}O_3$. AIP Conf Proc 1661:070005. https://doi.org/10.1063/1.4915383
62. Kumar M, Choudhary RJ, Phase DM (2015) Metastable magnetic state and exchange bias training effect in Mn-rich $YMnO_3$ thin films. J Phys D Appl Phys 48:125003. https://doi.org/10.1088/0022-3727/48/12/125003
63. Kajimoto R, Yoshizawa H, Shintani H, Kimura T, Tokura Y (2004) Magnetic structure of $TbMnO_3$ by neutron diffraction. Phys Rev B 70:012401. https://doi.org/10.1103/PhysRevB.70.012401
64. Sharma P, Kumar A, Qiu F, Cui J, Tan G (2024) Growth and characterization of perovskite $TbMnO_3$ single crystal: structural, vibrational and magnetic properties. J Alloys Compd. 990:174488. https://doi.org/10.1016/j.jallcom.2024.174488
65. Rovillain P, Cazayous M, Gallais Y, Sacuto A, Measson MA, Sakata H (2010) Magnetoelectric excitations in multiferroic $TbMnO_3$ by Raman scattering. Phys Rev B 81:054428. https://doi.org/10.1103/PhysRevB.81.054428
66. Pimenov A, Mukhin AA, Ivanov VY, Travkin VD, Balbashov AM, Loidl A (2006) Possible evidence for electromagnons in multiferroic manganites. Nat Phys 2:97–100. https://doi.org/10.1038/nphys212
67. Smolenskiĭ GA, Chupis IE (1982) Ferroelectromagnets. Sov Phys Usp 25:475. https://doi.org/10.1070/PU1982v025n07ABEH004570
68. Arima T (2007) Ferroelectricity induced by proper-screw type magnetic order. J Phys Soc Jpn 76:073702. https://doi.org/10.1143/JPSJ.76.073702

69. Sergienko IA, Dagotto E (2006) Role of the Dzyaloshinskii-Moriya interaction in multiferroic perovskites. Phys Rev B 73:094434. https://doi.org/10.1103/PhysRevB.73.094434
70. Lu C, Deniz H, Li X, Liu J-M, Cheong S-W (2016) Continuous magnetoelectric control in multiferroic $DyMnO_3$ films with twin-like domains. Sci Rep 6:20175. https://doi.org/10.1038/srep20175
71. Rahman MS, Ghose S, Hong L, Dhungana P, Fahami A, Gatabi JR, Rojas-Ramirez JS, Zakhidov A, Klie RF, Pandey RK, Ravi D (2016) Integration of $BiFeO_3/La_{0.7}Sr_{0.3}MnO_3$ heterostructures with III–V semiconductors for low-power non-volatile memory and multiferroic field effect transistors. J Mater Chem C 4:10386–10394. https://doi.org/10.1039/C6TC03386E
72. Tripathi PN, Ojha SK, Nazarov A (2020) Development of highly reliable $BiFeO_3/HfO_2$/Silicon gate stacks for ferroelectric non-volatile memories in IoT applications. J Mater Sci: Mater Electron 31:22107–22118. https://doi.org/10.1007/s10854-020-04713-9
73. Muneeswaran M, Gopiraman M, Dhanabalan SS, Giridharan NV, Akbari-Fakhrabadi A (2021) Multiferroic properties of rare earth-doped $BiFeO_3$ and their spintronic applications. In: Rajendran S, Qin J, Gracia F, Lichtfouse E (eds) Metal and metal oxides for energy and electronics. Environmental chemistry for a sustainable world, vol 55. Springer, Cham. https://doi.org/10.1007/978-3-030-53065-5_11
74. Man S, Leng X, Bai J, Kan S, Cui Y, Wang J, Xu L (2023) Enhancement of photoelectrochemical performance of $BiFeO_3$ by Sm^{3+} doping. Ceram Int 49:10255–10264. https://doi.org/10.1016/j.ceramint.2022.11.205
75. Chen G, Chen J, Pei W, Lu Y, Zhang Q, Zhang Q, He Y (2019) Bismuth ferrite materials for solar cells: current status and prospects. Mater Res Bull 110:39–49. https://doi.org/10.1016/j.materresbull.2018.10.011
76. Fei L, Hu Y, Li X, Song R, Sun L, Huang H, Gu H, Chan HLW, Wang Y (2015) Electrospun bismuth ferrite nanofibers for potential applications in ferroelectric photovoltaic devices. ACS Appl Mater Interfaces 7:3665–3670. https://doi.org/10.1021/acsami.5b00069
77. de Sousa R (2022) Multiferroics are a spintronics game changer. Physics 15:124. https://doi.org/10.1103/Physics.15.124
78. Pantel D, Alexe M (2012) Electroresistance effects in ferroelectric tunnel junctions. Phys Rev B 85:155324. https://doi.org/10.1103/PhysRevB.82.134105
79. Zubko P, Gariglio S, Gabay M, Ghosez P, Triscone J-M (2016) Interface physics in complex oxide heterostructures. Annu Rev Condens Matter Phys 7:387–423. https://doi.org/10.1146/annurev-conmatphys-062910-140445
80. Pulvirenti L, Lombardo C, Salmeri M, Bongiorno C, Mannino G, Presti FL, Cambria MT, Condorelli GG (2023) Self-assembled $BiFeO_3$@MIL-101 nanocomposite for antimicrobial applications under natural sunlight. Discover Nano 18:113. https://doi.org/10.1186/s11671-023-03883-9
81. Jalandhara D, Kumar S, Dalal J, Supreet SG, Kumar S, Badru R, Singh Y, Sharma SV, Kaushal S (2025) A multifunctional Co-doped $BiFeO_3$ nanocomposite: a promising candidate for photocatalytic degradation, antibacterial activity, and antioxidant applications. Mater Adv 6:641. https://doi.org/10.1039/D4MA01053A
82. Sasmal A, Patra A, Maity S, Pratihar S, Sen S (2022) Multiferroic $BiFeO_3$-based hydrophobic polymer composites for polarization rationalization-induced piezo-tribo hybrid energy harvesting and versatile self-powered mechanosensing. Sustain Energy Fuels 6:4652–4668. https://doi.org/10.1039/D2SE00947A
83. Mousania Y, Karimi S, Farmani A (2025) Designing high-performance SPR biosensor using hybrid structure of $BiFeO_3$ and 2D material for SARS–CoV–2 detection. Plasmonics. https://doi.org/10.1007/s11468-024-02516-w
84. Sagheer I, Ansar MT, Ramay SM, Huang H, Atiq S (2024) Efficient energy and memory storage capabilities in optimized $BiFeO_3/MnMoO_4/NiFe_2O_4$ triphasic composites for futuristic multistate devices. RSC Adv 14:29260–29270. https://doi.org/10.1039/D4RA05223D
85. Orudzhev F, Alikhanov N, Amirov A, Rabadanova A, Selimov D, Shuaibov A, Gulakhmedov R, Abdurakhmanov M, Magomedova A, Ramazanov S, Sobola D, Giraev K, Amirov A, Rabadanov K, Gadzhimagomedov S, Murtazali R, Rodionova V (2023) Porous hybrid PVDF/

$BiFeO_3$ smart composite with magnetic, piezophotocatalytic, and light-emission properties. Catalysts 13:874. https://doi.org/10.3390/catal13050874
86. Spaldin NA, Ramesh R (2019) Advances in magnetoelectric multiferroics. Nat Mater 18:203–212. https://doi.org/10.1038/s41563-018-0275-2
87. Zheng H et al (2004) Multiferroic $BaTiO_3$-$CoFe_2O_4$ nanostructures. Science 303:661–663. https://doi.org/10.1126/science.1094207
88. Cai R et al (2017) Multiferroic nanopatterned hybrid material with room temperature magnetic switching of the electric polarization. Adv Mater 29:1604604. https://doi.org/10.1002/adma.201604604
89. Poddar S et al (2018) Room-temperature magnetic switching of the electric polarization in ferroelectric nanopillars. ACS Nano 12:576–584. https://doi.org/10.1021/acsnano.7b07389
90. Tokura Y, Seki S, Nagaosa N (2014) Multiferroics of spin origin. Rep Prog Phys 77:076501. https://doi.org/10.1088/0034-4885/77/7/076501
91. Zhai J, Xing Z, Dong S, Li JF, Viehland D (2008) Magnetoelectric laminate composites: an overview. J Am Ceram Soc 91:351–358. https://doi.org/10.1111/j.1551-2916.2008.02259.x

Chapter 2
Synthesis Techniques of Perovskite Multiferroic Materials and Their Properties

Anjali Varshney, Sunil Chauhan, and Sangeeta Rawal

1 Overview

The class of chemicals with the general formula ABX_3 is called a perovskite material. It was first found in the Ural Mountains in 1839 by German geologist Gustav Rose [1, 2], who named it after the Russian mineralogist L. Perovski. In ABX_3, X represents an electronegative anion with ionic valence − 2, such as O^{2-}, S^{2-}, etc., while A and B correspond to the metal cations whose combined oxidation state typically add up to + 6, such as Li^+: Nb^{5+}; Sr^{2+}: Mn^{4+}; Ba^{2+}: Ti^{4+}; Bi^{3+}: Fe^{3+}; and La^{3+}: Fe^{3+} [3–6]. The perovskite materials can be oxides, halides, nitrides, sulfides, etc., and they can be found in a variety of forms, including powders and thin films [7–10]. The interest of researchers in the perovskite material stems from its ability to hold a different cation at the A- & B-sites alone or in combination with anions at the X-site [11, 12]. Bhalla et al. reported that each element of the periodic table can be employed at A- and B-sites because of its structure and adaptability [13]. The typical chemical formula is $A_2BB'O_6$, as the double perovskite compounds are made up of two simple perovskite structures, ABO_3 and $AB'O_3$. In these structures, B and B′ are typically transition metal cations, which can be ordered or disordered depending on their size, charge, and electronic configuration. To achieve rock salt ordering between these cations, half of the cations of the B-site are switched by another cation, B′, resulting in double perovskite compounds [14]. The fundamental perovskite crystal structure ABO_3 is represented in Fig. 1a, whereas the double perovskite structure made up

A. Varshney · S. Chauhan (✉) · S. Rawal
Department of Physics & Environmental Sciences, Sharda School of Engineering & Science, Sharda University, Greater Noida, Uttar Pradesh, India
e-mail: sunil.chauhan@sharda.ac.in

A. Varshney · S. Chauhan
Department of Physics & Environmental Sciences, Sharda School of Engineering & Science, Centre for Solar Cell and Renewable Energy, Sharda University, Greater Noida, Uttar Pradesh, India

M. Kumar (ed.), *Sustainable Perovskite Multiferroic Materials*, Materials Horizons: From Nature to Nanomaterials, https://doi.org/10.1007/978-981-96-7419-0_2

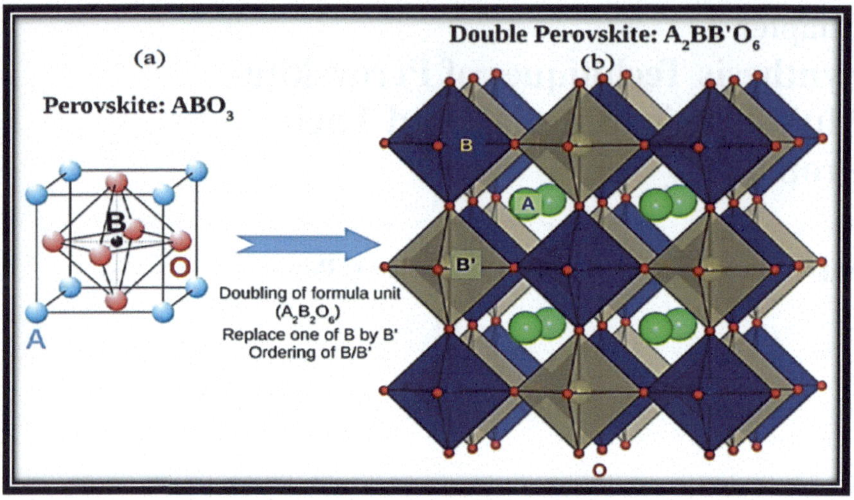

Fig. 1 **a** ABO_3: basic structure of a perovskite **b** $A_2BB'O_6$: the structure of double perovskites crystal [15]. Copyright, open access, MDPI publishing

of two simple perovskite structures, $A_2BB'O_6$, is shown in Fig. 1b. Transition metal elements (Mn, Co, Sc, Ni, Fe) are present at the B and B′ sites, while alkaline earth metals (Bi, Ca, Sr, Ba, etc.) or lanthanide elements (La) occupy the A-site [15]. For example, bismuth ferrite ($BiFeO_3$) deformed perovskite structure promotes multiferroic behavior, where ferroelectricity and antiferromagnetism coexist. $BiFeO_3$ is therefore a promising candidate for multiple applications.

There are two main types of perovskites: ideal and deformed. Materials that crystallize into shapes other than cubic ones are known as deformed perovskites. The X atoms create the corners of the octahedron, the B atoms form the core of the octahedral framework, and the A atoms form the center of the cube in a perfect perovskite with the Pm3‾m space group. Six X anions encircle the B cations in this structure, creating a BX_6 octahedron, whereas twelve X anions encircle the A cations, creating a cuboctahedral arrangement (AX_{12}). While in the case of deformed perovskites with non-cubic crystal structures are the result of perovskites frequently departing from the optimum shape. Changes in the cations at the A- and/or B-sites cause these distortions [16, 17], which change the bond lengths and angles, also affect the symmetry and volume of the unit cell.

The tolerance factor (t), given by Victor Moritz Goldschmidt [18], is one way that may be used to determine the stability of the perovskite material structure. Using the ionic radii of the cations and anions, the tolerance factor can be calculated using the formula below:

$$t = \frac{r_A + r_o}{\sqrt{2}(r_B + r_o)} \tag{1}$$

where r_O represents the ionic radii of the oxygen anion and r_A and r_B are the ionic radii of the A and B cations. The cation A always has a higher ionic radius than the cation B. When selecting elements to create the perfect perovskite material, choose the combination of A and B cations that depends on the tolerance factor. Equation (2) provides an alternative expression for Eq. (1) that may be applied to any temperature:

$$t = \frac{d_{A-O}}{\sqrt{2}(d_{B-O})} \tag{2}$$

where bond lengths for A–O and B–O are represented by d_{A-O} and d_{B-O}, respectively [19].

To the best of our knowledge, the perovskite material can hold various ions at its A- and B-sites. The tolerance factor varies as a result of variations in the cations at the A- and B-sites. The perovskite structure changes from a cubic to a non-cubic deformed structure as a result of the shift in tolerance factor. The tolerance factor should be between 0.88 and 1.09 for a stable perovskite [20]. The tolerance (t) factor of a perfect perovskite crystal is one. Generally, Perovskite materials exhibit a monoclinic or rhombohedral structure when t is less than 1, and a tetragonal or orthorhombic structure when t is greater than 1 [21].

The BO_6 octahedral lead is skewed from an optimum position due to distortion in the perovskite system, which results in a change or improvement in unit cell volume. The degree of distortion in the perovskite structure is thus gauged by the tolerance factor. These distortions can improve or change the properties of perovskites, making them suitable for a range of applications. The main requirements must be fulfilled in order to create a stable perovskite.

First, with a tolerance factor within the stable range, the average ionic radii of the A-site and B-site cations should be more than 0.90 Å and 0.51 Å, respectively [22, 23].

Second, the electro-neutrality of the perovskite structure must be maintained, meaning that the total positive charges of the cations at the A-site and B-site must balance the total negative charge contributed by the anions. The second need for the perovskite structure to retain electro-neutrality is that the cations at the A- and B-sites and the anions at the X-site have equal charges. $A^{1+}B^{5+}O_3$, $A^{4+}B^{2+}O3$, and $A^{3+}B^{3+}O_3$ are the typical charge distributions [16, 22].

The qualities needed for advanced applications are frequently absent from pure perovskite materials. Doping is frequently used to improve performance at the A- and/or B-sites. To produce materials with better or unique qualities, some of the original cations must be changed. $A_{1-x}A'_xBO_3$ for A-site doping, $AB_{1-y}B'_yO_3$ for B-site doping, or $A_{1-x}A'_xB_{1-y}B'_yO3$ for simultaneous doping at both sites are the general formulas for doped perovskites. For some applications, doping can improve performance by altering the ionic states, link lengths, and crystal structure. $BiFeO_3$ is produced utilizing a range of procedures, including hydrothermal methods, sol–gel processes, electrospinning, and solid-state reactions. Each synthesis process affects the material phase, purity, particle size, and morphology. Surface morphology was revealed by scanning electron microscopy (SEM), while X-ray diffraction (XRD) confirmed the phase, purity, and structural characteristics of BFO [23–27]. In conclusion, $BiFeO_3$ is a versatile perovskite material also known as the holy-grail BFO due to its deformed structure caused by its ionic arrangement and bonding properties. Its properties can be changed by controlled manufacture and doping, which makes it a valuable material for a range of technological applications that entirely rely on their synthesis method.

2 Perovskite Material Synthesis Methods

The synthesis route has a significant impact on the optical, chemical, and physical properties of perovskite materials. The appropriate synthesis technique needs to be chosen for the produced sample to possess the necessary properties. The synthesis procedures also affect the shape and crystal structure of the samples [28]. As an example, any of the perovskite materials that may be produced using a range of techniques that can be broadly divided into solid-state (ST), liquid-state (LT), and gas-state (GT) processes is bismuth ferrite ($BiFeO_3$).

Every technique has its benefits and is selected according to the material's intended shape. For instance, gas-state processes are utilized for thin films, liquid-state synthesis is best for producing nanomaterials, and solid-state synthesis is frequently employed for bulk materials. The use of solid-state synthesis techniques to $BiFeO_3$ (BFO) is the main emphasis of this research.

2.1 Solid-State Technique (ST)

One of the most popular methods for creating polycrystalline materials is the solid-state synthesis method, also known as the ceramic method, because it is commonly used to make ceramics. Using this technique, raw materials that do not mix chemically at normal temperature, such as oxides or carbonates, are mixed. A chemical process takes place on heating the mixture at high temperatures (700–1500 °C), which facilitates the creation of the required perovskite structure. Two variants of the solid-state synthesis method to the production of $BiFeO_3$ are described in Fig. 2.

2.1.1 The Mechanical Ball Milling Route

The solid-state technique for creating bulk perovskite materials is mechanical ball milling. The raw materials undergo manual processing, blending, ball milling, and grinding during the process. For instance, $BiFeO_3$ is synthesized by combining stoichiometric proportions of fundamental components like Bi_2O_3 and Fe_2O_3. To ensure even mixing, these items are first manually smashed with a mortar and pestle. After that, the mixture is ball milled for several hours at a moderate speed of 100–500 rpm

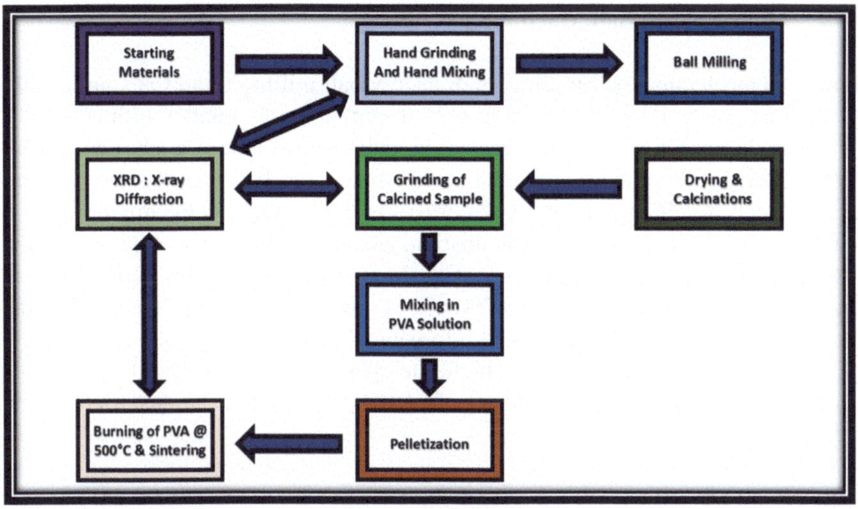

Fig. 2 Flow chart for the ball milling method used to synthesize oxide materials

in a planetary milling machine using a medium like acetone or alcohol. To encourage phase formation, the mixture is dried and calcined at temperatures between 600 and 800°C after milling [29–31]. X-ray diffraction (XRD) and other analytical methods are used to verify the purity of the $BiFeO_3$ phase in the calcined product. A great technique for creating particles smaller than a micrometer is the mechanical ball milling process shown in Figure 3.

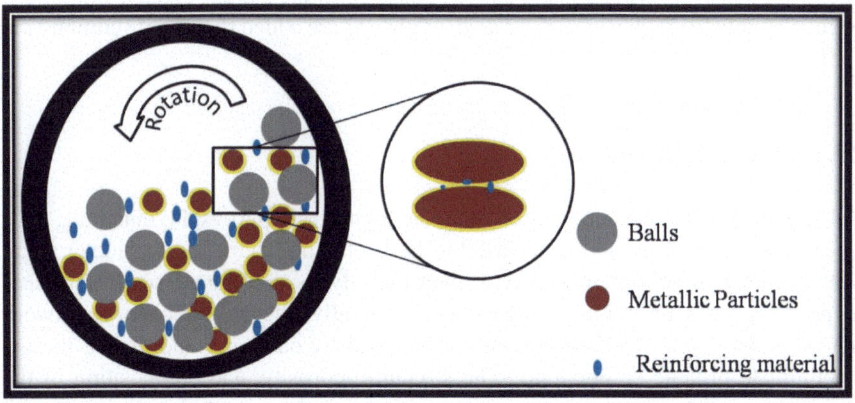

Fig. 3 Metallic nanocomposite ball milling process diagram [31]. Copyright, open access, MDPI publishing

2.1.2 High-Energy Ball Machining Method

Similar to mechanical ball milling, high-energy ball milling rotates at significantly greater rates (several hundred to thousands of rpm) and uses smaller milling balls. It can cause chemical reactions and decrease particles to nanometric levels; this process is particularly well-suited for creating nanoparticles. For $BiFeO_3$, zirconia balls are used to blend stoichiometric quantities of Bi_2O_3 and Fe_2O_3, and the mixture is then milled for 12–30 h. To avoid contamination, alcohol is frequently employed as the mixing medium. To maximize phase purity, the sample is dried at 90 °C after milling and subsequently annealed at 400–700 °C. To create dense $BiFeO_3$ ceramics, the resultant nanopowder can be formed into pellets using a binder such as polyvinyl alcohol (PVA) and then sintered at higher temperatures. To confirm the material's purity and phase stability after sintering, the XRD patterns are examined for the same [32, 33].

2.2 The Method of Liquid-State Synthesis (LT)

Apart from solid-state synthesis technique, one more popular technique exists for creating nanomaterials, especially oxide nanoparticles, is liquid-state synthesis. It makes use of raw chemicals that react at room temperature, such as nitrates, acetates, or oxalates. This procedure uses a different method, such as co-precipitation, sol–gel, hydrothermal, and auto-combustion, to create perovskite material, i.e., BFO. One of the method in this area is the auto-combustion technique. Its price and relative ease of usage make it suitable for the synthesis of perovskite nanoparticles. The first chemical, which are typically oxalates, acetates, or nitrates, is dissolved in de-ionized water. Organic fuels like urea, citric acid, or glycine are added to promote combustion [22].

2.2.1 Sol–Gel Method

Sol–gel chemical route can be used to create high-purity perovskite materials, such as ceramics, thin films, and nanoparticles. In this method, a metal alkoxide or metal salt precursor dissolves in a solvent, then hydrolyzes and condenses the mixture to produce a sol (colloidal suspension). Eventually, the sol becomes a network that resembles gel. The final product is obtained by calcining and drying the gel to eliminate organic components. The structure, porosity, and composition of the material can all be accurately controlled with this technique. Although it typically requires careful manipulation of precursors and stringent environmental conditions, it is perfect for creating thin films and nanostructured materials. This approach is time-consuming, difficult to manage, and may have instability problems [34]. For the sol–gel process to work, precursor solutions need to be mixed evenly with ultrasonic assistance for the intended result. The goal is to create homogenous, pure $BiFeO_3$

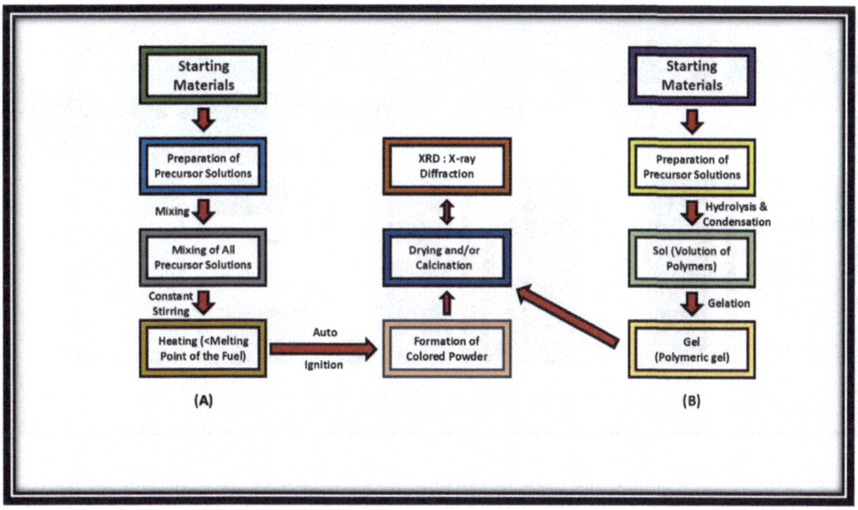

Fig. 4 Complete process of synthesis of perovskite material using **a** auto-combustion and **b** sol–gel

spherical nanoparticles by combining chelating agents for hydrolytic condensation to create gels. Utilizing the sol–gel method with ultrasonic assistance provides synthesized material with an average particle size of 60–80 nm. Nanomaterials synthesized by this method generally show good crystallinity and smaller microcrystalline size in comparison to the materials prepared using other conventional approaches. The complete synthesis of perovskite material via auto-combustion and sol–gel is represented in Fig. 4.

The sol–gel technique is described by Soltani and Entezari et al. [35] for BFO nanoparticles. The glycine-based sol–gel approach used by Cirkovi C. et al. to manufacture $BiFeO_3$ [36] particles is depicted in Fig. 5. First, a milky solution was made by dissolving bismuth nitrate (1.1 mmol) in 10.5 mL of distilled water, and a bright orange solution was produced by dissolving iron (III) nitrate (1 mmol) in 10 mL of distilled water. The complete process of synthesizing BFO via the sol–gel technique is illustrated in Fig. 5. To guarantee enough bismuth, the metal cations were combined in a 1.05:1 ratio (Bi: Fe). A vortex machine was used to mix the fluids to enhance dissolution. After mixing the two solutions and heating them to 80 °C for 15 min, 2.1 mol of glycine was added, and the mixture was stirred for a further 15 min at the same temperature. After adding 1.5 mL of 65% concentrated nitric acid, the liquid was evaporated for an hour at 150 °C on a hot plate. After completing a self-combustion process, the temperature was raised to 330 °C, producing a fluffy powder. This powder was annealed for one and a half hours at 550 °C in an oven, then obtained BFO's fine nanopowder as a result.

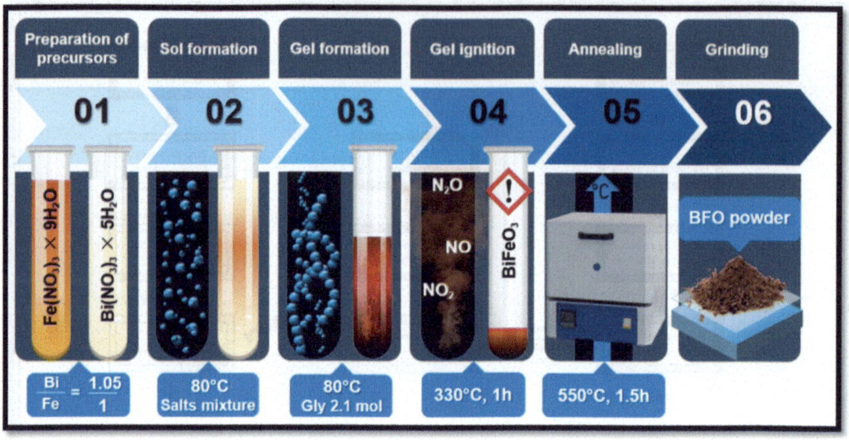

Fig. 5 Schematic illustration of sol–gel BFO nanoparticles synthesis [37]. Copyright, open access, MDPI publishing

2.2.2 Chemical Co-precipitation Techniques

Co-precipitation is a widely used method for producing metal oxide powders and nanomaterials. Precursor salts, like metal nitrates or sulfates, are dissolved in a solvent, usually water, in this process. A precipitating agent, like sodium hydroxide or ammonia, is then added to produce insoluble precipitates. To obtain the required material, the precipitates are filtered, cleaned, dried, and calcined. For getting better control over the size and composition of the particles via this method, which is also inexpensive and simple to utilize. It is often used to produce multicomponent oxides and nanoparticles. However, temperature, stirring speed, and pH all have a significant influence on the final product's quality and properties. The chemical co-precipitation method can result in incomplete and irregular $BiFeO_3$ crystals, despite its reputation for simplicity of usage and high yields [38]. Originally used to fabricate semiconductors, the co-precipitation process has significantly improved and offers helpful references for the synthesis of $BiFeO_3$.

In general, the following chemical equations can be used to depict the co-precipitation synthesis technique of $BiFeO_3$ nanoparticles:

- Formation of metal precursors:

$$Bi(NO_3)_3{\cdot}5H_2O + Fe(NO_3)_3{\cdot}9H_2O \rightarrow BiFe(NO_3)_6{\cdot}14H_2O \quad (3)$$

- Precipitation of metal oxides:

$$BiFe(NO_3)_6{\cdot}14H_2O + NaOH \rightarrow BiFeO_3\ (s) + NaNO_3 + 14H_2O\ (l) \quad (4)$$

- Calcination of metal oxides:

$$BiFeO_3\ (s) \rightarrow BiFeO_3\ (s)\ (calcined) \quad (5)$$

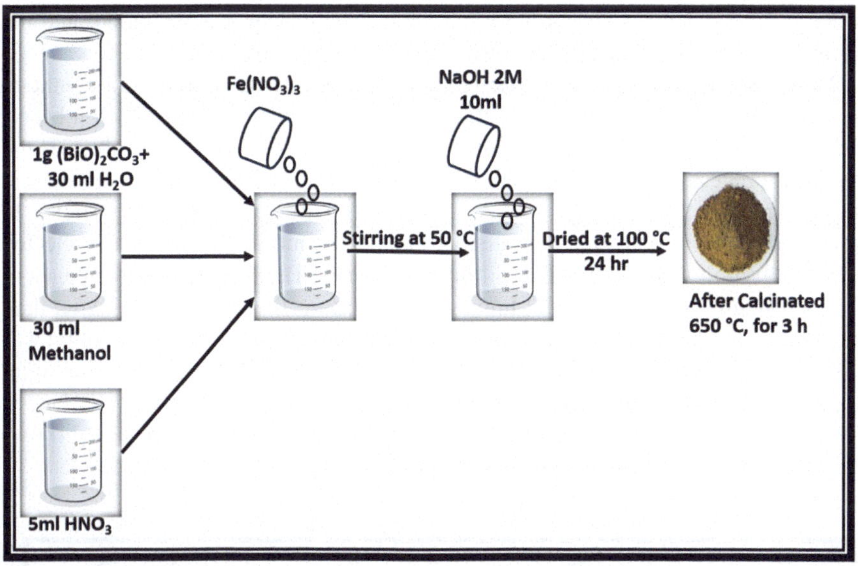

Fig. 6 Schematic diagram of co-precipitation approach to produce the BFO

High temperatures (500–650 °C) are typically used for the calcination process to eliminate any remaining water and produce crystalline $BiFeO_3$ nanoparticles.

One gram of bismuth (III) precursor—basic carbonate was mixed with 30 ml of distilled water to create a suspension, which was subsequently mixed with 30 ml of methanol. After stirring the raction for 1.5 hours, 5 mL of HNO_3 solution (ph=2) was added to dissolve the $(BiO)_2$. Furthermore, the temperature was raised to 50 °C by adding 5 g of $Fe(NO_3)_3$ to the reaction mass. Three hours were spent mixing the resultant reaction mass until it was completely homogenized, and 10 mL of NaOH was required for precipitation to occur. Figure 6 [39] illustrates the schematic diagram of the co-precipitation approach to produce BFO.

2.2.3 The Method of Auto-Combustion

A simple process, the auto-combustion approach allows for the simultaneous production of many nanoparticles. Controlling the size and shape of the generated catalysts is difficult, though, for a number of reasons [40–42]. In this method, the fuel helps the system in two ways; it forms stable complexes in addition to supplying heat. This procedure improves the creation of stable and uniform nanoparticles while preventing the precipitation of metal ions [43]. Using a glycine-assisted combustion route, clustered $BiFeO_3$ nanoparticles with a rhombohedral structure were successfully created. In this method, organic species were used as a reducing agent and NO_3^- as an oxidizing agent. Following a combustion reaction, the mixture is annealed in three steps: 350 °C for one hour, 500 °C for 6 h, and 600 °C for 1 h. After 180 min of exposure to visible light, around 80% of the oxfloxacin was degraded, and the estimated energy band gap of BFO was reported to be2.1 eV [44].

However, because of the intense reaction that occurs throughout the synthesis process, the kind and quantity of fuel, as well as the temperature at which it burns, can significantly affect the quality of the finished product [45]. The fuels utilized in the synthesis process have been compared in many reported articles. Sharmila et al. used the precursors Fe $(NO_3)_3$ $9H_2O$ and Bi $(NO_3)_3$ $5H_2O$ to create BFO nanoparticles. In a stoichiometric ratio, bismuth nitrate (0.2 M) and iron nitrate (0.2 M) were collected and dissolved in water with three milliliters of honey. The following percentages of excess bismuth were added to this homogenous mixture: 0%, 2%, and 5%. After that, the mixture was agitated for about two hours at room temperature. The solution was stirred for two hours before being dried in an oven set to 180 °C for the entire night. To produce the necessary $BiFeO_3$ nanoparticles, the dry powder was finally heated for 10 min at 450 °C and then annealed for two hours at 600 °C. Figure 7 [46] shows the entire BFO nanoparticle synthesis process.

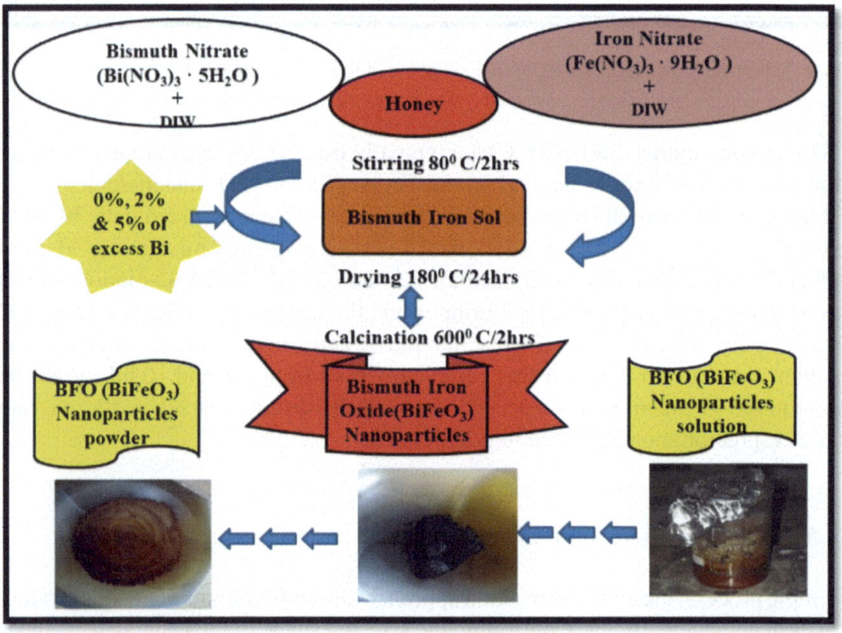

Fig. 7 Complete auto-combustion route is displayed for synthesis of $BiFeO_3$ [46]. Copyright, open access, MDPI publishing

2.2.4 Hydrothermal Technique (HT)

Hydrothermal synthesis is a popular technique for creating perovskite materials because this technique has exact control over the size, shape, and production of crystals. Using a sealed autoclave and high pressure and temperature, the hydrothermal technique produces materials in an aqueous solution. After the precursors have been dissolved in water, the mixture is heated above its boiling point to promote crystal formation in a regulated setting. With exact control over size, shape, and morphology, this technique is very effective in creating well-crystallized nanoparticles, nanorods, and other complicated structures. It is especially useful for materials that become unstable at high temperatures in open systems. But it requires expensive equipment and long processing times. Chien et al. [47] reported that the hydrothermal method is well known for creating catalysts with fully formed grains, little particle aggregation, tunable nanoparticle size, a uniform particle size distribution, morphology, and crystal structure [47, 48]. Also, this method involves reacting precursors in an aqueous solution at high temperatures and pressures, typically in a sealed autoclave. Because the hydrothermal method yields high-quality crystals with controlled stoichiometry, it is effective with a variety of perovskite materials. The advantage of hydrothermal synthesis over conventional solid-state methods is its capacity to produce materials at relatively low temperatures. Additionally, it enables the creation of nanostructured materials and the addition of dopants, both of which are essential for enhancing the functional properties of perovskites. For example, the multiferroic perovskite material Bismuth Ferrite ($BiFeO_3$) is commonly produced by hydrothermal methods.

For $BiFeO_3$, the hydrothermal synthesis route has been particularly helpful in producing nanoscale particles with multiple applications. To control pH and promote crystallization, precursors like iron and bismuth nitrate are commonly dissolved in an aqueous solution before a mineralizer like potassium hydroxide (KOH) is added. The reaction mixture is then heated in an autoclave for hours at temperatures ranging from 150 to 200 °C.

Hydrothermal synthesis is still a crucial method for developing research in electronics, renewable energy, and environmental applications since it can yield high-quality perovskite materials like $BiFeO_3$.

Because of the vessel's peculiarities, nanoparticle yields are lower and requires more time and energy to produce. Precursors based on bismuth-iron and similar precursors have a direct effect on the final product in the creation of pure $BiFeO_3$ nanoparticles. Additionally, the products possessed a 5 μm cubic microstructure made up of many tiny particles, as seen in Fig. 8 [48]. Furthermore, in comparison to alternative techniques, the produced $BiFeO_3$ nanorods showed superior cyclic stability of dye-degrading activity. Comprehensive information on the various methods utilized to manufacture $BiFeO_3$ is provided in Table 1.

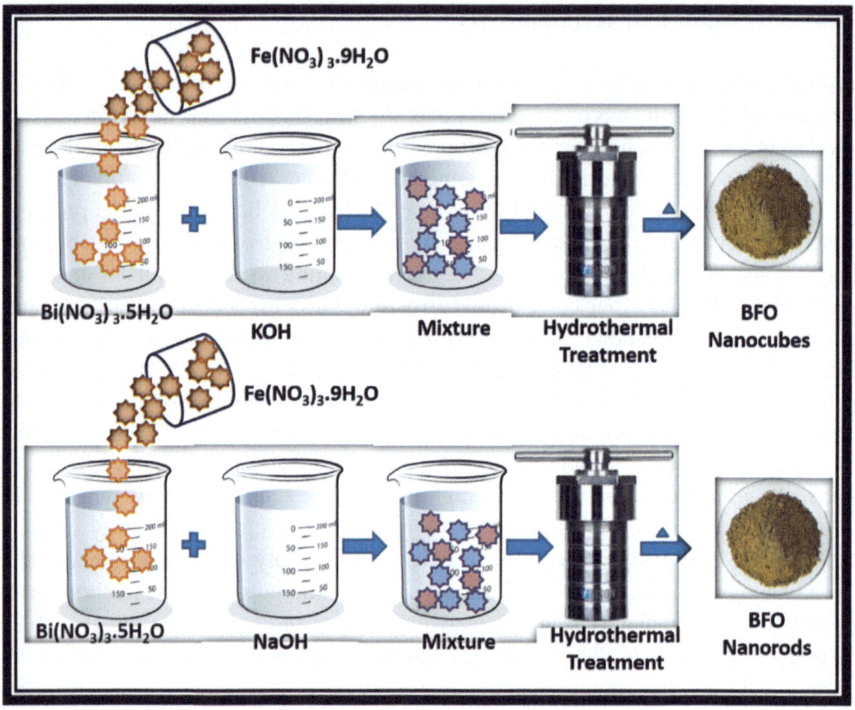

Fig. 8 Hydrothermal approach for synthesizing BFO nanocubes and nanorods

2.3 Method of Gas-State Synthesis (GT)

One crucial technique for producing nanoparticles from the bottom up is gas-state synthesis. Nanoparticles are created by joining smaller constituents, such as atoms or molecules. It makes use of equipment like lasers, plasmas, furnaces, and flames—all of which, in spite of their different designs, adhere to the same rules of reaction. By decreasing dispersion during synthesis, the method may create nanoparticles with a limited size distribution, which is one of its key benefits. Particle sizes may increase with increasing dispersions.

Thin films for a range of uses are also produced by gas-state synthesis. Numerous researchers have documented the synthesis of BFO thin films via sputtering, Chemical vapor deposition (CVD), Molecular beam epitaxy (MBE), and Pulse laser deposition (PLD) [72–77]. Chemical vapor deposition (CVD), Molecular beam epitaxy (MBE), laser ablation, DC and magnetron sputtering, thermal evaporation, and electron beam evaporation are common methods for producing thin films. For these techniques to provide high-quality samples with the desired properties, specific setups are needed.

The synthesis process can often be classified based on the temperature used. While certain materials are initially synthesized at intermediate temperatures (500–800

Table 1 Various routes for the synthesis of $BiFeO_3$

Synthesis process	Structure	References
Hydrothermal	Spherical	49
Hydrothermal	Cube	50
Hydrothermal	Polyhedral	51
Hydrothermal	–	52
Hydrothermal	Sphere	53
Co-precipitation	Rectangular	54
Co-precipitation	Irregular	55
Co-precipitation	Spherical	56
Hydrothermal	Spherical	57
Hydrothermal	Spherical	58
Hydrothermal	Spherical	59
Sol–gel	Spherical	60
Sol–gel	Spherical	61
Sol–gel	Spherical	62
Sol–gel	Spherical	63
Sol–gel	Spherical	64
Sol–gel	Spherical	65
Sol–gel	Polyhedral	66
Sol–gel	Irregular	67
Combustion	Irregular	68
Combustion	–	69
Solution combustion method		70
Solution combustion method		71

°C) and subsequently processed at higher temperatures, most materials are directly synthesized at their crystallization temperature under appropriate gas conditions. In order to enhance their qualities, some materials are also created at extremely low temperatures and then annealed at high temperatures. The applications of produced perovskite materials via GT are solar cells, semiconductor devices, capacitor dielectrics, optical and anticorrosion coatings, chemical reactor catalysts, and bio-implantable devices. The gas-state synthesis approach is anticipated to become even more important in the future due to its adaptability and the growing need for innovative materials in sectors including electronics, energy, and medicine.

2.4 Electrospinning Technique (ET)

Electrospinning is another widely used route for producing perovskite materials, particularly nanofibers. With this route, an electric field is used to pull a polymer solution comprising the constituents of the desired perovskite material into uniform nanofibers. When a high voltage is applied between the collector and the spinneret, which holds the solution, because of it, a charged jet is produced. This happens due to the electrostatic force is greater than the surface tension of the solution. As the jet travels toward the collector, the solvent evaporates, leaving behind solid nanofibers. These fibers can then go through a heat treatment known as calcination to get rid of any leftover polymer and promote the perovskite phase crystallization. For example, Bismuth Ferrite ($BiFeO_3$) can be produced by electrospinning a solution containing bismuth and iron nitrate. Calcination creates the $BiFeO_3$ phase, which yields nanofibers with enhanced multiferroic and photocatalytic properties due to its huge surface area and one-dimensional structure. Due to its many advantages for perovskite synthesis, including large surface area, scalability, and tight control over fiber morphology, electrospinning is a powerful approach for producing perovskite materials with specified properties for a range of applications [78–80]. The entire synthesis procedure for producing BFO nanofibers using electrospinning technology is depicted in Fig. 9.

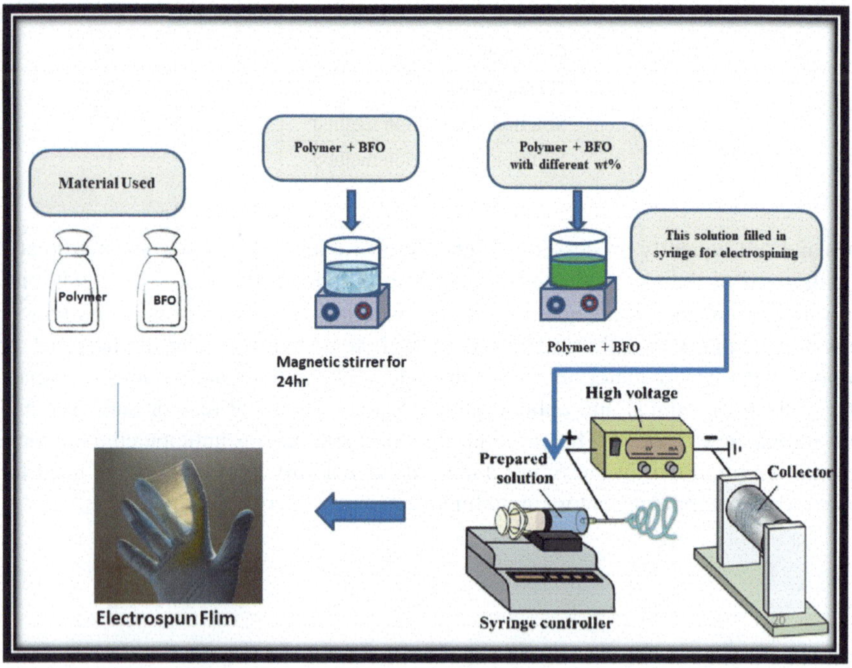

Fig. 9 Schematic diagram of BFO nanofibers synthesis through electrospinning route

3 Perovskite Material Properties

Perovskite materials are a family of substances with versatile physical and chemical properties that hold good promise for a variety of industrial and scientific applications. These characteristics are significantly influenced by their crystal structure and variable doping content, which allows them to proceed for further specific purposes. Here is a detailed analysis of their parameters given below.

3.1 Stability and Crystal Structure

Perovskite materials have good stability and capacity to embrace a wide range of cations and anions; the structure of this material enables the production of materials with a broad range of characteristics. A big cation (A) in the middle of a cubic lattice, a smaller cation (B) encircled by octahedra of anions (X), and the anions forming a three-dimensional framework make up the perovskite structure, represented by the formula ABX_3 as shown in Fig. 1.

3.1.1 Study Phase Purity by X-ray Diffraction

The phase purity of synthesized perovskite materials needs to be analyzed. Without understanding the phase and purity of the perovskite materials, no prediction can be made regarding their morphology. XRD is a practical technique for determining the phase of perovskite material. The XRD data can be used to identify the relative phase fractions of the different phases that are present in the synthesized material. The Rietveld refinement of the XRD pattern is used to find the lattice strain, unit cell volume, theoretical density, and crystallite size.

The synthesis conditions for the perovskite materials are additionally analyzed using the XRD technique. Compare the typical XRD pattern of $BiFeO_3$, which is made by the Solid State (ST), Sol–gel (SG), Hydrothermal (HT), and Electrospinning (ET) processes of the synthesized material in Fig. 10. The synthesized pure $BiFeO_3$ was structurally analyzed using the X-ray diffraction (XRD) technique. According to JCPDS card No. 71-2494 [81, 82], the diffraction peaks of the XRD patterns (012), (104), (110), (006), (202), (024), (116), (012), (018), (300), (208), (220), (131), (306), (128), (134), and (226) have been successfully used to index the rhombohedral crystal structure of perovskite with R3c space group. Table 2 displays the crystallite size and lattice characteristics of BFO nanoparticles.

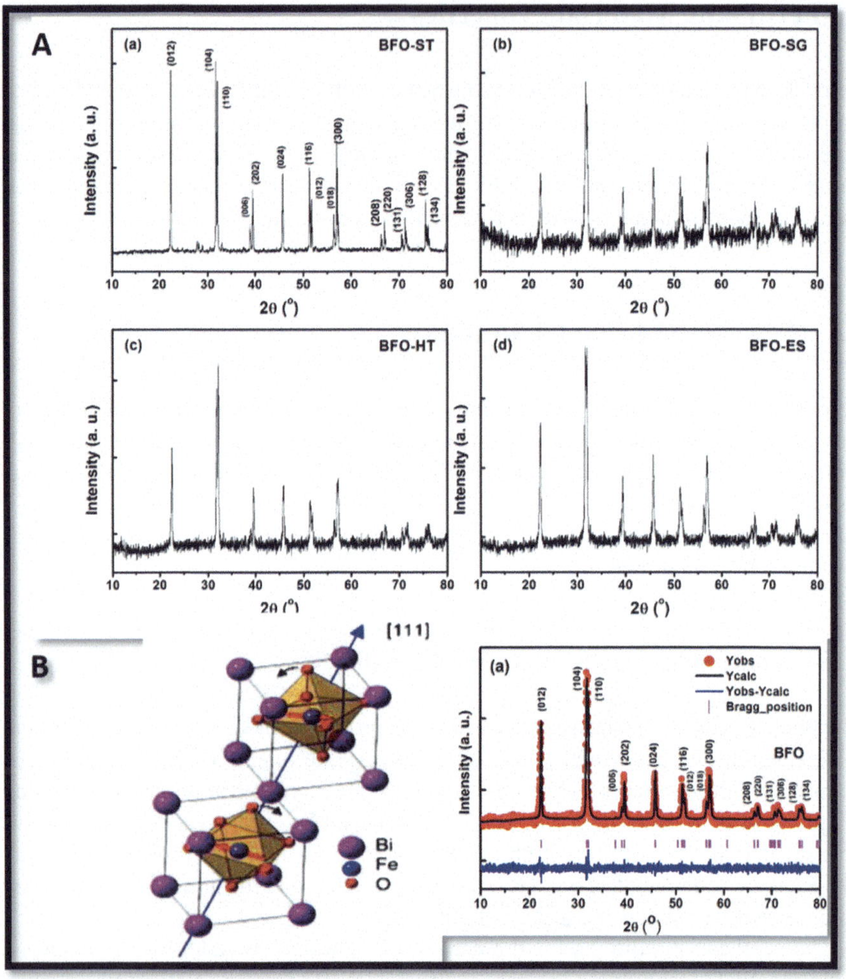

Fig. 10 XRD pattern of BFO which prepared using different techniques like solid state (ST), sol–gel (SG), hydothermal (HT), and electrospining (ES). The inset in **A** shows XRD pattern of BFO, which was synthesized using different techniques like solid state, sol–gel, electrospinning, and hydrothermal (a–d). **B** includes a crystal structure diagram with labeled elements: Bi, Fe, and O, and Rietveld refined XRD pattern of synthesized BFO Nanoparticles

3.1.2 Morphology Study of Perovskite Material

The morphology and microstructure of the perovskite materials can be investigated using scanning electron microscopy (SEM) and transmission electron microscopy (TEM), respectively. The morphology of bismuth ferrite ($BiFeO_3$) nanoparticles is commonly studied using SEM and TEM, which provide detailed information on particle size, shape, and surface characteristics. SEM image of the synthesized

Table 2 The crystallite size and lattice properties of BFO nanoparticles [83, 84]

Samples	Lattice parameters	Atom positions	*x*	*y*	*z*	*R*-factors (%)
BiFeO3	$a = 5.5771(1)$ (Å)	Bi	0.0	0.0	0.29662(9)	$R_{wp} = 4.47$
	$c = 13.8643(2)$(Å)	Fe	0.0	0.0	0.0188(1)	$R_p = 3.55$
R3c	$V = 373.465(8)$ (Å^3)	O	0.2319(24)	0.3405(21)	0.0833	$R_f = 3.92$
	Strain = 10 E-04 Crystallite size 88.7 nm					$R_{Bragg} = 2.73$

material (as shown in Fig. 11) depicts cubic, spherical, or irregular nanoparticles of 20–100 nm diameter, which completely depend on the circumstances and different synthesis techniques, that are shown in SEM pictures along with their dispersion and surface appearance. These images help reveal aggregation tendencies, which are ubiquitous due to the high surface energy of nanoscale $BiFeO_3$. Figure 11 shows several forms of bismuth ferrite compound (BFO) made using a variety of synthesis techniques, including solid state, sol–gel, hydrothermal, and electrospinning. Figure 11a shows cubic-like morphology Fig. 11b shows a particle-like shape, Fig. 11c shows an uneven sheets and particles, and Figure 11d shows a sheet-like shape with agglomerated nanoparticles.

Figure 12 shows TEM investigation demonstrating the atomic-scale internal structure and crystallinity of $BiFeO_3$ nanoparticles, which provides a more comprehensive analysis. High-resolution TEM images, which show characteristic lattice fringes that match the perovskite structure of $BiFeO_3$, are often used to validate the particles' crystalline shape.

The combined use of SEM and TEM provides a comprehensive understanding of the morphology and structural properties of $BiFeO_3$ nanoparticles, which are critical for optimizing their applications in photocatalysis, sensors, and energy devices.

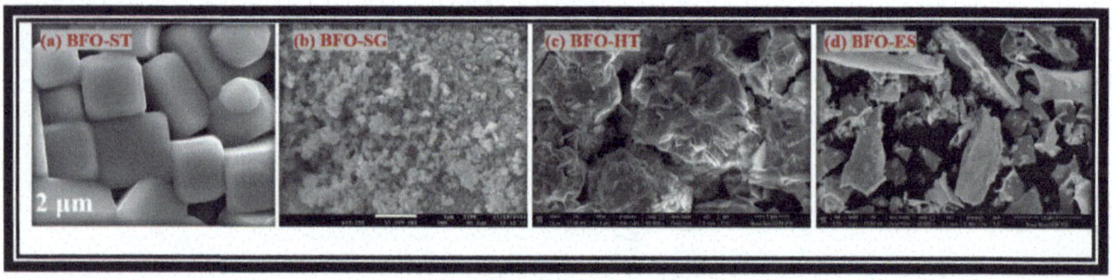

Fig. 11 SEM micrograph of BFO (**a–d**), which was prepared through different techniques like solid-state technique, sol–gel, hydrothermal, and electrospinning technique

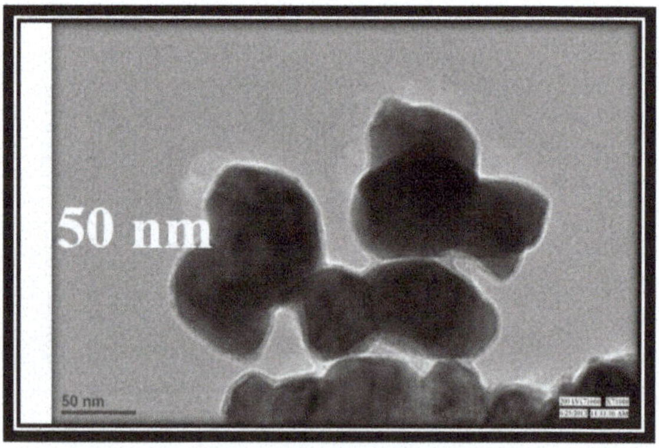

Fig. 12 TEM image of BFO nanoparticles

4 Properties of BFO

BFO is a multiferroic material that exhibits many properties like ferroelectricity, ferroelasticity, and ferro or antiferromagnetism. BFO shows Ferroelectric, Piezoelectric, Magnetic, Magnetoelectroelectric Coupling, Photovoltaic, and Photocatalysis characteristics, which have been discussed in detail below. These properties are also illustrated in Fig. 13.

4.1 Ferroelectrics

Bismuth Ferrite ($BiFeO_3$), a well-known multiferroic material, exhibits strong ferroelectric properties and a high spontaneous polarization (~ 100 μC/cm^2) along the [108] direction at room temperature. The ferroelectricity of the Bi^{3+} ions is due to the stereochemical activity of their 6s^2 lone pair electrons. $BiFeO_3$ can be utilized in non-volatile memory devices and high-performance capacitors due to its properties [74].

4.2 Piezoelectric

$BiFeO_3$ resembles significant piezoelectric activity due to its ferroelectric property, which means atomic locations within the crystal lattice vary when $BiFeO_3$ is

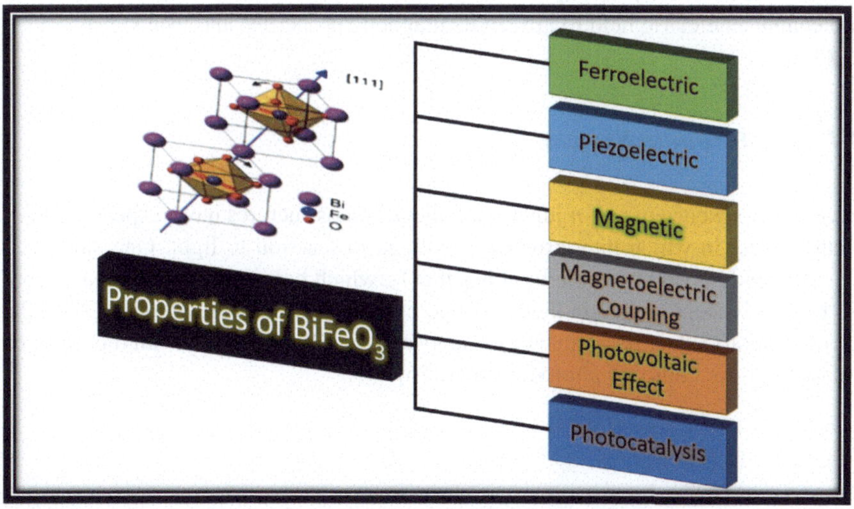

Fig. 13 Flowchart of BFO properties

subjected to a mechanical force, such as strain or pressure, which causes it to deform. Its piezoelectric coefficient (d33) ranges from 30 to 50 pC/N for bulk ceramics and can rise in nanoscale structures. Therefore, $BiFeO_3$ is attractive for usage in sensors, actuators, and energy harvesting systems [74].

4.3 Magnetism and Magnetoelectric Coupling

$BiFeO_3$ is a promising candidate for next-generation multifunctional devices, such as magnetoelectric sensors and transducers [74]. The magnetic behavior of materials such as $BiFeO_3$ is largely governed by the arrangement of Fe^{3+} spins within the crystal lattice. In these materials, the Fe^{3+} ions exhibit a specific spin alignment where adjacent spins are oriented in opposite directions. This antiparallel spin alignment is a characteristic feature of antiferromagnetic ordering. The reason for this behavior lies in the superexchange interaction between neighboring Fe^{3+} ions, which is mediated by oxygen anions. As a result of this interaction, the magnetic moments cancel each other out, leading to a net zero magnetization under ideal conditions. Therefore, the antiferromagnetic nature of the material arises because the Fe^{3+} spins are aligned so that adjacent spins are antiparallel. However, a small canting of spins leads to weak ferromagnetism because of the Dzyaloshinskii–Moriya (DM) interaction. $BiFeO_3$ is a *G*-type spin structure and a Néel temperature of about 643 K. In a G-type structure, each Fe^{3+} ion is surrounded by six nearest neighbors with oppositely aligned spins, resulting in a three-dimensional checkerboard-like arrangement of antiparallel spins. The coexistence of magnetic ordering and ferroelectricity makes BFO a promising candidate for multifunctional application, particularly in spintronic devices [85]. Due to its magnetoelectric coupling, $BiFeO_3$'s electric and magnetic orders are connected and enable an electric field to govern its magnetic properties and vice versa.

4.4 Photovoltaic System

Even in the absence of a *p–n* junction, $BiFeO_3$ nanostructures have a special photovoltaic effect in which they produce a voltage in reaction to light. One substance that efficiently absorbs visible light is $BiFeO_3$, which has a bandgap of ranges from 2.1 to 2.7 eV. Depolarization fields created by ferroelectric polarization improve the intrinsic photovoltaic performance of nanoscale thin films. $BiFeO_3$ can therefore be utilized in solar cells and photodetectors [76].

4.5 Photocatalysis

Because of their special qualities as a multiferroic material, $BiFeO_3$ nanoparticles have attracted a lot of interest in the field of photocatalysis. Under visible light, $BiFeO_3$ nanoparticles can function as photocatalysts because of their suitable band gap. The acceleration of a photoreaction in the presence of a catalyst, in which chemical processes are fueled by light energy, is known as photocatalysis. Ferroelectric polarization facilitates the small bandgap and effective charge separation of $BiFeO_3$, which are responsible for its visible light photocatalytic activity [86]. Each of these characteristics demonstrates the adaptability of $BiFeO_3$, providing a broad range of applications in electronics, energy, and the environment.

5 Applications of Perovskite Innovation

The exceptional potential of BFO materials makes them highly promising for the development of a wide range of uses with multiple and novel functionality. The researcher reported the applications of electronics, spintronics, and photonics devices [87–92]. The spontaneous polarization enhancement, photovoltaic effects, piezoelectric, switchable ferroelectric diode effects, and THz radiation properties of BFO nanomaterials have been discovered in recent studies. These properties may find application in next-generation, non-destructive memories, lead-free, actuators, spin valve devices, and ultrahigh-speed telecommunication devices [93–102]. Additionally, BFO nanoparticles have efficient visible light photocatalytic activity, indicating that they may be used as novel visible light-responsive photocatalysts for the breakdown of organic compounds. Since the antiferromagnetic and ferroelectric vectors are connected, the antiferromagnetic spins rotate when ferroelectric polarization is rotated with an external electric field. Materials research for thin film photovoltaic cells is essential for sustainable energy production because of the worldwide energy crisis. Bismuth ferrite ($BiFeO_3$) and other perovskite materials are the focus of a lot of research because of their remarkable qualities, which make them appropriate for a wide range of applications in various fields. Below is a list of some noteworthy applications.

5.1 Degradation by Photocatalysis

$BiFeO_3$ is used extensively as a photocatalyst in the environmental remediation industry [103]. Bismuth Ferrite is a promising material for photocatalytic degradation because of its qualities, which include photosensitivity at high excitation binding energy, small band gap, non-toxicity, and room temperature multiferroic features. $BiFeO_3$ is sensitive to a range of visible or solar light because of its narrow band gap energy. Furthermore, the capacity of photocatalysts for recycling is influenced by their multiferroic behavior [104–106].

5.2 Solar Energy

The problems with global warming and energy have become worse. Numerous studies have been conducted on green renewable energy [107–111]. The most fascinating area of contemporary research is the photovoltaic effect, which transforms light into electricity. The photovoltaic effect frequently involves two basic stages. First, input photons are absorbed by the device to produce an excited electron–hole pair that acts as a charge carrier. The driving force within the photoelectric device then moves the carriers to the electrode. $BiFeO_3$ is a ferroelectric material that has garnered a lot of interest lately because of its visible light photovoltaic and room temperature multiferroic properties, as well as possible applications related to them. The advantages of BFO films made by the chemical solution deposition method include their affordability and precise chemical component control. Ferroelectricity and other physical processes must be thoroughly understood to design and research highly efficient optoelectronic devices with multifunctional qualities, even though the photovoltaic capabilities of BFO-based materials have been thoroughly studied. The corresponding working mechanism of the ferroelectric polarization and piezo-phototronics in the photovoltaic La-doped BFO (BLFO)/ZnO heterojunction was reported by analyzing the energy band structure at the heterojunction. This allowed for a systematic clarification of the motivation behind the separation of photogenerated carriers. This study provides useful guidance and a plan for developing high-performance photoelectric systems by clearly demonstrating how the piezophototronic effect and ferroelectricity interact to alter the photovoltaic performance of heterojunctions [112]. The traditional sol–gel method was utilized to create both pure and doped BFO nanoparticles. It was found that doped BFO containing Cobalt had smaller crystallites than pure BFO. Co-doped BFO also enhanced its optical properties, which are connected to bandgap energy. It was shown that the bandgap energy was significantly affected by the structural change of transition metal Co-doped BFO from rhombohedral (R3c) to orthorhombic (Pn21a) symmetry. The energy bandgap of co-doped BFO is reduced from the range of 2.1–1.6 eV. Given that its reduced value is nearly equivalent to 1.5 eV, representing the maximum power conversion efficiency for a semiconductor-based photovoltaic solar cell, this composite may be applied in photovoltaic devices [113].

5.3 Treatment of Wastewater

Water is not just an essential economic resource but also an essential element of life [114]. Nonetheless, a sizable portion of the water is unsuitable for commercial or human use on surface of the earth. Furthermore, enormous volumes of wastewater are generated daily. Modern water treatment facilities may remove most of the toxins in these effluents, but the water may still include some hazardous ions and organic contaminants that are rising quickly. These impurities are removed from a solid's surface by chemisorption or physisorption. Because of their naturally high specific surface area, nanoparticles are being studied for their potential application in chemical adsorption. However, it can be challenging to remove these adsorbents from the purified liquid because of their small size. The best choices for the absorption of different substances are magnetic ferrite particles found in water sources [115–118]. Among the most widely used personal care and pharmaceutical products (PPCPs), Antibiotics are used to treat sick humans and animals. Because many antibiotics are hard to digest fully, the residues they leave behind are excreted into the environment through urine and feces. In several countries, including the US, China, and India, tetracycline was identified as the most frequently used antibiotic in animal feed. Over the past few decades, some technologies have been employed to remove and degrade tetracycline, and other researchers have thoroughly investigated this topic. The ceramic substance $BiFeO_3$ has exceptional magnetic properties, robust resistance to acids and alkalis, and the ability to effectively activate oxidants such as hydrogen peroxide. It is a good heterogeneous catalyst and a typical perovskite-type oxide (ABO_3). The physiochemical characteristics of a Co/BiFeO3 nanocomposite made using the Sol–gel method for the treatment of pharmaceutical wastewater were reported by Huixuan Li et al. [119].

6 Challenges and Future Scope

The multiferroic characteristics of BFO present both significant obstacles and exciting opportunities. Although BFO demonstrates ferromagnetism and ferroelectricity at ambient temperature, low coupling performance and the presence of antiferromagnetic configurations restrict its usefulness. The multiferroic behavior of BFO is largely dependent on its structure, particularly its distorted rhombohedral perovskite (R3c space group), whose shape and crystallinity affect its characteristics. X-ray diffraction (XRD) structural investigation shows that local strain increases with particle size. The ineffective coupling efficiency between ferroelectric and magnetic orders is a major hindrance that diminishes the usefulness of BFO for multifunctional applications, which affects both ferroelectric and magnetic properties. Furthermore, the magnetic performance of pure BFO is limited by its antiferromagnetic order, which calls for methods like doping, crystal structure modification, and composite structure creation to enhance these characteristics.

Stability and reusability are other problems that impact the long-term performance of BFO. The production of defects, such as oxygen vacancies and phase impurities, makes it challenging to achieve constant multiferroic behavior during repeated cycling. Another difficulty is achieving high-phase purity since secondary phases and imperfections like oxygen vacancies can harm performance. Optimizing crystallinity, grain size, and defect concentration can be achieved by regulating synthesis and sintering parameters, including calcination temperature and environment. Doping tactics have demonstrated the ability to get over these obstacles. By minimizing flaws and fine-tuning grain size, morphological control, especially through methods like sol–gel synthesis, is essential to improving the efficiency and quality of material. Furthermore, the microstructure and morphological differences of BFO have an impact on the efficiency constraints in its performance. The overall performance may be impacted by smaller grain or pore diameters that restrict access to perovskite sites. To overcome these obstacles and advance the practical application of BFO, synthesis methods must be optimized to grasp uniform crystallinity, improve coupling between ferroelectric and magnetic orders, and reduce defects.

7 Conclusion

This chapter provides information on the perovskite structure, including how stable it is and how it changes shape. It mainly talks about how perovskite materials can handle different kinds of cations at both the A- and B-sites. Perovskite materials have outstanding adaptability and the capacity to hold several cations, either singly or in combination, which makes them perfect for a broad range of applications across numerous sectors. This chapter talks about the different ways that perovskite materials can be made, focusing on the solid, liquid, and gas-state methods that are commonly used to make oxide powder forms of perovskite materials. These synthesis techniques greatly influence the structural, morphological, and optical properties of final materials. Bismuth ferrite ($BiFeO_3$), a perovskite material with interesting properties like multiferroicity (the presence of both ferroelectric and antiferromagnetic properties in a single phase), is one of the main examples used throughout the chapter. The production, structural characterization, and range of uses of perovskite materials are demonstrated using $BiFeO_3$ as an example. The versatility of $BiFeO_3$ perovskite material is explained by its multiferroic properties, which enable its use in solar energy and water treatment, and its photocatalytic activity, which breaks down organic contaminants and splits water to produce hydrogen. This chapter also briefly discusses characterization techniques for studying its morphology and surface. There are many uses for perovskite materials, including BFO perovskite material for energy storage and sensing devices. These materials have distinct emission characteristics, making them essential for developments in temperature sensors, LED technology, and biological research. The chapter emphasizes how the versatility and range of applications of perovskites, especially $BiFeO_3$, have made them indispensable components of numerous technological advancements.

Conflict of Interest Regarding the chapter, the authors disclose no conflicts of interest.

References

1. Roth RS (1957) Classification of Perovskite and other ABO_3-type compounds[1]. J Res Natl Bur Stand 58(2):75
2. Yashima M, Ali R (2009) Structural phase transition and octahedral tilting in the calcium titanate perovskite $CaTiO_3$. Solid State Ionics 180(2–3):120–126. https://doi.org/10.1016/j.ssi.2008.11.019
3. Weis RS, Gaylord TK (1985) Lithium niobate: summary of physical properties and crystal structure. Appl Phys A 37:191–203
4. Kwei GH, Lawson AC, Billinge SJL, Cheong SW (1993) Structures of the ferroelectric phases of barium titanate. J Phys Chem 97(10):2368–2377. https://doi.org/10.1021/j100112a043
5. Sakai H, Ishiwata S, Okuyama D, Nakao A, Nakao H, Murakami Y, Taguchi Y, Tokura Y (2010) Electron doping in the cubic perovskite $SrMnO_3$: isotropic metal versus chainlike ordering of Jahn-Teller polarons. Phys Rev B—Condensed Matter Mater Phys 82(18):180409. https://doi.org/10.1103/PhysRevB.82.180409
6. Wang Y, Zhu J, Zhang L, Yang X, Lu L, Wang X (2006) Preparation and characterization of perovskite $LaFeO_3$ nanocrystals. Mater Lett 60(13–14):1767–1770. https://doi.org/10.1016/j.matlet.2005.12.015
7. Kubicek M, Bork AH, Rupp JL (2017) Perovskite oxides–a review on a versatile material class for solar-to-fuel conversion processes. J Mater Chem A 5(24):11983–12000. https://doi.org/10.1039/C7TA00987A
8. Protesescu L, Yakunin S, Bodnarchuk MI, Krieg F, Caputo R, Hendon CH, Yang RX, Walsh A, Kovalenko MV (2015) Nanocrystals of cesium lead halide perovskites ($CsPbX_3$, X=Cl, Br, and I): novel optoelectronic materials showing bright emission with wide color gamut. Nano Lett 15(6):3692–3696. https://doi.org/10.1021/nl5048779
9. Sarmiento-Perez R, Cerqueira TF, Körbel S, Botti S, Marques MA (2015) Prediction of stable nitride perovskites. Chem Mater 27(17):5957–5963. https://doi.org/10.1021/acs.chemmater.5b02026
10. Kuhar K, Crovetto A, Pandey M, Thygesen KS, Seger B, Vesborg PC, Hansen O, Chorkendorff I, Jacobsen KW (2017) Sulfide perovskites for solar energy conversion applications: computational screening and synthesis of the selected compound $LaYS_3$. Energy Environ Sci 10(12):2579–2593. https://doi.org/10.1039/C7EE02702H
11. Kansara SB, Dhruv D, Kataria B, Thaker CM, Rayaprol S, Prajapat CL, Singh MR, Solanki PS, Kuberkar DG, Shah NA (2015) Structural, transport and magnetic properties of monovalent doped $La_{1-x}Na_xMnO_3$ manganites. Ceram Int 41(5):7162–7173. https://doi.org/10.1016/j.ceramint.2015.02.037
12. Pandey R, Kiran Pillutla R, Shankar U, Kumar Singh A (2013) Absence of tetragonal distortion in $(1-x)SrTiO_3$–xBi $(Zn_{1/2}Ti_{1/2})O_3$ solid solution. J Appl Phys 113(18). https://doi.org/10.1063/1.4804934
13. Bhalla AS, Guo R, Roy R (2000) The perovskite structure—a review of its role in ceramic science and technology. Mater Res Innovations 4(1):3–26. https://doi.org/10.1007/s100190000062
14. Saha-Dasgupta T (2020) Double perovskites with 3d and 4d/5d transition metals: compounds with promises. Mater Res Express 7(1):014003. https://doi.org/10.1088/2053-1591/ab6293
15. Rehman MU, Wang Q, Yu Y (2022) Electronic, magnetic and optical properties of double perovskite compounds: a first principle approach. Crystals 12(11):1597. https://doi.org/10.3390/cryst12111597
16. Ali Z, Ahmad I, Amin B, Maqbool M, Murtaza G, Khan I, Akhtar MJ, Ghaffor F (2011) Theoretical studies of structural and magnetic properties of cubic perovskites $PrCoO_3$ and $NdCoO_3$. Phys B: Condens Matter 406(20):3800–3804. https://doi.org/10.1016/j.physb.2011.06.065
17. Johnsson M, Lemmens P (2005) Crystallography and chemistry of perovskites. arXiv preprint cond-mat/0506606. https://doi.org/10.48550/arXiv.cond-mat/0506606

18. Goldschmidt VM (1926) Die gesetze der krystallochemie. Naturwissenschaften 14(21):477–485. https://doi.org/10.1007/BF01507527
19. Kumar D, Verma NK, Singh CB, Singh AK (2018) Evolution of structural characteristics of $Nd_{0.7}$ $Ba_{0.3}$ MnO_3 perovskite manganite as a function of crystallite size. In: AIP conference proceedings, vol 2009, no 1, Aug 2018. AIP Publishing. https://doi.org/10.1063/1.5052082
20. Duan R, Speyer RF, Alberta E, Shrout TR (2004) High Curie temperature perovskite $BiInO_3$–$PbTiO_3$ ceramics. J Mater Res 19(7):2185–2193. https://doi.org/10.1557/JMR.2004.0282
21. Kumar D, Singh AK (2019) Investigation of structural and magnetic properties of $Nd_{0.7}$ $Ba_{0.3}$ $Mn_{1\text{-}x}Ti_xO_3$ (x = 0.05, 0.15 and 0.25) manganites synthesized through a single-step process. J Magn Magn Mater 469:264–273. https://doi.org/10.1016/j.jmmm.2018.08.062
22. Atta NF, Galal A, El-Ads EH (2016) Perovskite nanomaterials–synthesis, characterization, and applications. Perovskite Mater Characterisation Prop Appl 107–151.https://doi.org/10.5772/61280
23. Porta P, De Rossi S, Faticanti M, Minelli G, Pettiti I, Lisi L, Turco M (1999) Perovskite-type oxides: I. Structural, magnetic, and morphological properties of $LaMn_{1-x}Cu_xO_3$, and $LaCo_{1-x}Cu_xO_3$ solid solutions with large surface area. J Solid State Chem 146(2):291–304. https://doi.org/10.1006/jssc.1999.8326
24. Chen M, Zhang H, Liu T, Jiang H, Chang A (2017) Preparation, structure and electrical properties of $La_{1-x}Ba_xCrO_3$ NTC ceramics. J Mater Sci: Mater Electron 28:18873–18878. https://doi.org/10.1007/s10854-017-7839-9
25. Yang J, Ma YQ, Zhang RL, Zhao BC, Ang R, Song WH, Sun YP (2005) Structural, transport, and magnetic properties in the Ti-doped manganites $LaMn_{1-x}Ti_xO_3$ ($0\leq x\leq 0.2$). Solid State Commun 136(5):268–272. https://doi.org/10.1016/j.ssc.2005.08.005
26. Kameli P, Salamati H, Heidarian A, Bahrami H (2009) Ferromagnetic insulating and reentrant spin glass behavior in Mg doped $La_{0.75}Sr_{0.25}MnO_3$ manganites. J Non-Cryst Solids 355(16–17):917–921. https://doi.org/10.1016/j.jnoncrysol.2009.04.035
27. Choudhary N, Verma MK, Sharma ND, Sharma S, Singh D (2020) Correlation between magnetic and transport properties of rare earth doped perovskite manganites $La_{0.6}R_{0.1}Ca_{0.3}MnO_3$ (R=La, Nd, Sm, Gd, and Dy) synthesized by Pechini process. Mater Chem Phys 242:122482. https://doi.org/10.1016/j.matchemphys.2019.122482
28. Ring TA (1996) Fundamentals of ceramic powder processing and synthesis. Elsevier. ISBN 0-12-588930-5
29. Lin LT, Chen JQ, Deng C, Tang L, Chen DJ, Meng JX, Cao LW (2015) Broadband near-infrared quantum-cutting by cooperative energy transfer in Yb^{3+}–Bi^{3+} co-doped $CaTiO_3$ for solar cells. J Alloy Compd 640:280–284. https://doi.org/10.1016/j.jallcom.2015.04.031
30. Pandey R, Tiwari A, Upadhyay A, Singh AK (2014) Phase coexistence and the structure of the morphotropic phase boundary region in (1–x) Bi $(Mg_{1/2}Zr_{1/2})$ O_3–$xPbTiO_3$ piezoceramics. Acta Mater 76:198–206. https://doi.org/10.1016/j.actamat.2014.05.023
31. Joy J, Krishnamoorthy A, Tanna A, Kamathe V, Nagar R, Srinivasan S (2022) Recent developments on the synthesis of nanocomposite materials via ball milling approach for energy storage applications. Appl Sci 12(18):9312. https://doi.org/10.3390/app12189312
32. Zhang L, Xu Z, Cao L, Yao X (2007) Synthesis of BF–PT perovskite powders by high-energy ball milling. Mater Lett 61(4–5):1130–1133. https://doi.org/10.1016/j.matlet.2006.06.069
33. Upadhyay A, Pandey R, Singh AK (2017) Origin of ferroelectric P-E loop in cubic compositions and structure of poled $(1–x)Bi(Mg_{1/2}Zr_{1/2})_{3–x}PbTiO_3$ piezoceramics. J Am Ceram Soc 100(4):1743–1750. https://doi.org/10.1111/jace.14735
34. Kim JK, Kim SS, Kim WJ (2005) Sol–gel synthesis and properties of multiferroic $BiFeO_3$. Mater Lett 59(29–30):4006–4009. https://doi.org/10.1016/j.matlet.2005.07.050
35. Soltani T, Entezari MH (2013) Sono-synthesis of bismuth ferrite nanoparticles with high photocatalytic activity in degradation of Rhodamine B under solar light irradiation. Chem Eng J 223:145–154. https://doi.org/10.1016/j.cej.2013.02.124
36. Ćirković J, Radojković A, Golić DL, Tasić N, Čizmić M, Branković G, Branković Z (2021) Visible-light photocatalytic degradation of mordant blue 9 by single-phase $BiFeO_3$ nanoparticles. J Environ Chem Eng 9(1):104587. https://doi.org/10.1016/j.jece.2020.104587

37. Petrukhin D, Salnikov V, Nikitin A, Sidane I, Slimani S, Alberti S, Peddis D, Omelyanchik A, Rodionova V (2024) Effect of bismuth ferrite nanoparticles on physicochemical properties of polyvinylidene fluoride-based nanocomposites. J Compos Sci 8(8):329. https://doi.org/10.3390/jcs8080329
38. Li L, Liu Y, Zhang S, Liang M, Li F, Yuan Y (2020) Enhanced mineralization of bisphenol A by eco-friendly $BiFeO_3$–MnO_2 composite: performance, mechanism and toxicity assessment. J Hazard Mater 399:122883. https://doi.org/10.1016/j.jhazmat.2020.122883
39. Berbentea A, Ciopec M, Duteanu N, Negrea A, Negrea P, Nemeş NS, Pascu B, Svera P, Ianăşi C, Duda Seiman DM, Muntean D, Boeriu E (2024) Advanced photocatalytic degradation of cytarabine from pharmaceutical wastewaters. Toxics 12(6), 405. https://doi.org/10.3390/toxics12060405
40. Farhadi S, Zaidi M (2009) Bismuth ferrite ($BiFeO_3$) nanopowder prepared by sucrose-assisted combustion method: a novel and reusable heterogeneous catalyst for acetylation of amines, alcohols and phenols under solvent-free conditions. J Mol Catal A: Chem 299(1–2):18–25. https://doi.org/10.1016/j.molcata.2008.10.013
41. Kumar A, Kumar A, Krishnan V (2020) Perovskite oxide-based materials for energy and environment-oriented photocatalysis. ACS Catal 10(17):10253–10315. https://doi.org/10.1021/acscatal.0c02947
42. Vasei HV, Masoudpanah SM, Adeli M, Aboutalebi MR (2018) Solution combustion synthesis of ZnO powders using CTAB as fuel. Ceram Int 44(7):7741–7745. https://doi.org/10.1016/j.ceramint.2018.01.202
43. Nersisyan HH, Lee JH, Ding JR, Kim KS, Manukyan KV, Mukasyan AS (2017) Combustion synthesis of zero-, one-, two-and three-dimensional nanostructures: current trends and future perspectives. Prog Energy Combust Sci 63:79–118. https://doi.org/10.1016/j.pecs.2017.07.002
44. Gupta G, Kansal SK, Umar A, Akbar S (2023) Visible-light driven excellent photocatalytic degradation of ofloxacin antibiotic using $BiFeO_3$ nanoparticles. Chemosphere 314:137611. https://doi.org/10.1016/j.chemosphere.2022.137611
45. Pourgolmohammad B, Masoudpanah SM, Aboutalebi MR (2017) Synthesis of $CoFe_2O_4$ powders with high surface area by solution combustion method: effect of fuel content and cobalt precursor. Ceram Int 43(4):3797–3803. https://doi.org/10.1016/j.ceramint.2016.12.027
46. Sharmila M, Mani RJ, Parvathiraja C, Kader SA, Siddiqui MR, Wabaidur SM, Islam MA, Lai WC (2022) Visible light photocatalyst and antibacterial activity of BFO (bismuth ferrite) nanoparticles from honey. Water 14(10):1545. https://doi.org/10.3390/w14101545
47. Chien SWC, Ng DQ, Kumar D, Lam SM, Jaffari ZH (2022) Investigating the effects of various synthesis routes on morphological, optical, photoelectrochemical and photocatalytic properties of single-phase perovskite $BiFeO_3$. J Phys Chem Solids 160:110342. https://doi.org/10.1016/j.jpcs.2021.110342
48. Rajitha B, Rao KV, Suvarna RP (2020) Synthesis of multiferroic $BiFeO_3$ microcrystals for photocatalytic activity and stability performance. Mater Today: Proc 26:126–129. https://doi.org/10.1016/j.matpr.2019.06.325
49. Chen D, Niu F, Qin L, Wang S, Zhang N, Huang Y (2017) Defective $BiFeO_3$ with surface oxygen vacancies: Facile synthesis and mechanism insight into photocatalytic performance. Sol Energy Mater Sol Cells 171:24–32. https://doi.org/10.1016/j.solmat.2017.06.021
50. Li Y, Wang XT, Zhang XQ, Li X, Wang J, Wang CW (2020) New hydrothermal synthesis strategy of nano-sized $BiFeO_3$ for high-efficient photocatalytic applications. Physica E 118:113865. https://doi.org/10.1016/j.physe.2019.113865
51. Kolivand A, Sharifnia S (2021) Enhanced photocatalytic hydrogen evolution from water splitting by Z-scheme CdS/$BiFeO_3$ heterojunction without using sacrificial agent. Int J Energy Res 45(2):2739–2752. https://doi.org/10.1002/er.5966
52. Dutta V, Sharma S, Raizada P, Hosseini-Bandegharaei A, Kaushal J, Singh P (2020) Fabrication of visible light active $BiFeO_3$/CuS/SiO_2 Z-scheme photocatalyst for efficient dye degradation. Mater Lett 270:127693. https://doi.org/10.1016/j.matlet.2020.127693

53. Si YH, Xia Y, Li YY, Shang SK, Xiong XB, Zeng XR, Zhou J (2018) Enhanced visible-active photocatalytic behaviors observed in Mn-doped $BiFeO_3$. Mod Phys Lett B 32(17):1850185. https://doi.org/10.1142/S0217984918501853
54. Chaiwichian S, Wetchakun K, Kangwansupamonkon W, Wetchakun N (2017) Novel visible-light-driven $BiFeO_3$-Bi_2WO_6 nanocomposites toward degradation of dyes. J Photochem Photobiol A 349:183–192. https://doi.org/10.1016/j.jphotochem.2017.09.034
55. Samran B, Chaiwichian S (2019) Highly enhanced photoactivity of $BiFeO_3/Bi_2WO_6$ composite films under visible light irradiation. Physica B 575:411683. https://doi.org/10.1016/j.physb.2019.411683
56. Liu Y, Wang H, Qian L, Zhao X, Yao L, Wang J, Xing X, Mo G, Chen Z, Wu Z (2020) Bismuth–iron-based precursor: preparation, phase composition, and two methods of thermal treatment. RSC Adv 10(35):20713–20723. https://doi.org/10.1039/D0RA00177E
57. Jiang J, Zou J, Anjum MN, Yan J, Huang L, Zhang Y, Chen J (2011) Synthesis and characterization of wafer-like $BiFeO_3$ with efficient catalytic activity. Solid State Sci 13(9):1779–1785. https://doi.org/10.1016/j.solidstatesciences.2011.07.008
58. Wang X, Mao W, Zhang J, Han Y, Quan C, Zhang Q, Yang T, Yang J, Li XA, Huang W (2015) Facile fabrication of highly efficient g-$C_3N_4/BiFeO_3$ nanocomposites with enhanced visible light photocatalytic activities. J Colloid Interface Sci 448:17–23. https://doi.org/10.1016/j.jcis.2015.01.090
59. Papadas IT, Subrahmanyam KS, Kanatzidis MG, Armatas GS (2015) Templated assembly of $BiFeO_3$ nanocrystals into 3D mesoporous networks for catalytic applications. Nanoscale 7(13):5737–5743. https://doi.org/10.1039/C5NR00185D
60. An J, Zhu L, Zhang Y, Tang H (2013) Efficient visible light photo-Fenton-like degradation of organic pollutants using in situ surface-modified $BiFeO_3$ as a catalyst. J Environ Sci 25(6):1213–1225. https://doi.org/10.1016/S1001-0742(12)60172-7
61. Arora M, Sati PC, Chauhan S, Singh H, Yadav KL, Chhoker S, Kumar M (2013) Structural, magnetic and optical properties of $Bi_{1-x}Dy_xFeO_3$ nanoparticles synthesized by sol–gel method. Mater Lett 96:71–73. https://doi.org/10.1016/j.matlet.2012.12.114
62. Wang X, Lin Y, Ding X, Jiang J (2011) Enhanced visible-light-response photocatalytic activity of bismuth ferrite nanoparticles. J Alloy Compd 509(23):6585–6588. https://doi.org/10.1016/j.jallcom.2011.03.074
63. Arya G, Yogiraj J, Negi NS, Shah J, Kotnala RK (2017) Observation of enhanced multiferroic, magnetoelectric and photocatalytic properties in Sm-Co codoped $BiFeO_3$ nanoparticles. J Alloy Compd 723:983–994. https://doi.org/10.1016/j.jallcom.2017.06.325
64. Guo Y, Pu Y, Cui Y, Hui C, Wan J, Cui C (2017) A simple method using citric acid as the template agent to improve photocatalytic performance of $BiFeO_3$ nanoparticles. Mater Lett 196:57–60. https://doi.org/10.1016/j.matlet.2017.03.023
65. Neogi S, Ghosh R (2020) Origin of irreversible to reversible transition in acetone detection for Y-doped $BiFeO_3$ perovskite. J Appl Phys 128(14). https://doi.org/10.1063/5.0023131
66. Bagwaiya T, Khade P, Reshi HA, Bhattacharya S, Shelke V, Kaur M, Debnath AK, Muthe KP, Gadkari SC (2018) Investigation on gas sensing properties of Ag doped $BiFeO_3$. In: AIP conference proceedings, vol 1942, no 1, April 2018. AIP Publishing. https://doi.org/10.1063/1.5028910
67. Xu H, Xu J, Wei J, Zhang Y (2020) Fast response isopropanol sensing properties with sintered $BiFeO_3$ nanocrystals. Materials 13(17):3829. https://doi.org/10.3390/ma13173829
68. Asefi N, Masoudpanah SM, Hasheminiasari M (2019) Photocatalytic performances of $BiFeO_3$ powders synthesized by solution combustion method: the role of mixed fuels. Mater Chem Phys 228:168–174. https://doi.org/10.1016/j.matchemphys.2019.02.059
69. Bhoi YP, Nayak AK, Gouda SK, Mishra BG (2018) Photocatalytic mineralization of carbendazim pesticide by a visible light active novel type-II $Bi_2S_3/BiFeO_3$ heterojunction photocatalyst. Catal Commun 114:114–119. https://doi.org/10.1016/j.catcom.2018.06.018
70. Azam A, Jawad A, Ahmed AS, Chaman M, Naqvi AH (2011) Structural, optical and transport properties of Al^{3+} doped $BiFeO_3$ nanopowder synthesized by solution combustion method. J Alloy Compd 509(6):2909–2913. https://doi.org/10.1016/j.jallcom.2010.11.153

71. Iorgu AI, Maxim F, Matei C, Ferreira LP, Ferreira P, Cruz MM, Berger D (2015) Fast synthesis of rare-earth (Pr^{3+}, Sm^{3+}, $Eu^{3+}$and Gd^{3+}) doped bismuth ferrite powders with enhanced magnetic properties. J Alloy Compd 629:62–68. https://doi.org/10.1016/j.jallcom.2014.12.108
72. Vergari M, Osmani S, Basagni A, Scattolin E, Rea M, Gasparotto A, Rizzi GA, Gavioli L (2025) Influence of surface iron content on the photoelectrochemical properties of BiFeO films deposited via SCBD. Available at SSRN 5021353. https://doi.org/10.1016/j.matchemphys.2025.130600
73. Catalan G, Scott JF (2009) Physics and applications of bismuth ferrite. Adv Mater 21(24):2463–2485. https://doi.org/10.1002/adma.200802849
74. Wang JB, Neaton JB, Zheng H, Nagarajan V, Ogale SB, Liu B, Viehland D, Vaithyanathan V, Schlom DG, Waghmare UV, Spaldin NA, Ramesh R (2003) Epitaxial $BiFeO_3$ multiferroic thin film heterostructures. Science 299(5613):1719–1722. https://doi.org/10.1126/science.1080615
75. Zhao T, Scholl A, Zavaliche F, Lee K, Barry M, Doran A, Cruz MP, Chu YH, Ederer C, Spaldin NA, Das RR, Kim DM, Baek SH, Eom CB, Ramesh R (2006) Electrical control of antiferromagnetic domains in multiferroic $BiFeO_3$ films at room temperature. Nat Mater 5(10):823–829. https://doi.org/10.1038/nmat1731
76. Browne ND (2018) Nanoscale stress-induced conducting states in functional oxides. Doctoral dissertation, Queen's University Belfast
77. Seidel J, Martin LW, He Q, Zhan Q, Chu YH, Rother A, Hawkridge ME, Maksymovych P, Yu P, Gajek M, Balke N, Kalinin SV, Gemming S, Wang F, Catalan G, Scott JF, Spaldin NA, Orenstein J, Ramesh R (2009) Conduction at domain walls in oxide multiferroics. Nat Mater 8(3):229–234. https://doi.org/10.1038/nmat2373
78. Yang YC, Liu Y, Wei JH, Pan CX, Xiong R, Shi J (2014) Electrospun nanofibers of p-type $BiFeO_3$/n-type TiO_2 hetero-junctions with enhanced visible-light photocatalytic activity. https://doi.org/10.1039/C4RA04258A
79. Yuan X, Li X, Liu Y, Li C (2017) Fabrication and property of Nd-doped $BiFeO_3$ nanofibers by electrospinning. AIP Adv 7(7). https://doi.org/10.1063/1.4986531
80. Xie SH, Li JY, Proksch R, Liu YM, Zhou YC, Liu YY, Ou Y, Lan LN, Qiao Y (2008). Nanocrystalline multiferroic $BiFeO_3$ ultrafine fibers by sol–gel based electrospinning. Appl Phys Lett 93(22). https://doi.org/10.1063/1.3040010
81. Nadeem M, Khan W, Khan S, Husain S, Singh F, Ansari A, Shukla DK, Ahad A, Chakradhary VK, Akhtar MJ (2020) Structural, optical and enhanced multiferroic properties of La/Cr co-substituted $BiFeO_3$ nanostructures. J Mater Sci: Mater Electron 31:11177–11194. https://doi.org/10.1007/s10854-020-03666-3
82. Mukherjee A, Chakrabarty S, Kumari N, Su WN, Basu S (2018) Visible-light-mediated electrocatalytic activity in reduced graphene oxide-supported bismuth ferrite. ACS Omega 3(6):5946–5957. https://doi.org/10.1021/acsomega.8b00708
83. Chauhan S, Kumar M, Yousuf A, Rathi P, Sahni M, Singh S (2021) Effect of Na/Co co-substituted on structural, magnetic, optical and photocatalytic properties of $BiFeO_3$ nanoparticles. Mater Chem Phys 263:124402. https://doi.org/10.1016/j.matchemphys.2021.124402
84. Chauhan S, Kumar M, Chhoker S, Katyal SC (2016) A comparative study on structural, vibrational, dielectric and magnetic properties of microcrystalline BiFeO3, nanocrystalline $BiFeO_3$ and core–shell structured $BiFeO_3$ @ SiO_2 nanoparticles. J Alloy Compd 666:454–467. https://doi.org/10.1016/j.jallcom.2016.01.116
85. Choi T, Lee S, Choi YJ, Kiryukhin V, Cheong SW (2009) Switchable ferroelectric diode and photovoltaic effect in $BiFeO_3$. Science 324(5923):63–66. https://doi.org/10.1126/science.1168636
86. García-Sancho C, Cecilia JA, Mérida-Robles JM, González JS, Moreno-Tost R, Infantes-Molina A, Maireles-Torres P (2018) Effect of the treatment with H_3PO_4 on the catalytic activity of Nb_2O_5 supported on Zr-doped mesoporous silica catalyst. Case study: Glycerol dehydration. Appl Catal B 221:158–168. https://doi.org/10.1016/j.apcatb.2017.09.016

87. Zhu LF, Liu Q, Zhang BP, Cen ZY, Wang K, Li JJ, Bai Y, Wang XH, Li JF (2018) Temperature independence of piezoelectric properties for high-performance $BiFeO_3$–$BaTiO_3$ lead-free piezoelectric ceramics up to 300 °C. RSC Adv 8(62):35794–35801. https://doi.org/10.1039/C8RA07553K
88. Kariya K, Yoshimura T, Ujimoto K, Fujimura N (2019) Quantitative analysis of the direct piezoelectric response of bismuth ferrite films by scanning probe microscopy. Sci Rep 9(1):19727. https://doi.org/10.1038/s41598-019-56261-w
89. Li J, Guo L, Yang Y, Rao Y, Gan G, Zhang H (2019) Bi_2O_3 adjusting equivalent permeability and permittivity of M-type barium ferrite for antenna substrate application. Mater Res Express 6(5):056113. https://doi.org/10.1088/2053-1591/ab069c
90. Fantechi E, Innocenti C, Zanardelli M, Fittipaldi M, Falvo E, Carbo M, Shullani V, Di Cesare ML, Ghelardini C, Ferretti AM, Ponti A, Sangregorio C, Ceci P (2014) A smart platform for hyperthermia application in cancer treatment: cobalt-doped ferrite nanoparticles mineralized in human ferritin cages. ACS Nano 8(5):4705–4719. https://doi.org/10.1021/nn500454n
91. Sakar M, Balakumar S, Saravanan P, Jaisankar SN (2013) Manifestation of weak ferromagnetism and photocatalytic activity in bismuth ferrite nanoparticles. In: AIP conference proceedings, vol 1512, no 1, Feb 2013, pp 228–229. American Institute of Physics. https://doi.org/10.1063/1.4790994
92. Maleki H, Haselpour M, Fathi R (2018) The effect of calcination conditions on structural and magnetic behavior of bismuth ferrite synthesized by co-precipitation method. J Mater Sci: Mater Electron 29:4320–4326. https://doi.org/10.1007/s10854-017-8379-z
93. Sakar M, Balakumar S, Saravanan P, Jaisankar SN (2013) Annealing temperature mediated physical properties of bismuth ferrite ($BiFeO_3$) nanostructures synthesized by a novel wet chemical method. Mater Res Bull 48(8):2878–2885. https://doi.org/10.1016/j.materresbull.2013.04.008
94. Gore SK, Jadhav SS, Tumberphale UB, Shaikh SM, Naushad M, Mane RS (2017) Cation distribution, magnetic properties and cubic-perovskite phase transition in bismuth-doped nickel ferrite. Solid State Sci 74:88–94. https://doi.org/10.1016/j.solidstatesciences.2017.10.009
95. Mathur P, Thakur A, Singh M (2008) Effect of nanoparticles on the magnetic properties of Mn–Zn soft ferrite. J Magn Magn Mater 320(7):1364–1369. https://doi.org/10.1016/j.jmmm.2007.11.008
96. Punia P, Naagar M, Chalia S, Dhar R, Ravelo B, Thakur P, Thakur A (2021) Recent advances in synthesis, characterization, and applications of nanoparticles for contaminated water treatment—a review. Ceram Int 47(2):1526–1550. https://doi.org/10.1016/j.ceramint.2020.09.050
97. Thakur A, Chevalier A, Mattei JL, Queffelec P (2010) Low-loss spinel nanoferrite with matching permeability and permittivity in the ultrahigh frequency range. J Appl Phys 108(1). https://doi.org/10.1063/1.3455875
98. Rana K, Thakur P, Sharma P, Tomar M, Gupta V, Thakur A (2015) Improved structural and magnetic properties of cobalt nanoferrites: influence of sintering temperature. Ceram Int 41(3):4492–4497. https://doi.org/10.1016/j.ceramint.2014.11.143
99. Thakur A, Kumar P, Thakur P, Rana K, Chevalier A, Mattei JL, Queffelec P (2016) Enhancement of magnetic properties of $Ni_{0.5}Zn_{0.5}Fe_2O_4$ nanoparticles prepared by the co-precipitation method. Ceram Int 42(9):10664–10670. https://doi.org/10.1016/j.ceramint.2016.03.173
100. Thakur P, Gahlawat N, Punia P, Kharbanda S, Ravelo B, Thakur A (2022) Cobalt nanoferrites: a review on synthesis, characterization, and applications. J Supercond Novel Magn 35(10):2639–2669. https://doi.org/10.1007/s10948-022-06334-1
101. Pathania A, Thakur P, Trukhanov AV, Trukhanov SV, Panina LV, Lüders U, Thakur A (2019) Development of tungsten doped Ni–Zn nano-ferrites with fast response and recovery time for hydrogen gas sensing application. Results Phys 15:102531. https://doi.org/10.1016/j.rinp.2019.102531

102. Chahar D, Taneja S, Thakur P, Thakur A (2020) Remarkable resistivity and improved dielectric properties of Co–Zn nanoferrites for high frequency applications. J Alloy Compd 843:155681. https://doi.org/10.1016/j.jallcom.2020.155681
103. Chakraborty S, Pal M (2018) Highly efficient novel carbon monoxide gas sensor based on bismuth ferrite nanoparticles for environmental monitoring. New J Chem 42(9):7188–7196. https://doi.org/10.1039/c8nj01237g
104. Sahni M, Mukhopadhyay S, Mehra RM, Chauhan S, Sati PC, Kumar M, Singh M, Kumar N (2021) Effect of Yb/Co co-dopants on surface chemical bonding states of $BiFeO_3$ nanoparticles with promising photocatalytic performance in dye degradation. J Phys Chem Solids 152:109926. https://doi.org/10.1016/j.jpcs.2020.109926
105. Chauhan S, Kumar M, Chhoker S, Katyal SC, Jewariya M, Suma BN, Kunte G (2015) Structural modification and enhanced magnetic properties with two phonon modes in Ca–Co codoped $BiFeO_3$ nanoparticles. Ceram Int 41(10):14306–14314. https://doi.org/10.1016/j.ceramint.2015.07.062
106. Kebede MT, Devi S, Dillu V, Chauhan S (2022) Influence of novel Cd–Ni co-substitution on structural, magnetic, optical and photocatalytic properties of $BiFeO_3$ nanoparticles. J Alloy Compd 894:162552. https://doi.org/10.1016/j.jallcom.2021.162552
107. Mathur P, Thakur A, Singh M (2008) Low temperature synthesis of $Mn_{0.4}Zn_{0.6}In_{0.5}Fe_{1.5}O_4$ nanoferrite for high-frequency applications. J Phys Chem Solids 69(1):187–192, https://doi.org/10.1016/j.jpcs.2007.08.014
108. Punia P, Dhar R, Ravelo B, Trukhanov AV, Panina LV, Thakur P, Thakur A (2021) Microstructural, optical and magnetic study of Ni–Zn nanoferrites. J Supercond Nov Magn 34(8):2131–2140. https://doi.org/10.1007/s10948-021-05967-y
109. Thakur A, Mathur P, Singh M (2007) Controlling the properties of manganese-zinc ferrites by substituting In^{3+} and Al^{3+} ions. Z Phys Chem 221(6):837–845. https://doi.org/10.1524/zpch.2007.221.6.837
110. Thakur P, Chahar D, Taneja S, Bhalla N, Thakur A (2020) A review on MnZn ferrites: synthesis, characterization and applications. Ceram Int 46(10):15740–15763. https://doi.org/10.1016/j.ceramint.2020.03.287
111. Thakur A, Thakur P, Hsu JH (2014) Structural, magnetic and electromagnetic characterization of In^{3+} substituted Mn-Zn nanoferrites. Z Phys Chem 228(6–7):663–672. https://doi.org/10.1515/zpch-2014-0477
112. Zhang Y, Yang L, Zhang Y, Ding Z, Wu M, Zhou Y, Diao C, Zheng H, Wang X, Wang ZL (2020) Enhanced photovoltaic performances of La-doped bismuth ferrite/zinc oxide heterojunction by coupling piezo-phototronic effect and ferroelectricity. ACS Nano 14(8):10723–10732. https://doi.org/10.1021/acsnano.0c05398
113. Rhaman MM, Matin MA, Hakim MA, Islam MF (2021) Bandgap tuning of samarium and cobalt co-doped bismuth ferrite nanoparticles. Mater Sci Eng B 263:114842. https://doi.org/10.1016/j.mseb.2020.114842
114. Chakraborty S, Pal M (2019) Enhanced and ultrafast dye contaminated wastewater remediation using molybdenum doped bismuth ferrite nanoparticles. Beilstein Arch 20. https://doi.org/10.3762/bxiv.2019.103.v1
115. Punia P, Bharti MK, Dhar R, Thakur P, Thakur A (2022) Recent advances in detection and removal of heavy metals from contaminated water. ChemBioEng Rev 9(4):351–369. https://doi.org/10.1002/cben.202100053
116. Mathur P, Thakur A, Singh M, Harris G (2008) Preparation and characterization of $Mn_{0.4}Ni_x Zn_{0.6-x} Fe_2O_4$ soft spinel ferrites for low and high frequency applications by citrate precursor method. Zeitschrift Für Physikalische Chemie 222(4):621–633. https://doi.org/10.1524/zpch.2008.5265
117. Mathur P, Thakur A, Lee JH, Singh M (2010) Sustained electromagnetic properties of Ni–Zn–Co nanoferrites for the high-frequency applications. Mater Lett 64(24):2738–2741. https://doi.org/10.1016/j.matlet.2010.08.056
118. Thakur A, Sharma P, Thakur P, Sharma A, Thakur A, Ravelo B, Hsu JH (2015) Effect of high spin ions on microstructural, optical, magnetic and electrical properties of hydrothermally

prepared Ni–Mg nanoferrites. Int J Mod Phys B: Condens Matter Phys; Stat Phys; Appl Phys, 29(24). https://doi.org/10.1142/S0217979215501830

119. Li J, Guo L, Yang Y, Rao Y, Gan G, Zhang H (2019) Bi_2O_3 adjusting equivalent permeability and permittivity of M-type barium ferrite for antenna substrate application. Mater Res Express 6(5):056113. https://doi.org/10.1088/2053-1591/ab069c

Chapter 3
Impact of Doping, Solid Solution, and Composite Formation Mechanisms on Sustainable Perovskite Multiferroic Materials

Rongli Gao and Wei Cai

1 Introduction

Multiferroic materials are one of new type of multifunctional materials, which have two or more orders simultaneously such as ferroelectricity, ferromagnetism, and ferroelasticity [1–3]. The spontaneous stress or deformation of ferroelasticity is usually accompanied by the appearance of ferroelectricity [4]. Therefore, multiferroic materials generally refer to magnetoelectric materials that have both ferroelectric order and ferromagnetic order coupling. In order to improve the storage capacity and speed, we need to prepare new types of magnetoelectric storage devices, and multiferroic materials have potential applications in electronic components due to their magnetoelectric coupling effect. The study of the magnetoelectric coupling performance of multiferroic materials is a broad new field in the current physical and chemical properties of materials and quantum mechanics.

At present, multiferroic materials with both ferroelectricity and ferromagnetism, which are widely studied, have a unique magnetoelectric coupling effect (MEC effect), that is, the magnetic properties can be affected when an external electric field is applied on the material. Similarly, an external magnetic field will also affect the electrical properties of the material. Ferroelectromagnetic multifunctional materials are a type of multiferroic materials, generally referring to materials that have both ferroelectricity (or antiferroelectricity) and ferromagnetism (or antiferromagnetism) within a certain temperature range [5]. In summary, multiferroic materials

R. Gao (✉) · W. Cai
School of Materials and New Source, Chongqing University of Science and Technology, Chongqing, China
e-mail: gaorongli2008@163.com

W. Cai
e-mail: caiwei_cqu@163.com

M. Kumar (ed.), *Sustainable Perovskite Multiferroic Materials*, Materials Horizons: From Nature to Nanomaterials, https://doi.org/10.1007/978-981-96-7419-0_3

refer to systems that simultaneously exhibit ferromagnetic and ferroelectric or ferroelastic orders within a certain temperature range. In recent years, some new functions such as magnetoelectric effect and magnetodielectric effect have emerged due to the coexistence and coupling of ferromagnetic and ferroelectric orders, which contain various ordered coupling mechanisms and promising broad application prospects, and have attracted much attention [6].

Single-phase multiferroic materials refer to single-phase composite materials that possess two or more properties of ferroelectric order, ferromagnetic order, and ferroelastic order under certain conditions (temperature). We generally consider materials that possess both ferromagnetic and ferroelectric properties to be single-phase multiferroic materials. In these materials, their polarization can be directly changed by applying an external magnetic field due to their intrinsic magnetoelectric coupling effect, resulting in a fast response speed and making them more suitable for memory fabrication. Therefore, the magnetoelectric coupling effect of single-phase materials has become one of the hot spots in recent years. Currently, the commonly used single-phase multiferroic materials include RMn_2O_5 (R = Yb, Tm), $RMnO_3$ (R = Dy, Gd), $BiMnO_3$, $BiCrO_3$, $BiFeO_3$, etc. [7–11]. Due to the fact that the ferroelectric Curie temperature or antiferromagnetic Neel temperature of most single-phase multiferroic materials is lower than room temperature, they can only exhibit certain magnetoelectric effects at low temperatures, but it is difficult to achieve strong magnetoelectric coupling at room temperature, which limits their practical applications. Due to the weak magnetoelectric coupling effect of single-phase multiferroic materials, strong magnetoelectric performance cannot be achieved, which makes them potentially unsuitable for various functional devices. There are also some unavoidable problems in the preparation process, and their equipment requirements are high. Therefore, although there are many single-phase multiferroic materials with magnetoelectric effects, most of them are difficult to apply. Therefore, we need to study composite multiferroic materials to solve the defects of single-phase multiferroic materials.

Composite multiferroic materials are materials prepared by combining two or more single-phase multiferroic materials together, thereby improving their magnetoelectric coupling coefficient and polarization and magnetization sensitivity. Compared with single-phase multiferroic materials, multiferroic composite materials can generate stronger magnetoelectric coupling by selecting materials with ferromagnetic and ferroelectric ordering Curie temperatures higher than room temperature, and combining the two phases in different proportions, thereby improving the magnetoelectric coupling coefficient. The magnetoelectric coupling effect refers to the ability to induce magnetic polarization under an external electric field, while inducing changes in the polarization strength and dielectric constant of materials when an external magnetic field was applied [12–14]. Therefore, this type of material is widely used in new generation sensors, information storage devices, and spintronics devices, becoming a hot topic in the research field of new functional materials [15]. And this kind of composite is based on Suchtelen's product properties proposed in 1972, which achieves greater magnetoelectric coupling performance through the combination of ferromagnetic and ferroelectric phases, and is widely used in microelectronics technology.

$BiFeO_3$ is maybe currently almost the only single-phase multiferroic material that shows both high antiferromagnetic ordering (T_N ~ 643 K) and high ferroelectric ordering (T_C ~ 1103 K), which is far above room temperature, making it a promising multiferroic material for future widespread applications. However, during the research process, it was found that $BiFeO_3$ also has many shortcomings, such as the presence of a large number of oxygen vacancies during preparation, high leakage conductivity, weak ferroelectric polarization, and its own spatially modulated spiral magnetic structure, which causes magnetic moments to cancel each other out and no magnetism to be observed macroscopically. These to some extent limit its application prospects [16]. In summary, the large leakage current, weak magnetic properties, and difficulty in synthesizing a series of identical materials seriously affect the magnetoelectric coupling performance of BFO materials, greatly hindering their application prospects [11]. Doping is one of the effective ways to improve the magnetoelectric coupling performance of BFO. The rare earth elements used for A-site doping mainly include Eu, Nd, La, Gd, Pr, Dy, etc. [17–24]. Therefore, doping with rare earth ions can reduce the leakage current of BFO and improve its ferroelectric properties. The use of rare earth doping with small ionic radii can better enhance the magnetic properties of BFO, while a large difference in ionic radii can lead to significant lattice distortion, thereby enhancing its magnetic properties, but the effect on magnetic enhancement is limited [25–27]. The transition group elements used for B-site doping mainly include Ni, Cr, Zn, Mn, Co, etc. [28–30]. B-site doping can also disrupt the antiferromagnetic helical structure of BFO, thereby enhancing its magnetism. For example, Xie Yan et al. reduced the leakage current density of BFO samples through Ti doping, and the residual polarization intensity increased with the increase in Ti doping content [31]. Huang Zhengbei et al. effectively increased the magnetization of BFO ceramics by doping Sm element and found that doping La can effectively suppress grain growth [32].

$BaTiO_3$ is the most commonly used perovskite material, which has high dielectric constant and low dielectric loss, as well as good chemical stability and ferroelectricity, and has a wide range of applications in electronic components. Therefore, $BaTiO_3$ is known as the pillar of the electronic ceramic industry [33]. The dielectric constant of $BaTiO_3$ is usually high (~ 2000 or above at 1 k Hz); therefore, $BaTiO_3$-based ceramics or films are widely used in the manufacture of electronic components such as sensor, transducer, spintronic devices, multilayer ceramic capacitors (MLCC), and so forth [34–36]. At different temperatures, the structure of barium titanate undergoes significant changes, resulting in varying properties exhibited. When the temperature is larger than 130 °C, the crystal structure of barium titanate ceramics undergo a transition from the ferroelectric phase to the paraelectric phase, with a peak appearing on the dielectric constant curve. Above the Curie temperature, the structure of barium titanate is cubic. The temperature of the tetragonal phase remains stable between 10 and 130 °C. In the tetragonal phase, the ratio of lattice constant (c/a) decreases with the decrease in particle size [36]. Therefore, if the grain size can be changed by doping, sintering parameters and by controlling the preparation method, the dielectric and ferroelectric properties of ceramics can be altered. Research has found that the dielectric constant increases with decreasing grain size [37]. In recent years, due to the higher chemical stability of Zr^{4+}compared to Ti^{4+}, Ba (ZrxTi1-x)O_3 (BZT) has gained great attentions because it shows excellent dielectric properties and piezoelectric property. Usually, the temperature stability has also be improved. As a consequent, it has been chosen to make ceramic capacitors [38]. In addition, the substitution of zirconium at the Ti position is an effective method to reduce the Curie temperature, which exhibits some interesting characteristics in the dielectric properties of $BaTiO_3$ ceramics [39]. As is well known, at the Curie temperature TC, the properties of BZT ferroelectric phase transition undergo drastic changes with increasing Zr content. At higher Zr content, BZT ceramics exhibit a wider dielectric constant temperature curve near TC, which is due to the uneven distribution of Zr ions at the Ti position and mechanical stress in the grains [38].

Spinel-structured ferrites have increasingly important scientific value in the microwave industry due to their low cost, ease of preparation, and excellent electrical and magnetic properties. Among various ferrite materials, spinel-type cobalt ferrite $CoFe_2O_4$ is a well-known hard magnetic material, which has been extensively studied due to its high coercivity, moderate saturation magnetization, significant chemical stability and mechanical hardness [40, 41], as well as the largest magnetostriction among spinel ferrites. Combining with ferroelectricity produces magnetoelectric effects [42]. Adding a small amount of foreign ions such as lithium ions [43], aluminum ions [44], and lanthanum ions [45] to cobalt ferrite can improve its magnetoelectric properties.

These materials are very suitable for use in magnetoelastic sensors, actuators, and recently, as magnetoelastic components in composite multiferroics. The replacement of nonmagnetic ions at any position can alter the interaction between a–a or b–b and a–b, which may lead to significant changes in their physical properties. Therefore, by doping nonmagnetic ions into cobalt ferrite, its properties, including electrical and magnetic properties, can be adjusted or controlled. Numerous experimental studies have shown that partially replacing cobalt with other ions (such as zinc and cadmium) in cobalt ferrite can increase the saturation magnetization to a certain extent. Scholars have shown that doping Zn can greatly improve its magnetoelectric properties [46].

Due to the good ferroelectricity of $BiFeO_3$, high dielectric constant and low dielectric loss of $BaTiO_3$, and good ferromagnetism of $CoFe_2O_4$, doping can theoretically better compensate for its own defects. Therefore, we will investigate the magnetoelectric properties of $BiFeO_3$, $BaTiO_3$, and $CoFe_2O_4$ materials by combining them according to different compositions. Previous studies [47] reported the introduction of $CoFe_2O_4$ into $0.7BiFeO_3$–$0.3BaTiO_3$ to form composite materials for increased magnetization. Related studies have shown that $0.7BiFeO_3$–$0.3BaTiO_3$–0.3wt% MnO_2 (Mn BF BT) ceramics exhibit excellent ferroelectric properties [48]. Yang et al. found that BFO-BTO and YIG phases can coexist in composite materials without any second phase.

With the continuous development and progress of scientific research, the preparation methods of composite ceramics are becoming increasingly advanced, the processes are becoming more mature, and the equipment requirements are constantly increasing. Among them, solid-phase method, hydrothermal method, and sol–gel method are more mature methods for preparing composite ceramics. However, different preparation methods have their own advantages and disadvantages. We need to combine the properties of the materials themselves to find the most suitable preparation scheme, which not only reduces the preparation cost but also seeks a simple method to ensure the superiority of their magnetoelectric properties.

Due to their excellent magnetoelectric coupling performance at room temperature and simultaneous ferroelectricity and ferromagnetism, multiferroic materials have been increasingly used in our daily lives. With the continuous growth of information, the storage capacity of information is facing challenges. We can use the magnetoelectric coupling performance of multiferroic materials to prepare new types of magnetoelectric storage devices, which is expected to improve the current storage situation.

This paper uses solid-phase method to prepare bismuth ferrite–cobalt ferrite–barium titanate composite ceramics. By adjusting different sintering temperatures and composition examples, the relationship between microstructure and composition and the relationship between microstructure and magnetoelectric coupling performance are explored, laying a solid foundation for the research of new information storage devices. The main research contents are as follows: First of all, $Bi_{0.9}Sm_{0.1}Fe_{0.85}Ti_{0.15}O_3$ composite particles were prepared by solid-phase method. Secondly, $BaTi_{0.85}Zr_{0.15}O_3$ and $Co_{0.7}Zn_{0.3}Fe_2O_4$ particles were prepared by solid-phase method. Subsequently, $Bi_{0.9}Sm_{0.1}Fe_{0.85}Ti_{0.15}O_3$), $Co_{0.7}Zn_{0.3}Fe_2O_4$, and $BaTi_{0.85}Zr_{0.15}O_3$ particles were mixed in the formation of $(1 - x)(Bi_{0.9}Sm_{0.1}Fe_{0.85}Ti_{0.15}O_3)$—$x\,(0.2Co_{0.7}Zn_{0.3}Fe_2O_4$–$0.8BaTi_{0.85}Zr_{0.15}O_3)$ with $x =$ 0.5, 0.6, 0.7, 0.8, 0.9. Finally, the corresponding composite ceramics were sintered. The effect of different compositions on the properties was investigated.

2 Experimental Process

2.1 Raw Materials

All the main raw materials used in this paper were purchased from Chengdu Kelong, China. Bi_2O_3(99.5%a), Sm(99.5%a), Fe_2O_3(98.5%), La_2O_3(99.5%), ZrO_2(99.5%), ZnO(99%), TiO_2(98%), CoO(98.5%), and BaO(99%) were used without further purification.

2.2 Preparation Process

(1) Synthesis of $Bi_{0.9}Sm_{0.1}Fe_{0.85}Ti_{1.5}O_3$ Powders

Bi_2O_3, Sm_2O_3, Fe_2O_3, and TiO_2 were weighted according to different molar concentrations of $1 - x$ ($Bi_{0.9}Sm_{0.1}Fe_{0.85}Ti_{0.15}O_3$) ($x = 0.5$, 0.6, 0.7, 0.8, 0.9). The raw materials were added into a ball mill for mixing and thorough grinding (a small amount of alcohol or distilled water was added into the mortar and grind for more than 1 h; ball mill for 12 h). The powder obtained by grinding is sintered in a box furnace or tube furnace at a temperature of 3 °C per minute to 700 °C for 8 h. The sintered powder was milled again by using the same grinding process as mentioned above. The powder obtained by grinding is sintered in a box furnace or tube furnace at a temperature of 3 °C per minute to 800 °C for 4 h. The sintered powder was milled again by using the same grinding process as the above step.

(2) Preparation Process of $Co_{0.7}Zn_{0.3}Fe_2O_4$

The raw materials, including CoO, ZnO, and Fe_2O_3, were weighed according to the molar concentrations of $BaTi_{0.85}Zr_{0.15}O_3$. The raw materials were put into a mortar (or ball mill if of high quality) for mixing and thorough grinding (add a small amount of alcohol or distilled water to the mortar and grind for more than 1 h; ball mill for 12 h). The powder obtained by grinding is sintered in a box furnace or tube furnace at a temperature of 3 °C per minute to 800 °C for 8 h. Grind the sintered powder again using the same grinding process as previous step mentioned. The powder obtained by grinding is sintered in a box furnace or tube furnace at a temperature of 3 °C per minute to 900 °C for 4 h. Grind the sintered powder again using the same grinding process as the above step.

(3) Preparation Process of $0.8BaTi_{0.85}Zr_{0.15}O_3$ Materials

BaO, TiO_2, and ZrO_2 were weighed according to the molar concentrations of $BaTi_{0.85}Zr_{0.15}O_3$. The raw materials were put into a mortar (or ball mill if of high quality) for mixing and thorough grinding (add a small amount of alcohol or distilled water to the mortar and grind for more than 1 h; ball mill for 12 h). The powder obtained by grinding is sintered in a box furnace or tube furnace at a temperature of 3 °C per minute to 1000 °C for 8 h. Grind the sintered powder again using the same grinding process as step two. The powder obtained by grinding is sintered in a box furnace or tube furnace at a temperature of 3 °C per minute to 1100 °C for 4 h. Grind the sintered powder again using the same grinding process as step two.

(4) Preparation Process of $(1 - x)$ $(Bi_{0.9}Sm_{0.1}Fe_{0.85}Ti_{0.15}O_3)$—$x$ $(0.2Co_{0.7}Zn_{0.3}Fe_2O_4$–$0.8BaTi_{0.85}Zr_{0.15}O_3)$ Magnetoelectric Composite Ceramics

First, mix and grind the required $Bi_{0.9}Sm_{0.1}Fe_{0.85}Ti_{1.5}O_3$, $Co_{0.7}Zn_{0.3}Fe_2O_4$, and $BaTi_{0.85}Zr_{0.15}O_3$ powders. Add an appropriate amount of liquid paraffin or PVA (usually 5–7% by mass) to the ground powder and stir. Then, place it in an oven at a temperature of 80 °C and keep it warm for 4–5 min. After that, take it out and stir for 2–3 min. Repeat this process 4–5 times. Weigh 6–8% polyvinyl alcohol and slowly add the dried powder and stir. Stir for 2–5 min and place the powder in an oven at 80–100 °C (maximum temperature not exceeding 100 °C) for 5 min. Repeat this step until the agglomerated powder particles are uniformly sized and easily dispersed. Weigh 0.4–0.5 g (depending on the type of material, with a tablet thickness of about 1 mm) of powder and evenly spread it in the mold to ensure that the strength of the pressed green body is basically consistent at all points. Apply pressure as slowly as possible before the pressure gauge moves, so that the gas in the powder is expelled. After running the meter, apply pressure within the range of 10–15 Mpa and hold for 10 min. When demolding, the bottom mold cannot be quickly removed to avoid rapid air influx and cracking of the green body. When demolding, do not rotate the mold to remove it, as it may cause scratches or deformation inside the mold. Excluding organic compounds (such as polyvinyl alcohol) in the ceramic green body, the highest extrusion temperature selected is 500 °C due to the boiling point of liquid paraffin being around 320 °C. However, since the formed ceramic pieces are relatively dense, the extrusion time can be appropriately extended during the extrusion process, that is, the insulation time can be increased during the insulation process. Raise the temperature to 100 °C in a box furnace for 1 h, hold for 1 h, and then raise the temperature to 400 °C after 15 h, hold for 5 h. After another 5 h, heat up to1200 °C, hold for 5 h, and then cool naturally to room temperature.

In order to investigate the crystal structures, X-ray diffraction (XRD, D/max 2500, Rigaku, Japan) was used. The radiation source is Cu target, and the wavelength λ of CuK_{α} is 0.154081 nm. The sweeping speed is 0.1 degree per second, and the scanning step is 0.02°. To characterize the surface image and the grain size as well as the relative density, electron microscope (FE-SEM, JSM-7800F, JEOL, Japan) was used. Before measured, the surface of the prepared ceramics was polished, and then, it was sintered at 600 °C in order to make the grain boundary more obvious. The dielectric constant and dielectric loss were determined by using a LCR instrument analyzer (HP4980A, Agilent, USA). The dielectric constant was calculated according to the following equation:

$$\varepsilon_r = \frac{Cd}{\varepsilon_0 S} \tag{1}$$

Here, ε_r, C, d, and ε_0 are the relative dielectric constant, the capacitance (F), the thickness (m) of the ceramics, and the vacuum dielectric constant (8.85×10^{-12} F/m), and S is the effective electrode area on the sample. In this chapter, ferroelectric system (TF2000, aix–ACCT Inc., Germany) was used to measure the ferroelectric hysteresis loops, from which the ferroelectric parameters such as remanent polarization (P_r) and coercive electric field (E_c) can be obtained. Magnetic properties were carried out by using a vibrating sample magnetometer (VSM). The magnetoelectric (ME) coupling effect was measured according to the P-E curves with and without applying external magnetic field.

3 Results and Discussions

3.1 Crystal Structure

Figure 1 shows the XRD patterns of (1 − x) ($Bi_{0.9}Sm_{0.1}Fe_{0.85}Ti_{0.15}O_3$)—$x$ ($0.2Co_{0.7}Zn_{0.3}Fe_2O_4$–$0.8BaTi_{0.85}Zr_{0.15}O_3$) magnetoelectric composite ceramics with different compositions (x = 0.5, 0.6, 0.7, 0.8, 0.9). From the figure shown above, it can be seen that compared with the XRD phase standard card of bismuth ferrite, cobalt ferrite, and barium titanate, the basic characteristic peaks can roughly match. From the diffraction peaks of all samples, it can be seen that the crystallinity of all samples tends to be good, and all diffraction peaks are aligned with the standard card. As x increases, the ferroelectric peak $Bi_{0.9}Sm_{0.1}Fe_{0.85}Ti_{0.15}O_3$ will gradually decrease, while the ferromagnetic peak $0.2Co_{0.7}Zn_{0.3}Fe_2O_4$–$0.8BaTi_{0.85}Zr_{0.15}O_3$ will increase. When $x = 0.5$ and $x = 0.9$, heterogeneous phases $Ba_2Ti_9O_{20}$ and Ti_5O_9 will be generated, respectively. This may be due to the abundance of ferroelectric and ferromagnetic phases, uneven mixing, and insufficient reaction. In addition, because of the high temperature, some ions such as Ba^{+2}, Bi^{+3}, Ti^{+2}, Fe^{+3}, and other ions will move and intermigration among $Bi_{0.9}Sm_{0.1}Fe_{0.85}Ti_{0.15}O_3$, $Co_{0.7}Zn_{0.3}Fe_2O_4$ and $0.8BaTi_{0.85}Zr_{0.15}O_3$. Therefore, some impurities may be formed.

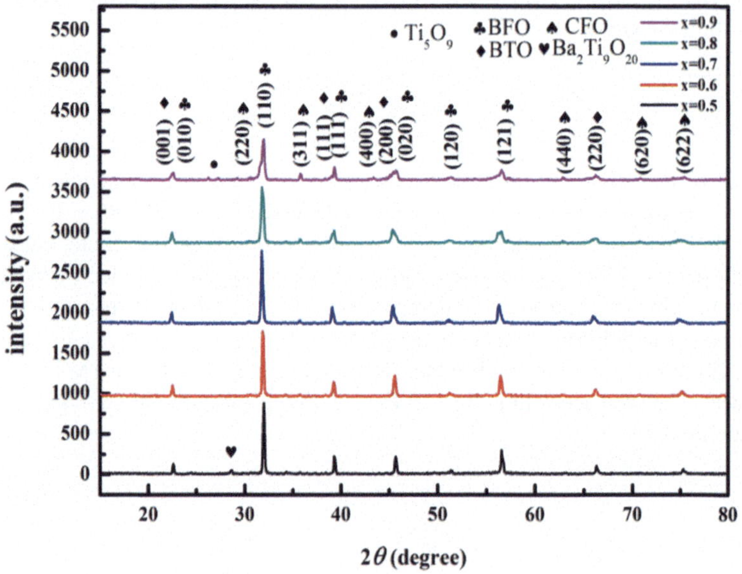

Fig. 1 XRD patterns of (1 − x) ($Bi_{0.9}Sm_{0.1}Fe_{0.85}Ti_{0.15}O_3$)—$x$ ($0.2Co_{0.7}Zn_{0.3}Fe_2O_4$–$0.8BaTi_{0.85}Zr_{0.15}O_3$) magnetoelectric composite ceramics with different x

3.2 Surface Morphology

Figure 2 shows SEM images of composite ceramics with different compositions (x = 0.5, 0.6, 0.7, 0.8, 0.9). From the graph, it can be seen that as x increases, the surface pores of the sample gradually increase, while the average grain size first increases and then decreases. The reason is that in the sintering process, some PVA was volatilized, and the pores will be caused. When $x = 0.9$, not only is the average grain size small, but there are also many pores on the surface of the sample, resulting in an increase in leakage current and poor ferroelectric performance.

3.3 Dielectric Properties

Figure 3a shows the dielectric constant as a functions of frequency of the composite ceramics with different compositions (x = 0.5, 0.6, 0.7, 0.8, 0.9). From the graph, it can be seen that as x increases, the dielectric constant gradually increases. This may be because as x increases, the grain size decreases and there are more grain boundaries, resulting in increased polarization and dielectric constant. When $x = 0.9$, the dielectric constant is highest at low frequencies and lowest at high frequencies. This may be because when $x = 0.9$, there are more micropores on the sample surface, which affects its electron displacement polarization at high frequencies, resulting in the lowest dielectric constant. The sample with $x = 0.8$ also has more micropores on its surface, so the dielectric constant of $x = 0.8$ is relatively small.

Figure 3b is the frequency dependence of dielectric loss of composite ceramics with different compositions (x = 0.5, 0.6, 0.7, 0.8, 0.9). From the graph, it can be seen that as x increases, the dielectric loss also gradually increases. When $x = 0.9$, its dielectric loss is abnormally high. This may be because as x increases, the surface pores of the sample increase and the grain size decreases, resulting in more impurities and defects, causing heat generation and energy loss during the polarization process.

Figure 4a and b show the dielectric temperature curves measured at frequencies of 100 Hz and 2000 kHz, respectively. Comparing the two graphs, it can be seen that as the frequency increases, its dielectric constant decreases. As the temperature increases, the dielectric constant of the sample gradually increases. This is because the degree of noncoincidence between the positive and negative charge centers of the original unit cell increases with temperature, that is, the spontaneous polarization intensity gradually increases. According to the relevant derivation formulas of ferroelectric phase transition physics, the dielectric constant gradually increases with temperature. However, the temperature curve has not reached its highest point, so its Curie temperature cannot be determined.

Figure 5 illustrates temperature dependence of dielectric constant of the prepared composite ceramics with different compositions. From the dielectric temperature diagrams of different composition examples, it can be seen that the samples will exhibit dielectric loss peaks within a certain temperature range. This is because the

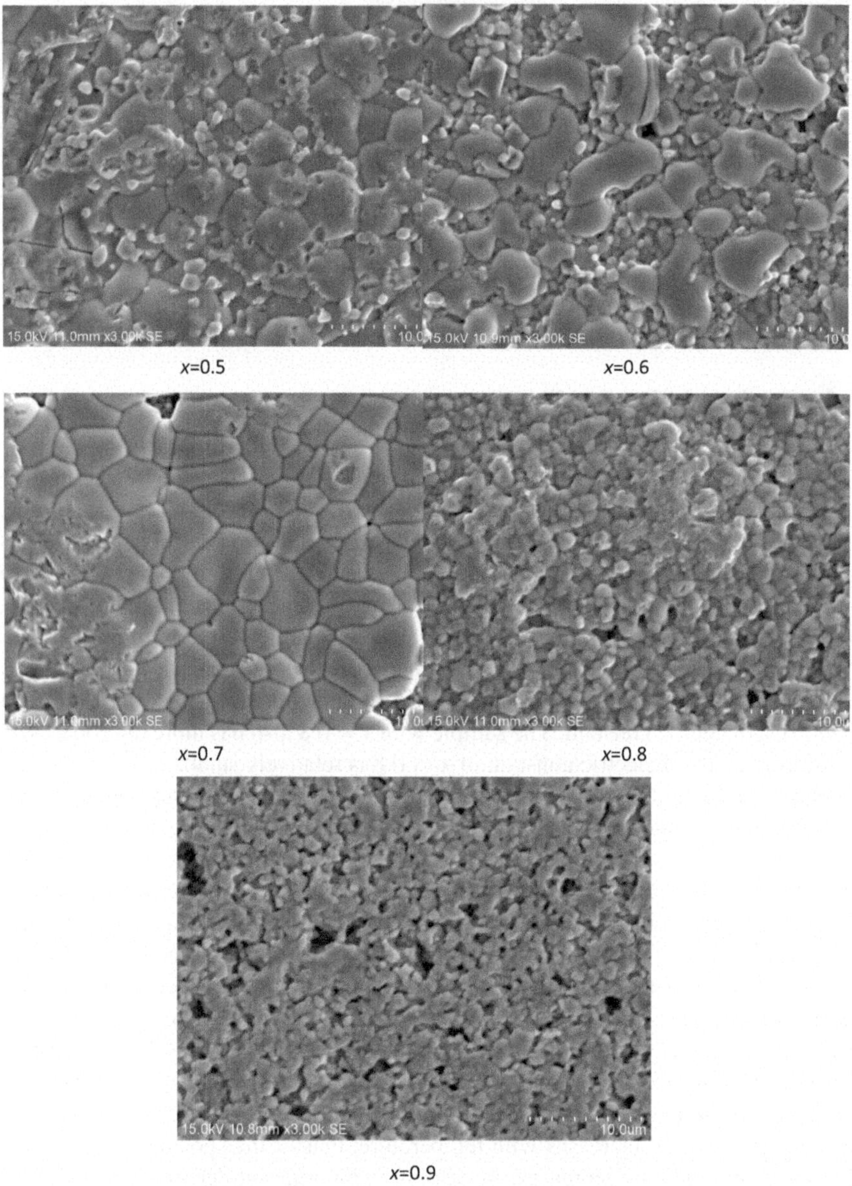

Fig. 2 SEM images of (1 − x) ($Bi_{0.9}Sm_{0.1}Fe_{0.85}Ti_{0.15}O_3$)—$x$ ($0.2Co_{0.7}Zn_{0.3}Fe_2O_4$–$0.8BaTi_{0.85}Zr_{0.15}O_3$) magnetoelectric composite ceramics with different compositions: $x = 0.5$, $x = 0.6$, $x = 0.7$, $x = 0.8$, $x = 0.9$

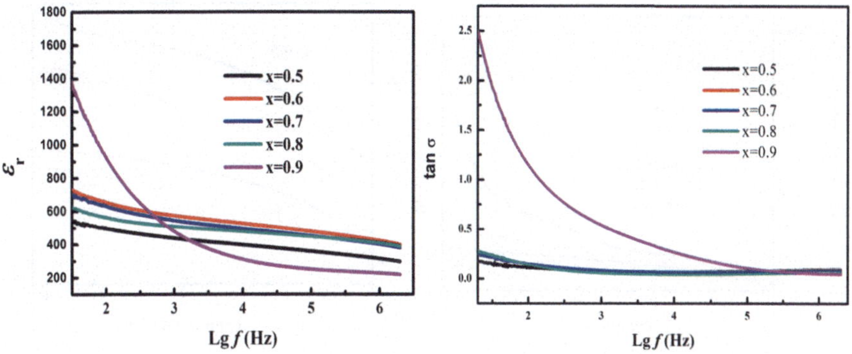

Fig. 3 Dielectric constant (**a**) and loss (**b**) as a function of frequency of $(1-x)$ $(Bi_{0.9}Sm_{0.1}Fe_{0.85}Ti_{0.15}O_3)$—$x$ $(0.2Co_{0.7}Zn_{0.3}Fe_2O_4–0.8BaTi_{0.85}Zr_{0.15}O_3)$ magnetoelectric composite ceramics with different compositions

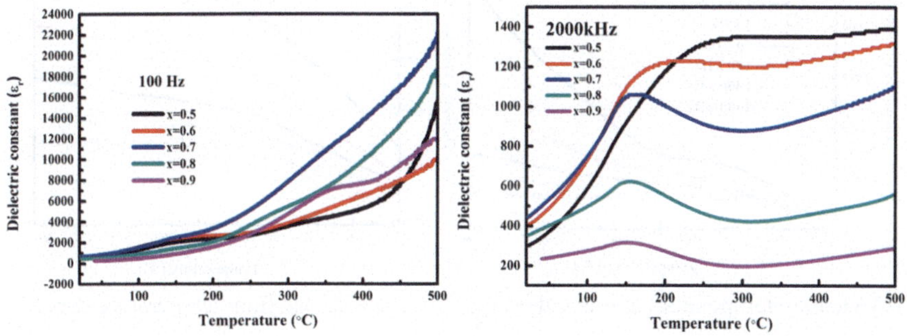

Fig. 4 Dielectric temperature of $(1-x)$ $(Bi_{0.9}Sm_{0.1}Fe_{0.85}Ti_{0.15}O_3)$—$x$ $(0.2Co_{0.7}Zn_{0.3}Fe_2O_4–0.8BaTi_{0.85}Zr_{0.15}O_3)$ magnetoelectric composite ceramics with different compositions at 100 Hz (**a**) and 2000 kHz (**b**)

polarization time of several polarization mechanisms is different, and when they cannot keep up with the changes in the electric field, relaxation polarization will occur, leading to the appearance of relaxation peaks. This relaxation peak is mainly caused by the charge, which may be due to the increase in grain boundaries caused by the small grain size, resulting in an increase in charge.

Figure 6 show the dielectric loss diagrams of different composition examples at different frequencies. From these graphs, it can be seen that the loss of the sample increases with temperature, and the higher the frequency, the greater the loss. This is because as the temperature of the dielectric increases, the heat generation also increases, resulting in increased losses. However, the dielectric does not experience severe heat generation at high frequencies, so the losses at high frequencies are reduced. From the graph dielectric loss of different samples measured at a frequency of 10 kHz, it can be seen that as x increases, the loss first decreases and then increases.

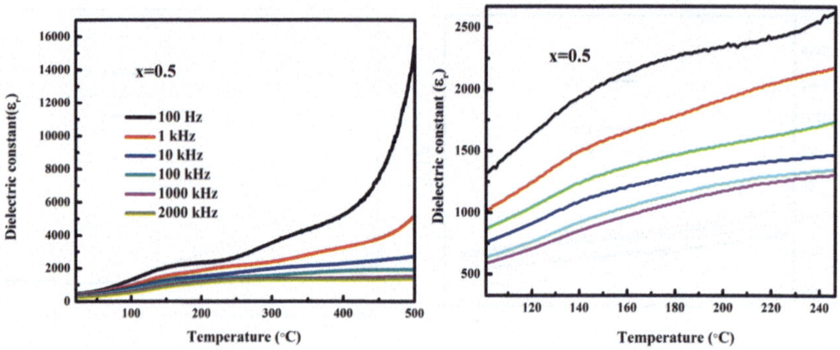

(a) Medium temperature curve with x=0.5; (b) Enlarged Medium temperature curve with x=0.5

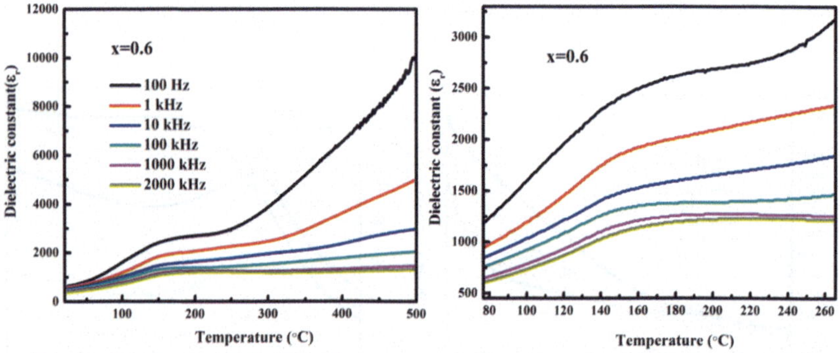

(c) Medium temperature curve with x=0.6; (d) Enlarged Medium temperature curve with x=0.6

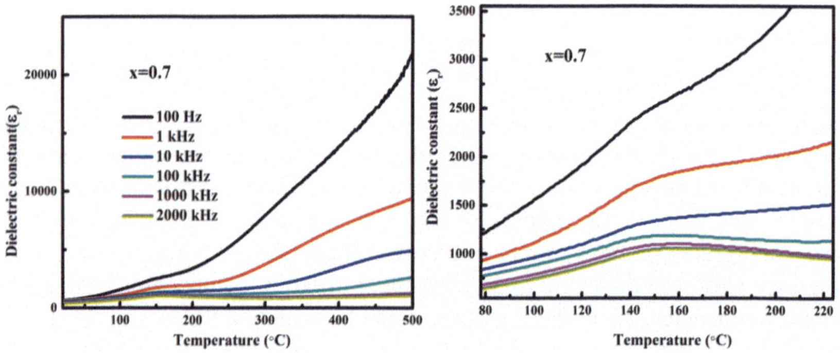

(e) Medium temperature curve with x=0.7; (f) Enlarged Medium temperature curve with x=0.7

Fig. 5 Temperature dependence of dielectric constant of the prepared $(1-x)$ $(Bi_{0.9}Sm_{0.1}Fe_{0.85}Ti_{0.15}O_3)$—$x$ $(0.2Co_{0.7}Zn_{0.3}Fe_2O_4–0.8BaTi_{0.85}Zr_{0.15}O_3)$ magnetoelectric composite ceramics with different compositions

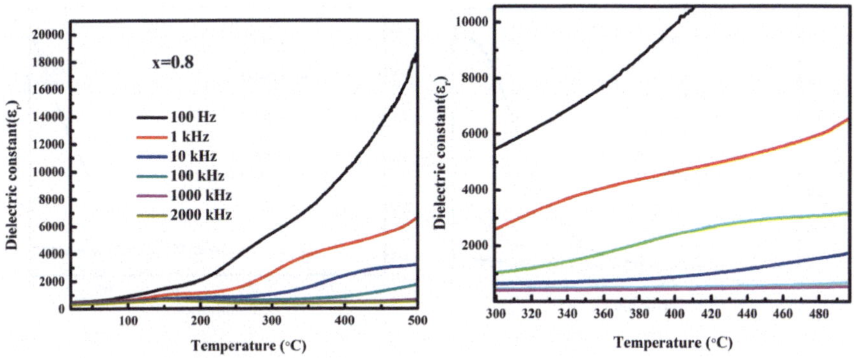

(g) Medium temperature curve with x=0.8; (h) Enlarged Medium temperature curve with x=0.8

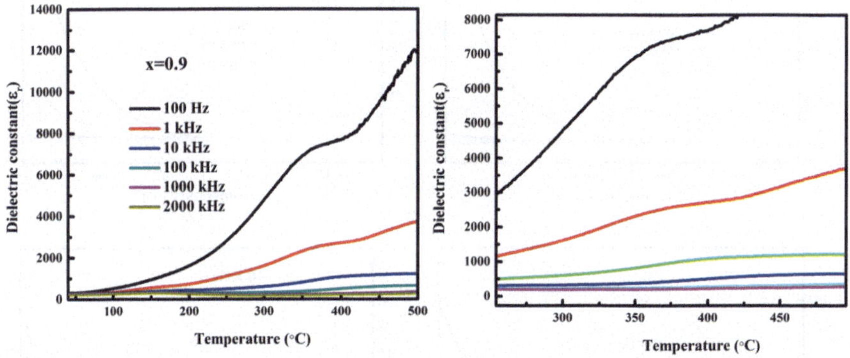

(i) Medium temperature curve with x=0.9; (j) Enlarged Medium temperature curve with x=0.9

Fig. 5 (continued)

When $x = 0.9$, the loss is the highest. This may be because when $x = 0.9$, the surface holes are the most, the leakage current is the highest, and the heat generation increases, resulting in an increase in loss.

3.4 *Ferroelectric Properties*

Figure 7a shows the hysteresis loops of composite ceramics with different compositions measured at a hysteresis loop of 80 kV/cm and a frequency of 2 kHz. From the graph, it can be seen that as x increases, the ferroelectric phase gradually decreases, and the residual polarization strength also gradually decreases. When $x = 0.9$, the

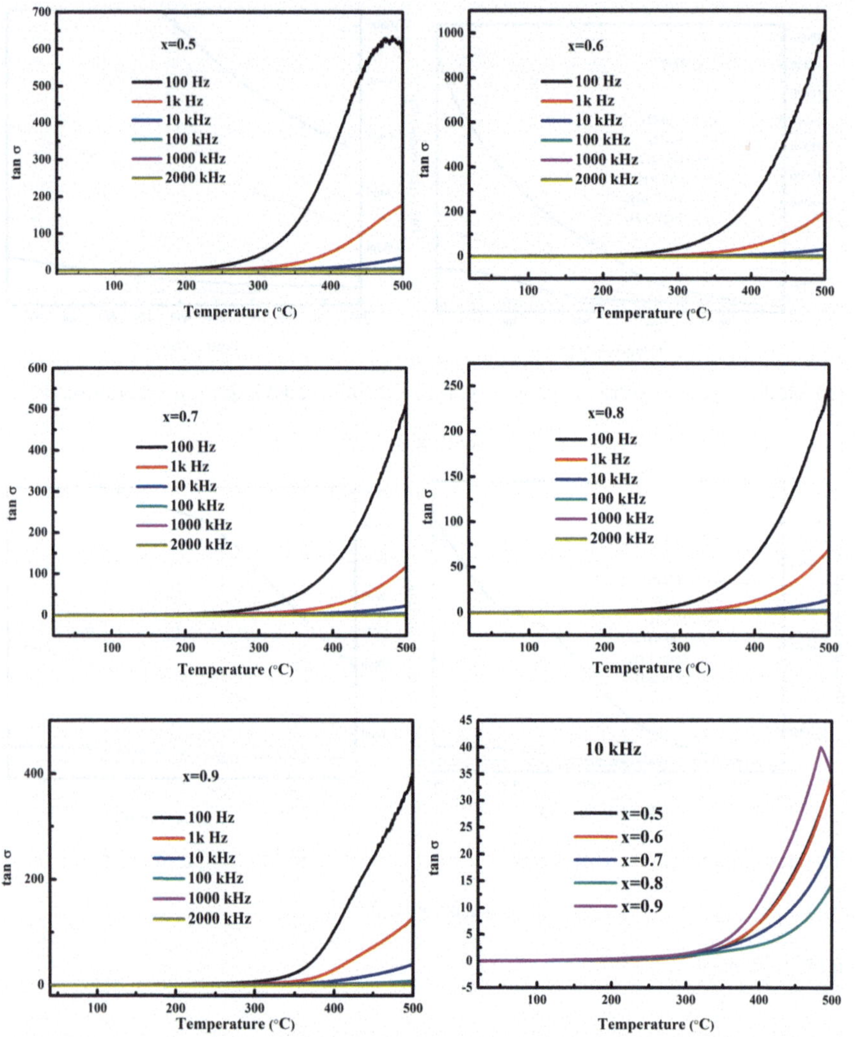

Fig. 6 Temperature dependence of dielectric loss of the prepared (1 − *x*) ($Bi_{0.9}Sm_{0.1}Fe_{0.85}Ti_{0.15}O_3$)—*x* ($0.2Co_{0.7}Zn_{0.3}Fe_2O_4$–$0.8BaTi_{0.85}Zr_{0.15}O_3$) magnetoelectric composite ceramics with different compositions and dielectric loss at 10 kHz for different compositions

residual polarization intensity reaches its maximum. From the SEM images, it can be seen that the main reason for the excessive leakage current is due to the large number of pores on the sample surface, which weakens the polarization ability. The polarization ability of materials is directly proportional to the amount of electricity, which is closely related to the current. In polarization, there are three types of currents: polarization current, leakage current, and conduction current, which contribute to

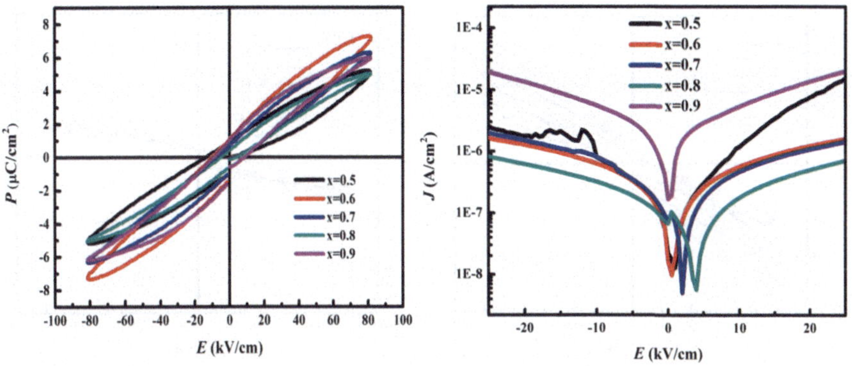

Fig. 7 **a** Hysteresis loop of (1 − *x*) ($Bi_{0.9}Sm_{0.1}Fe_{0.85}Ti_{0.15}O_3$)—*x* ($0.2Co_{0.7}Zn_{0.3}Fe_2O_4$–$0.8BaTi_{0.85}Zr_{0.15}O_3$) magnetoelectric composite ceramics with different composition examples; **b** leakage current of composite ceramics with different composition examples

polarization. And what is truly closely related to polarization ability is the magnitude of polarization current. So what we really need is a small leakage current and a large polarization current.

Figure 7b shows the leakage current measurements of composite ceramics with different compositions under a voltage of 1 kV. From the graph, it can be seen that as x increases, the leakage current first decreases and then increases. When $x = 0.8$, the leakage current is the smallest. This may be because as x increases, its BFO content gradually decreases, so its volatilization also decreases during sintering, resulting in a decrease in its leakage current. And $x = 0.9$ reaches the maximum leakage current. In summary, when $x = 0.6$, its leakage current is relatively small, and the residual polarization strength is also relatively large, so its ferroelectric performance is optimal when $x = 0.6$.

Figure 8 shows hysteresis loops of (1 − *x*) ($Bi_{0.9}Sm_{0.1}Fe_{0.85}Ti_{0.15}O_3$)—*x* ($0.2Co_{0.7}Zn_{0.3}Fe_2O_4$–$0.8BaTi_{0.85}Zr_{0.15}O_3$) magnetoelectric composite ceramics with different compositions tested at temperatures of 30, 50, 70, 90, 110, and 130 °C with an electric field strength of 50 kV/cm and a frequency of 2000 Hz. From the graph, it can be seen that the residual polarization intensity of all samples gradually increases with the increase in temperature. This is because as the temperature increases, it destroys the noncoincidence of positive and negative charges in the unit cell, enhances its polarization ability, and thus increases its residual polarization strength.

Figure 9 shows hysteresis loops obtained by testing different compositions at different voltages and frequencies. From these graphs, it can be seen that the electric field strength of certain hysteresis loops cannot be increased at a frequency of 4 kHz, which may be due to the high frequency and short time, resulting in excessive resistance. As the frequency increases, the time decreases, the number of electric domains turning along the electric field decreases, and the residual polarization strength decreases. As the electric field strength increases, the number of electric

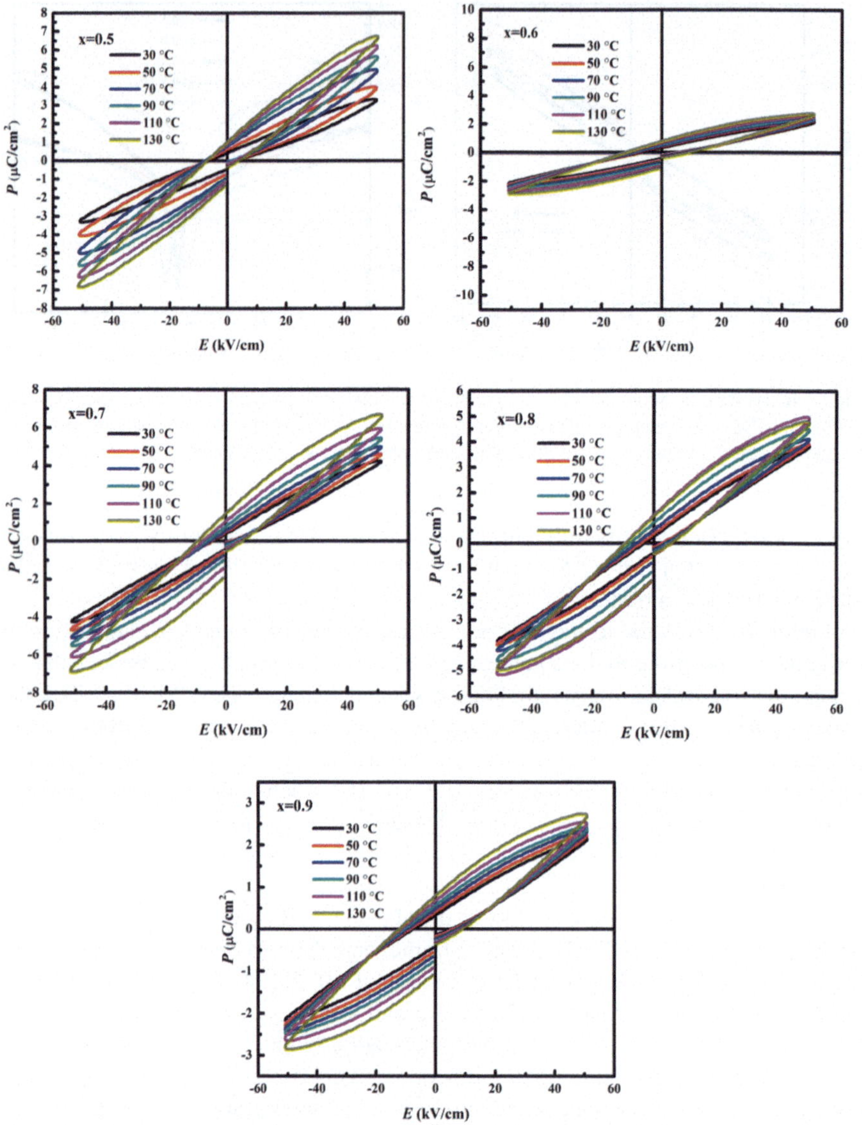

Fig. 8 Hysteresis loops of (1 − *x*) ($Bi_{0.9}Sm_{0.1}Fe_{0.85}Ti_{0.15}O_3$)—*x* ($0.2Co_{0.7}Zn_{0.3}Fe_2O_4$–$0.8BaTi_{0.85}Zr_{0.15}O_3$) magnetoelectric composite ceramics with different compositions tested at temperatures of 30, 50, 70, 90, 110, and 130 °C with an electric field strength of 50 kV/cm and a frequency of 2000 Hz

domains turning along the electric field increases, resulting in an increase in residual polarization strength and better ferroelectricity.

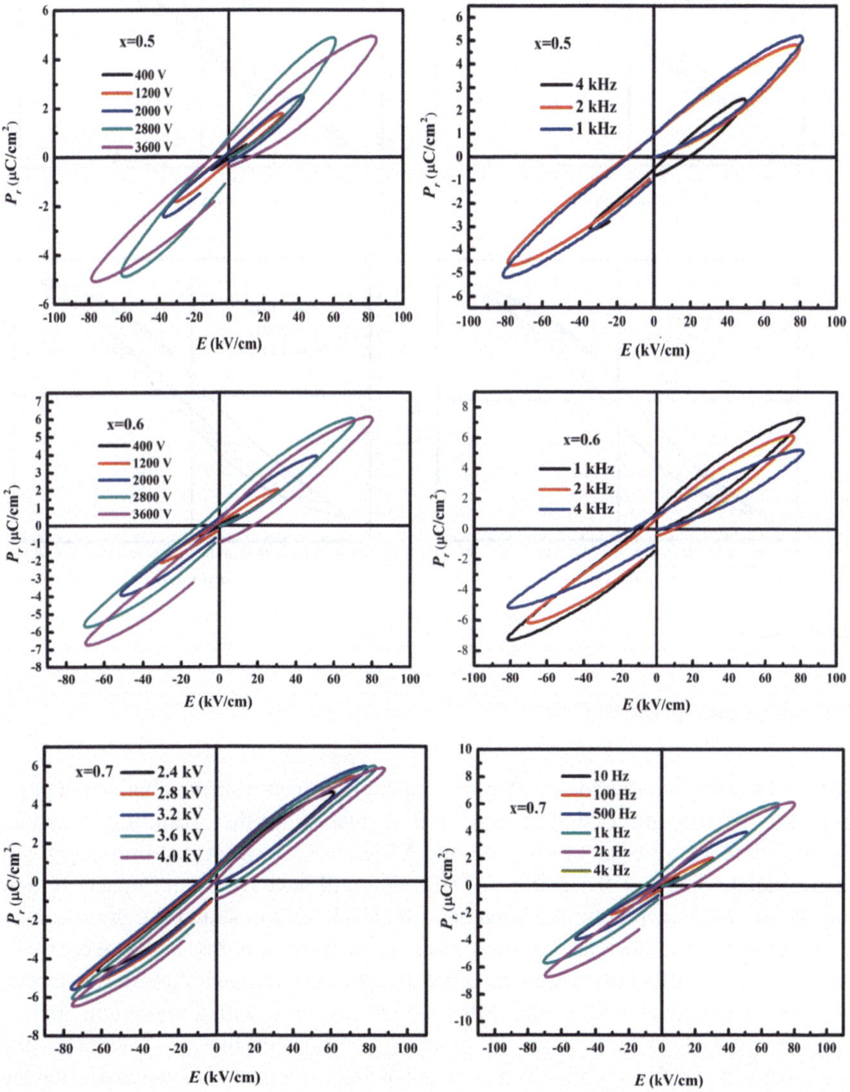

Fig. 9 Hysteresis loop at different frequencies and voltage of the prepared (1 − *x*) ($Bi_{0.9}Sm_{0.1}Fe_{0.85}Ti_{0.15}O_3$)—*x* ($0.2Co_{0.7}Zn_{0.3}Fe_2O_4$–$0.8BaTi_{0.85}Zr_{0.15}O_3$) magnetoelectric composite ceramics

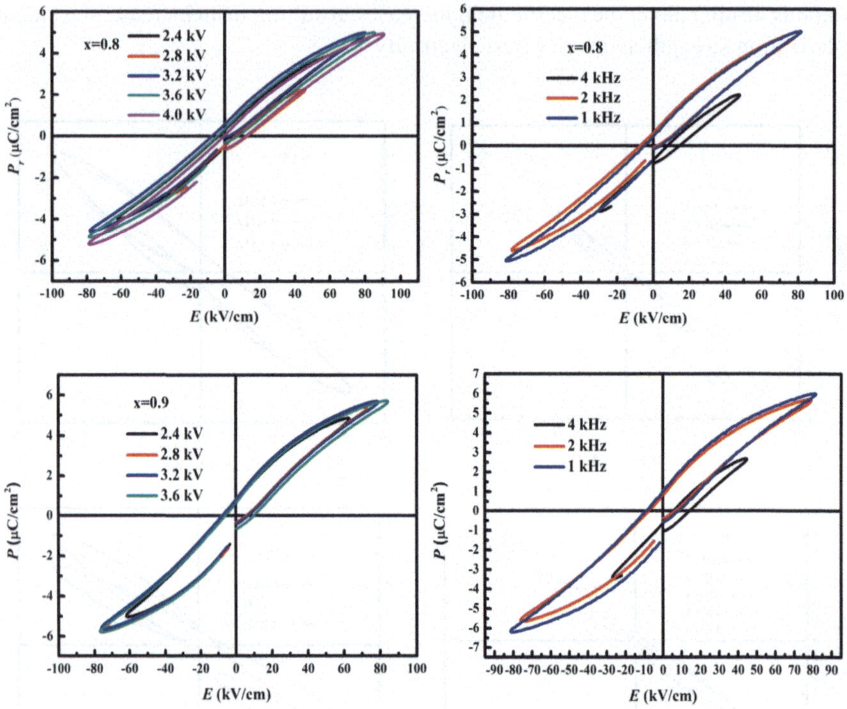

Fig. 9 (continued)

3.5 *Ferromagnetism*

Different materials sintered at different temperatures have varying changes in grain size, and different grains will be squeezed against each other, resulting in certain impurity defects or pores. From SEM and XRD, it can be seen that impurities will be generated when $x = 0.5$ and $x = 0.9$, which will have a certain impact on their magnetism. As x increases, the content of BaTiO3 increases, and the grain cannot grow, resulting in a decrease in grain size, an increase in pores, and a decrease in grain size, which affects the magnetic properties. Figure 10a and b show the hysteresis loops of composite ceramics with different compositions, with a maximum applied magnetic field of 25 kOe. From the graph, it can be seen that as x increases, the residual magnetization increases, the saturation magnetization also increases, and the magnetism is enhanced. When x increases from 0.5 to 0.9, the maximal magnetization increases from 0.3 emu/g to about 17 emu/g. This may be because as x increases, the CFO component of its ferromagnetic phase increases, resulting in enhanced magnetic properties. It can be inferred that the larger x, the stronger the magnetism. At the same time, the values of resident magnetization show weak variation, indicating that all the prepared composite ceramics show soft magnetization behavior.

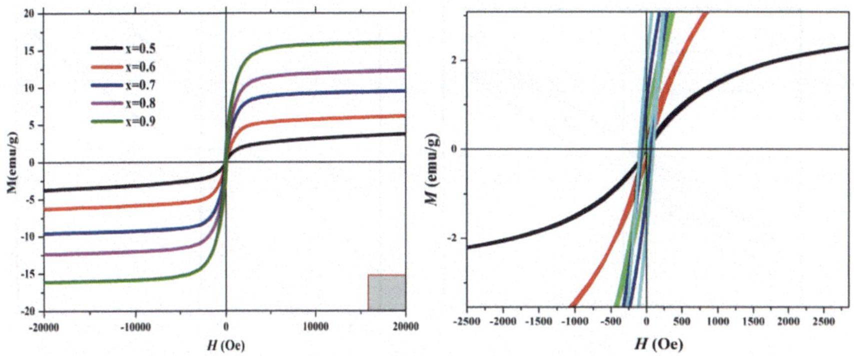

Fig. 10 **a** Hysteresis loop of (1 − *x*) ($Bi_{0.9}Sm_{0.1}Fe_{0.85}Ti_{0.15}O_3$)—*x* ($0.2Co_{0.7}Zn_{0.3}Fe_2O_4$–$0.8BaTi_{0.85}Zr_{0.15}O_3$) magnetoelectric composite ceramics with different compositions

3.6 Magnetoelectric Coupling Performance

In order to investigate the MEC effect of ($Bi_{0.9}Sm_{0.1}Fe_{0.85}Ti_{0.15}O_3$)—($0.2Co_{0.7}Zn_{0.3}Fe_2O_4$–$0.8BaTi_{0.85}Zr_{0.15}O_3$) magnetoelectric composite ceramics, P-E curves were measured before and after applying a magnetic field, and finally, the curves were also measured after canceling the external magnetic field so as to discuss whether the variations of P-E curves were affected by the magnetic field or not, as shown in Fig. 11. Indeed, it can be found that the combination of bismuth ferrite, barium titanate, and cobalt ferrite exhibits certain magnetoelectric coupling properties due to their ferroelectric phase and ferromagnetic phase. In order to investigate the magnetoelectric coupling performance of the composite ceramic, this paper measured the hysteresis loop of the samples with $x = 0.7$ by applying a magnetic field and compared it with the samples without a magnetic field. Figure 11a and b show the hysteresis loops of $x = 0.7$ composite ceramics with an external magnetic field of 8 mT, an electric field strength of 50 kV/cm, and a frequency of 2 kHz. From the figure, it can be seen that when no magnetic field is applied, the hysteresis loop is relatively thin and weak, with a minimum residual polarization intensity of 0.8422 C/cm^2. When a magnetic field is applied, due to the magnetoelectric coupling effect, the external magnetic field will cause changes in electrical properties, resulting in an increase in residual polarization intensity to 1.3045 C/cm^2. When the magnetic field is removed, the electric domains caused by its magnetic field cannot keep up with the changes in the magnetic field, so they are not completely flipped, and the remaining polarization strength is 1.0094 C/cm^2. So, it can be seen that bismuth ferrite–barium titanate–cobalt ferrite composite ceramics have certain magnetoelectric coupling properties at room temperature.

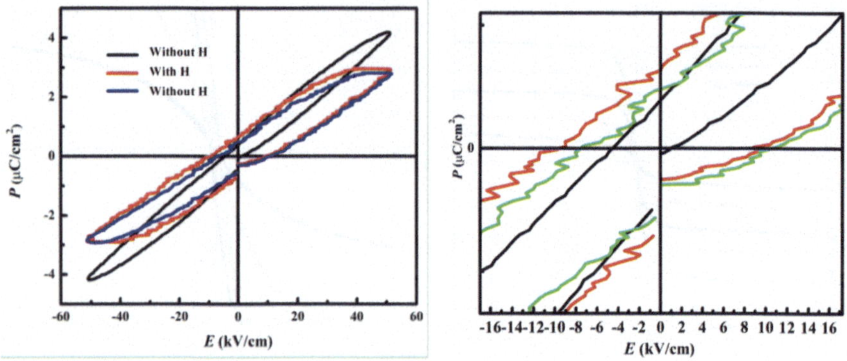

Fig. 11 **a** Hysteresis curves of (1 − x) ($Bi_{0.9}Sm_{0.1}Fe_{0.85}Ti_{0.15}O_3$)—$x$ ($0.2Co_{0.7}Zn_{0.3}Fe_2O_4$–$0.8BaTi_{0.85}Zr_{0.15}O_3$) magnetoelectric composite ceramics with different compositions measured by external magnetic fields. **b** Enlarged hysteresis curves of (1 − x) ($Bi_{0.9}Sm_{0.1}Fe_{0.85}Ti_{0.15}O_3$)—$x$ ($0.2Co_{0.7}Zn_{0.3}Fe_2O_4$–$0.8BaTi_{0.85}Zr_{0.15}O_3$) magnetoelectric composite ceramics with different compositions measured by external magnetic fields

4 Conclusion

This paper uses solid-phase method to prepare bismuth ferrite–barium titanate–cobalt ferrite composite ceramics. During the experiment, different composite ceramic samples were obtained by setting five different composition examples (1 − x) ($Bi_{0.9}Sm_{0.1}Fe_{0.85}Ti_{0.15}O_3$)—$x$ ($0.2Co_{0.7}Zn_{0.3}Fe_2O_4$–$0.8BaTi_{0.85}Zr_{0.15}O_3$), x = 0.5, 0.6, 0.7, 0.8, 0.9). The effects of sintering temperature and composition on the crystal structure, microstructure, and other properties of the bismuth ferrite–barium titanate–cobalt ferrite composite ceramics were analyzed. The sintering temperature has a significant impact on the phase structure, ferroelectricity, surface morphology, and dielectric properties of the composite ceramic. Although different components do not affect the phase composition of bismuth ferrite–barium titanate–cobalt ferrite composite ceramics, the intensity of XRD diffraction patterns may vary. Different components have a significant impact on surface morphology, electrical properties, and magnetic properties. When x increases, the grain size first increases and then decreases, and the dielectric and ferroelectric properties also improve significantly. However, the maximum loss and leakage current occur when $x = 0.9$. As the temperature increases, the residual polarization intensity of all samples increases. As x increases, the more $CoFe_2O_4$ there is, and the magnetic properties gradually increase.

This composite ceramic exhibits certain magnetoelectric coupling properties at room temperature. When subjected to an external magnetic field, the electric domains follow the deflection, polarization is enhanced, and all hysteresis loops undergo certain changes. The residual polarization strength increases, and ferroelectricity is enhanced.

References

1. Ao H, Mu HW, Chen C et al (2024) Impact of $PbZr_{0.5}Ti_{0.5}O_3$ particle size on the ferroelectric and magnetoelectric properties of $Mn_{0.5}Zn_{0.5}Fe_2O_4$-$PbZr_{0.5}Ti_{0.5}O_3$ multiferroic fluids. Adv Eng Mater 2401697. https://doi.org/10.1002/adem.202401697
2. Gao RL, Chen C, Ren KJ et al (2024) Enhanced magnetoelectric properties of flexible $CoFe_2O_4$/PVDF composite films with different $CoFe_2$O4 particle concentrations. Mater Today Chem 42:102386. https://doi.org/10.1016/j.mtchem.2024.102386
3. Chen C, Ao H, Ding YW et al (2024) Settlement stability, electrical properties, and magnetoelectric properties of $Mn_{0.5}Zn_{0.5}Fe_2O_4$–$PbZr_{0.5}Ti_{0.5}O_3$ multiferroic fluids modulating by particle sizes. Mater Today Commun 40:110145. https://doi.org/10.1016/j.mtcomm.2024.110145
4. Chen C, Mu HW, Zhu YL et al (2024) Study on the magnetoelectric coupling properties of $Mn_{0.5}Zn_{0.5}Fe_2O_4$–$PbZr_yTi_{1-y}O_3$ multiferroic fluids with different polarization intensities. Mater Sci Eng B 304:117337. https://doi.org/10.1016/j.mseb.2024.117337
5. Wu H, Xu RC, Zhou C et al (2022) Effect of core size on the magnetoelectric properties of $Cu_{0.8}Co_{0.2}Fe_2O_4$@$Ba_{0.8}Sr_{0.2}TiO_3$ ceramics. J Phys Chem Solids 160:110314. https://doi.org/10.1016/j.jpcs.2021.110314
6. Wu H, Ao H, Li WC et al (2021) Improvement of magnetoelectric coupling effect in $Ba_{0.8}Sr_{0.2}TiO_3$–$Co_{0.5}Cu_{0.5}Fe_2O_4$ multiferroic fluids by tuning the composition. Mater Today Chem 21:100511. https://doi.org/10.1016/j.mtchem.2021.100511
7. Gao RL, Lei TY, Fu CL et al (2021) Photovoltaic effect in Sr-doped $BiFeO_3$ thin films. Ferroelectrics 582(1):70–79. https://doi.org/10.1080/00150193.2021.1951037
8. Gao RL, Fu CL, Cai W et al (2016) Electric control of the Hall effect in Pt/$Bi_{0.9}La_{0.1}FeO_3$ bilayers. Sci Rep 6:20330. https://doi.org/10.1038/srep20330
9. Ruan J, Li C, Yuan Z et al (2016) Four-state non-volatile memory in a multiferroic spin filter tunnel junction. Appl Phys Lett 109(25):252903
10. Gao RL, Cai W, Chen G et al (2016) Enhanced ferroelectric photovoltaic effect based on converging depolarization field. Mater Res Bull 84:93–98. https://doi.org/10.1016/j.materresbull.2016.07.031
11. Hu J, Chen L, Nan C et al (2016) Multiferroic heterostructures integrating ferroelectric and magnetic materials. Adv Mater 28(1):15–39
12. Gao RL, Wang ZH, Chen G et al (2018) Influence of core size on the multiferroic properties of $CoFe_2O_4$@$BaTiO_3$ core shell structured composites. Ceram Int 44:S84–S87. https://doi.org/10.1016/j.ceramint.2018.08.234
13. Li CY, Xu RC, Gao RL et al (2020) Structure, dielectric, piezoelectric, antiferroelectric and magnetic properties of $CoFe_2O_4$–$PbZr_{0.52}Ti_{0.48}O_3$ composite ceramics. Mater Chem Phys 249:123144. https://doi.org/10.1016/j.matchemphys.2020.123144
14. Wu H, Zeng ZX, Xing SL et al (2021) Microstructure, magnetodielectric and multiferroic properties of $xCo_{0.8}Cu_{0.2}Fe_2O_4$–y($_{0.8}BaTiO_3$–0.2$BiAlO_3$) composite ceramics. Adv Eng Mater 2100410. https://doi.org/10.1002/adem.202100410
15. Ren KJ, Ding YW, Chen C et al (2024) Synthesis of a magneto electric fluid based on a core shell architecture of Cobalt ferrite and Barium Titanate. Mater Res Bull 113170. https://doi.org/10.1016/j.materresbull.2024.113170
16. Li H, Ding YW, Ren KJ et al (2024) Effect of rapid/slow annealing routes on the magnetic and photoelectric properties of $BiFeO_3$/$CoFe_2O_4$ multilayer thin films. J Alloy Compd 989(25):174203. https://doi.org/10.1016/j.jallcom.2024.174203
17. Gao RL, Chen YS, Sun JR et al (2012) Complex transport behavior accompanying domain switching in $La_{0.1}Bi_{0.9}FeO_3$ sandwiched capacitors. Appl Phys Lett 101:152901. https://doi.org/10.1063/1.4757987
18. Li ZX, Wang ZH, Gao RL et al (2018) Dielectric, ferroelectric and magnetic properties of $Bi_{0.78}La_{0.08}Sm_{0.14}Fe_{0.85}Ti_{0.15}O_3$ ceramics prepared at different sintering conditions. Process Appl Ceram 12(4):394–402. https://doi.org/10.1063/1.4757987

19. Ji C, Fan T, Chen G et al (2021) Influence of sintering method on microstructure, electrical and magnetic properties of $BiFeO_3$–$BaTiO_3$ solid solution ceramics. Materials Today Chemistry 20:100419. https://doi.org/10.1016/j.mtchem.2020.100419
20. Deng XL, Zeng ZX, Gao RL et al (2020) Study of structural, optical and enhanced multiferroic properties of Ni doped BFO thin films synthesized by sol-gel method. J Alloy Compd 831(5):1–9. https://doi.org/10.1016/j.jallcom.2020.154857
21. Wang ZH, Gao RL, Chen G et al (2020) Dielectric, ferroelectric and magnetoelectric properties of in-situ synthesized $CoFe_2O_4$/$BaTiO_3$ composite ceramics. Ceram Int 46(7):9154–9160. https://doi.org/10.1016/j.ceramint.2019.12.165
22. Gao RL, Bai L, Xu ZY et al (2018) Electric-field induced magnetization rotation in magnetoelectric multiferroic fluids. Adv Electron Mater 4(6):1800030. https://doi.org/10.1002/aelm.201800030
23. Xu RC, Zhang SL, Wang FQ et al (2019) The study of microstructure, dielectric and multiferroic properties of (1–x) $Co_{0.8}Cu_{0.2}Fe_2O_4$–x$Ba_{0.6}Sr_{0.4}TiO_3$ composites. J Electron Mater 48(1):386–400. https://doi.org/10.1007/s11664-018-6718-3
24. Zhou D, Gang J, Zheng Y et al (2011) Electrophoretic deposition of $BaTiO_3$/$CoFe_2O_4$ multiferroic composite films. Appl Surf Sci 257(17):7621–7626. https://doi.org/10.1016/j.apsusc.2011.03.141
25. Li T, Wang H, Hu Z et al (2016) Temperature dependence of magnetoelectric coupling in $La_{0.7}Sr_{0.3}MnO_3$ /$BaTiO_3$ layered heterostructure with various volume fraction. Thin Solid Films 616:1–5. https://doi.org/10.1016/j.tsf.2016.07.071
26. Kumar AS, Lekha CSC, Vivek S et al (2016) Multiferroic and magnetoelectric properties of $Ba_{0.85}Ca_{0.15}Zr_{0.1}Ti_{0.9}O_3$–$CoFe_2O_4$ core–shell nanocomposite. J Magn Magn Mater 418:294–299. https://doi.org/10.1016/j.jmmm.2016.02.065
27. Pithan C, Hennings D, Waser R (2005) Progress in the synthesis of nanocrystalline $BaTiO_3$ powders for MLCC. Int J Appl Ceram Technol 2(1):1–14. https://doi.org/10.1111/j.1744-7402.2005.02008.x
28. Battisha IK, Hamad ABA, Mahani RM (2009) Structure and dielectric studies of nano-composite Fe_2O_3: $BaTiO_3$ prepared by sol–gel method. Phys B Phys Condens Matter 404(16):2274–2279
29. Sadhana K, Krishnaveni T, Praveena K et al (2008) Microwave sintering of nanobarium titanate. Scripta Mater 59(5):495–498
30. Khamman O, Yimnirun R, Laosiritaworn Y et al (2009) Effects of vibro-milling time and calcination on phase formation and particle size of lead zirconate nanopowders. Mater Lett 35(1):177–180
31. Li X, Shih WH (2010) Size effects in barium titanate particles and clusters. J Am Ceram Soc 80(11):2844–2852. https://doi.org/10.1111/j.1151-2916.1997.tb03202.x
32. Tsege EL, Hong HTV, Atabaev TS et al (2016) Effect of Er^{3+} and Yb^{3+} co-doping on the performance of a ZnO-based DSSC. J Korean Phys Soc 68(12):1381–1389
33. Tang XG, Wang J, Wang XX et al (2004) Effects of grain size on the dielectric properties and tunabilities of sol–gel derived $Ba(Zr_{0.2}Ti_{0.8})O_3$ ceramics. Solid State Commun 131(3):163–168
34. Zhang SW, Zhang H, Zhang BP et al (2009) Dielectric and piezoelectric properties of $(Ba_{0.95}Ca_{0.05})(Ti_{0.88}Zr_{0.12})O_3$ ceramics sintered in a protective atmosphere. J Eur Ceram Soc 29(15):3235–3242. https://doi.org/10.1016/j.jeurceramsoc.2009.06.034
35. Cheng FX, Jia JT, Xu ZG et al (1999) Microstructure, magnetic, and magneto-optical properties of chemical synthesized Co–RE (RE=Ho, Er, Tm, Yb, Lu) ferrite nanocrystalline films. J Appl Phys 86(5):2727
36. Kitamoto Y, Kantake S, Shirasaki F et al (1999) Co ferrite films with excellent perpendicular magnetic anisotropy and high coercivity deposited at low temperature. J Appl Phys 85(8):4708–4710. https://doi.org/10.1063/1.370455
37. Selim MS, Turky G, Shouman MA et al (1999) Effect of Li_2O doping on electrical properties of $CoFe_2O_4$. Solid State Ionics 120(1–4):173–181. https://doi.org/10.1016/S0167-2738(99)00008-9

38. Ata AMAE, Attia SM, Meaz TM (2004) AC conductivity and dielectric behavior of $CoAl_xFe_xO$. Solid State Sci 6(1):61–69
39. Kahn ML, Zhang ZJ (2001) Synthesis and magnetic properties of $CoFe_2O_4$ spinel ferrite nanoparticles doped with lanthanide ions. Appl Phys Lett 78(23):3651–3653
40. Hou YH, Zhao YJ, Liu ZW et al (2011) First-principles investigations of Zn (Cd) doping effects on the electronic structure and magnetic properties of $CoFe_2O_4$. J Appl Phys 109(7):64–71
41. Liu XH, Xu Z, Wei XY, et al (2010) Ferroelectric, ferromagnetic, and magnetoelectric characteristics of $0.9(0.7BiFeO_3–0.3BaTiO_3)–0.1CoFe_2O_4$ ceramic composite. J Am Ceram Soc 93(10):2975–2977
42. Liu XH, Xu Z, Qu SB et al. (2008) Ferroelectric and ferromagnetic properties of Mn-doped $0.7BiFeO_3–0.3BaTiO_3$ solid solution. Ceram Int 34(4):797–801
43. Ruan J, Li C, Yuan Z et al (2016) Four-state non-volatile memory in a multiferroic spin filter tunnel junction. Appl Phys Lett 109(25):252903. https://doi.org/10.1063/1.4972786
44. Khomeriki R, Chotorlishvili L, Tralle I et al (2017) Positive-negative birefringence in multiferroic layered metasurfaces. Nano Lett 16(11):5–56. https://doi.org/10.1021/acs.nanolett.6b03936
45. Hu JM, Chen LQ, Nan CW (2016) Multiferroic heterostructures integrating ferroelectric and magnetic materials. Adv Mater 28(1):15–39. https://doi.org/10.1002/adma.201502824
46. Peters JJP, Apachitei G, Beanland R et al (2016) Polarization curling and flux closures in multiferroic tunnel junctions. Nat Commun 7:13484. https://doi.org/10.1038/ncomms13484
47. Annapureddy V, Kim M, Palneedi H et al (2016) Low-loss piezoelectric single-crystal fibers for enhanced magnetic energy harvesting with magnetoelectric composite. Adv Energy Mater 24(24):24–35. https://doi.org/10.1002/aenm.201601244
48. Luo F, Li Z, Chen J et al (2021) High piezoelectric properties in $0.7\ BiFeO_3–0.3\ BaTiO_3$ ceramics with MnO and MnO_2 addition. J Eur Ceram Soc 42(3):954–964. https://doi.org/10.1016/j.jeurceramsoc.2021.10.052

Chapter 4
Room-Temperature Perovskite Multiferroic Material Bismuth Ferrite ($BiFeO_3$): A Highly Investigated Material Among Various Perovskite Multiferroic Materials

Anjali Varshney, Sunil Chauhan, Manish Kumar, Oscar Raymond Herrera, and Subhash Sharma

1 Overview

When scientists started examining the presence of several ferroic properties like as ferromagnetism, ferroelasticity, and ferroelectricity—in a single material in the middle of the twentieth century, the idea of multiferroic materials was born. Even though ferroelectricity (found in Rochelle salt in 1920) and ferromagnetism (first seen in the nineteenth century) were well-established fields by the 1950s, scientists were starting to speculate about materials that may display both qualities concurrently at same time [1]. A special family of materials known as multiferroic materials is defined as having several ferroic orders, usually ferroelectricity, ferromagnetism,

A. Varshney · S. Chauhan (✉)
Department of Physics & Environmental Sciences, Sharda School of Engineering & Science, Sharda University, Greater Noida, Uttar Pradesh, India
e-mail: sunil.chauhan@sharda.ac.in

Centre for Solar Cell and Renewable Energy, Department of Physics, Sharda School of Basic Sciences and Research, Sharda University, Greater Noida, Uttar Pradesh, India

M. Kumar
Experimental Research Laboratory, Department of Physics, ARSD College, University of Delhi, New Delhi, India

O. R. Herrera · S. Sharma
Centro de Nanociencias y Nanotecnología, Universidad Nacional Autónoma de México, Ensenada, Baja California, México

S. Sharma
Conahcyt-Centro de Nanociencias y Nanotecnología, Universidad Nacional Autónoma de México, Ensenada, Baja California, México

M. Kumar (ed.), *Sustainable Perovskite Multiferroic Materials*, Materials Horizons: From Nature to Nanomaterials, https://doi.org/10.1007/978-981-96-7419-0_4

and ferroelasticity, all within the same phase [1]. Separately, ferromagnetism is the property of a material having a spontaneous magnetic moment that can be aligned with the help of an applied magnetic field, whereas ferroelectricity is the property of a material having a spontaneous electric polarization that may reversed by an external applied electric field [2]. The coexistence of these characteristics in a material offers intriguing prospects for the advance of multifunctional applications and devices, wherein electric fields exert control over magnetic states and magnetic fields impact electric polarization, a phenomenon referred to as magnetoelectric coupling [3]. Due to its potential, this creates energy-efficient devices. Before the identification and thorough study of material like bismuth ferrite ($BiFeO_3$), which presented the possibility of real-world applications in the early 2000s, the concept of multiferroic materials was mostly theoretical [4].

Since the conditions necessary for ferroelectricity and magnetism to coexist in the same material are frequently incompatible, multiferroics are generally uncommon. Materials exhibiting ferroelectricity typically have an off-centered core atom, which produces a spontaneous dipole moment—a characteristic common to non-magnetic materials [5]. On the other hand, materials having partially full d or f orbitals are often magnetized due to the generation of magnetic moments by electron spins. Finding materials with both qualities is difficult due to this incompatibility. However, due to their capacity to support both electric and magnetic ordering through lattice distortions and electron interactions, complex oxides—especially those with perovskite structures—have emerged as top contenders for multiferroic behavior [6].

One of the most important properties of multiferroics is their capability to alter both their magnetic and electric polarization characteristics by applying a magnetic field [7]. This phenomenon has the potential to greatly enhance the design of spintronic devices, which store and process data using electron spin in addition to charge [8]. Devices with greater memory densities, less power consumption, and quicker speeds than existing technologies might result from the combination of spintronics with multiferroicity [9]. Multiferroic materials are being investigated for applications in sensors, actuators, and energy harvesting devices in addition to spintronics. Their capacity to react to many stimuli at once may result in very sensitive and effective technologies [10]. They are also appealing for usage in future electronics because they are in nonvolatile memory, where data is kept even when the power is switched off [3].

A small number of scientists have examined nanostructured ferroics about possible multiferroics. Sunil's group [11, 12], for example, considered self-assembled BFO.

As a result, BFO research is essential to the overall development of multiferroic materials. The number of publications on this topic is trending steadily upward, which shown in Fig. 1a, b. In Fig. 1b, there is a tendency toward a steadily rising number of publications on "$BiFeO_3$". According to the statistical data, there has been a growing interest in BFO-based materials. Also, Fig. 1a displays statistical data on peer-reviewed papers from 2000 to 2024 that were found using the keywords $BiFeO_3$ base material for piezo-photocatalytic activity, where one can see that research increased per year due to its fast degradation properties. Most studies on BFO-based materials have focused on their ferroelectric, magnetic, magnetoelectric, piezoelectric, and more recently, photovoltaic capabilities, according to a more thorough analysis of the data. Figure 2 presents statistical data from refereed articles about the features listed above from 2000 to 2024. Research on "ferroelectric" and "magnetic" materials predominated over investigations on BFO-based materials.

The magnetoelectric, piezoelectric, and, more recently, photovoltaic characteristics have also received good interest. Due to the wide range of adjustable shapes and characteristics, thin films account for almost half of all articles on BFO ceramic bulks, nanostructures, and thin films published in peer-reviewed journals between 2000 and 2024.

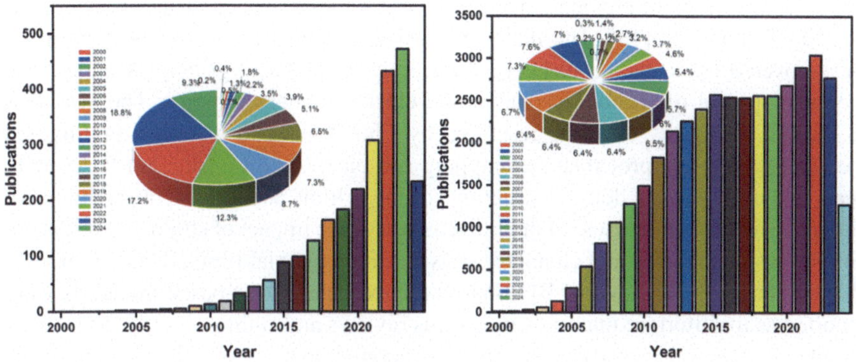

Fig. 1 **a** BFO-based material for piezo-photocatalytic activity in peer-reviewed journals throughout the 2000–2024 period, and **b** Multiferroic and BFO papers publications during the same period

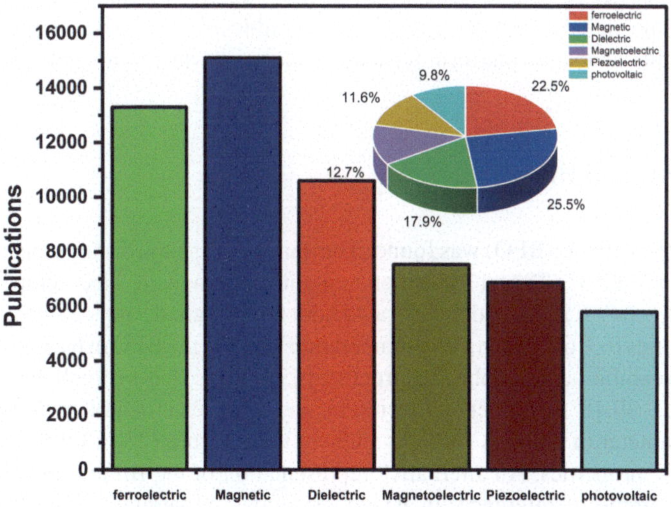

Fig. 2 Statistics of peer-reviewed works about BFO properties from 2000 to 2024

In the last 24 years, several research teams have provided summaries of the developments of BFO-based materials from various perspectives and at various stages of development, This has significantly aided in the advancement of research and produced some significant discoveries [13]. For instance, Ramesh et al. review [14] focused on the role of multiferroic BFO in ferromagnetism and electric field regulation. Scott et al. were reported, Bismuth ferrite's basic physics and possible uses in spintronics and memory devices, which might be handled both magnetically and electrically [15, 16]. An explanation of the preparation techniques (such as oxide precursors and wet chemistry) and associated characteristics of BFO-based ceramics was given by Silva et al. [17]. The piezoelectric and ferroelectric characteristics of BFO doped with rare-earth elements and made via liquid-phase sintering were emphasized by Yuan et al. [18]. The main processing concerns for BFO-based ceramics were covered by Rojac et al. [19], along with the physical principles underlying the development of secondary phases and electrical characteristics. The synthesis of $BiFeO_3$ thin films was first described by Chu et al. [20], who also highlighted the use of X-ray reciprocal space mapping and piezoresponse force microscopy to analyze the crystal and domain structures of these films and shed light on the ferroelectric switching processes. In their discussion of the impact of strain and thickness on the crystal structure and characteristics of BFO epitaxial films, Sando et al. [21] also emphasized the range of BFO-based device types. BFO-based materials were reported as multiferroic materials in various reviews and summaries [22–55]. BFO-based materials were mostly taken into consideration in a few specific characteristics or as a specific class of multiferroic materials in the majority of the prior evaluations [56–62]. Furthermore, each of the earlier review studies on BFO was mostly limited to a single kind of BFO. There are not many studies that discuss the various BFO forms in detail and offer comparisons between them. Utilizing both solar energy and environmental mechanical energies, these biocompatible and non-toxic composites may offer a readily adaptable platform for large-scale wastewater treatment applications.

2 Holy-Grail $BiFeO_3$

In the 1950s, $BiFeO_3$ (BFO) was found. This material displayed a ferroelectric Curie temperature 830 °C (T_C ~ 1103 K) and an antiferromagnetic Neel temperature of 370 °C (T_N ~ 647 K), which demonstrated both ferromagnetic and antiferromagnetic characteristics [63, 64]. At ambient temperature (25 °C), the BFO in bulk with a space group R3c exhibits a rhombohedral structure [65] and also has a rhombohedral angle, around 89.3–89.48° at ambient temperature, and a ferroelectric polarization-causing lattice parameter of 3.965 Å [66]. At ambient temperature (25 °C), BFO displayed multiferroic properties. An alternative representation of the $BiFeO_3$ unit cell is the hexagonal frame of reference, in which the c-axis is parallel to the diagonal of the cube, as in [001] hexagonal [67] pseudocubic. It was proven that the thermal expansion coefficient may not be isotropic or linear when the values ranged from 6.5 $\times 10^{-6}$ to 13 $\times 10^{-6}$ K^{-1} [68]. An important structural consideration is the rotation angle of the oxygen octahedra. In an ideal cubic perovskite structure, this angle is 0°, indicating no tilting of the octahedra and a perfect alignment of the ions based on their sizes. Here is the formula to determine if the ions line up with the perovskite unit cell: $(r_{Bi} + r_O)/l$; here r represents the ionic radii, and l indicates the octahedral length.

Bismuth ferrite, the only naturally occurring magnetoelectric substance, is termed BFO, or $BiFeO_3$. The nanoparticles of bismuth ferrite (BFO) have a noteworthy photocatalytic activity because of their huge surface area, narrow energy band gap, and stable crystal structure [69]. $BiFeO_3$ (BFO) crystallizes in a distorted perovskite structure with the general formula ABO_3, where Bi^{3+} occupies the A-site at the cube corners, Fe^{3+} occupies the B-site at the body center, and O^{2-} ions are located at the face centers, forming corner-sharing FeO_6 octahedra. In this structure, A and B represent cations, while O represents oxygen anions. The perovskite structure of bismuth ferrite is intriguing because of its strong magnetic characteristics, photocatalytic activity, and multiferroic qualities. At room temperature alone, BFO reflects both the strong ferroelectric effect and the magnetic effect [70]. Numerous changeable features, such as photochemical cells [71], non-volatile memory [72], optics [73], and thin-film capacitors [72], are included in these broad properties. BFO has a low band gap when compared to other semiconductors like TiO_2, making it a capable photocatalyst in the visible light spectrum. Another benefit of BFO material is that it blocks 44% of solar radiation into the visible light range [74]. At ambient temperature, $BiFeO_3$, or bismuth ferrite, is a model system for single-phase multiferroics that has been extensively studied due to its high residual polarization intensity of 150 $\mu C/cm^2$, high Curie temperature (830 °C), and high Neel temperature 370 °C. However, due to its weak liner magnetoelectric interaction and spiral spin structure with a periodicity of 62 nm, one form of G-type antiferromagnetic material BFO, it is difficult to utilize in multifunctional devices [75]. Because the BFO nanoparticles are smaller than 62 nm, it has recently been discovered that they display reasonably significant ferromagnetism [76]. All the factors are displayed in Table 1.

Figure 3 illustrates the all key parameters of BFO.

Because of its special blend of multiferroic characteristics, such as ferroelectricity, magnetism, and piezoelectricity, bismuth ferrite ($BiFeO_3$) is a very promising material for various room-temperature uses. One of the properties of this material is photocatalytic activity.

3 Factors Influencing $BiFeO_3$'s Photocatalytic Activity

The bulk photovoltaic effect led to the initial discovery of bismuth ferrite material [85]. Because of its photovoltaic nature, it has uses. Additionally, BFO materials are widely used in photocatalytic hydrogen production and organic pollutant degradation in presence of visible light spectrum (sunlight) [86]. Using BFO, the Rhodamine B

Table 1 All parameters of BFO multifunctional material

Property	Value	References
Crystal system[a]	Trigonal $a_{rh} = 5.6364$ Å, $\alpha_{rh} = 59.348°$ $a_{hex} = 5.6364$ Å, $c_{hex} = 13.8688$ Å $a_{pc} = 3.965$ Å, $\alpha_{pc} =$ 89.35°	[64–66, 68]
Space group	R3c	
Tolerance factor[b]	0.89	
Density	8.40 g/cm^3	
Thermal expansion	Nonlinear and anisotropic	[77]
Ferroelectricity	$T_c = 820$–830 °C	[78–83]
Polarization	70–100 μC/cm^2	[78–84]
Piezoelectricity	$D_{33} = 15$–60 pm/V	[82, 83]
Dielectric constant	$\varepsilon_r = 30$	[84]
Magnetic (antiferromagnetic)	Canted G-type antiferromagnetic $T_N = 370$ °C	[75–77]

a-Rhombohedral, hexagonal, and pseudocubic systems are the names given to a trigonal crystal structure

b-Fe^{3+} (FeO_6) octahedral and eight-coordinated Bi^{3+} (BiO_8) polyhedral have high spin according to the ionic radius ratio rule from polarization calculations based on ionic locations in the crystal structure, thin-film, and single crystal experimental data

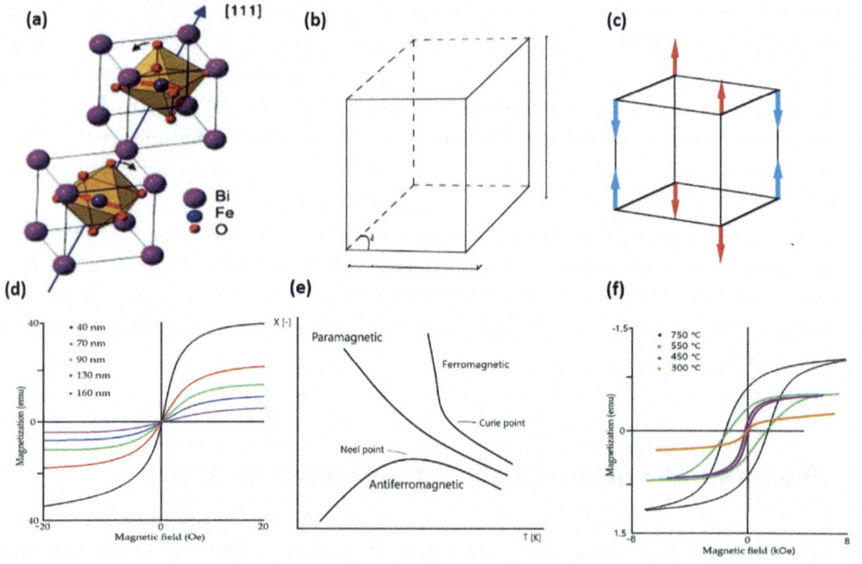

Fig. 3 **a** Bismuth ferrite unit in the perovskite structure. **b** Rhombohedral crystallographic structure. **c** G-type magnetization order. **d** Hysteresis loop of various particle sizes of BFO material. **e** Temperature-dependent magnetic hysteresis loop of BFO material. **f** Magnetic susceptibility versus temperature curve. Copyright, Open Access, MDPI publishing

dye in industrial effluent can be broken down in 35 min of direct sunshine irradiation [87]. Moreover, BFO nanoparticles were employed in the photoelectrochemical (PEC) method to produce hydrogen [88, 89].

Numerous variables, including particle size, surface shape, and dopant type, can significantly alter a material's photocatalytic activity and capacity to evolve hydrogen. Two changes are commonly employed to improve the photocatalytic activity of materials: doping [90] and sensitization [91].

3.1 Impact on the Surface of BFO

It is well understood that the surface shape of a material has a major impact on its photocatalytic activity. To create a material with high photocatalytic activity, form control is essential. When compared to other dominating various crystal facets, Fei et al. revealed that rods and pills with a highly showing lattice plane (111) had an improved visible light sensitivity [92, 93]. Additionally, it is discovered that the surface shape and surface area of bismuth ferrite nanoparticles exhibit unique influences when exposed to visible light. Research indicates that bismuth ferrite nanoparticles have superior photocatalytic activity in comparison with bulk bismuth ferrite material [94]. Similarly, Cai et al. reported changes in photocatalytic activities as a result of the surface area variation, and they found that the catalysis sequence under visible light was cube greater than plate and then sheet [95].

Additionally, it has been demonstrated that hexagon-shaped BFO nanoparticles exhibited superior characteristics than the spherical perovskite type [96]. About the surface area morphology effect, variations in surface can correspond to variety energy levels, such as valence bands (VB) and conduction bands (CB) from various crystal lattices, which improves the surfaces' surface qualities and amplifies photocatalytic reactions.

3.2 Heterojunctional Effect

In general, an interaction including electron and hole interaction in separate layers of two solid-state materials is referred to as a heterojunction. This procedure has two primary components; a donor semiconductor (typically with a higher conduction band) and an acceptor semiconductor (with a lower conduction band). This band alignment facilitates charge separation, which significantly enhances the photocatalytic activity by suppressing the recombination of photogenerated electron–hole pairs [97–99]. Fe_2O_3 and BFO create a heterojunction in bismuth ferrite. Despite having distinct light absorption ranges, visible light can excite BFO and Fe_2O_3, producing electrons and holes in the VB and CB, respectively. Electrons and holes migrate from the VB and CB of BFO and Fe_3O_3 to the corresponding bands of their respective alternate molecules because of the difference in their potential of energy bands. The movement and transfer of holes and electrons increase the material's capacity for photocatalysis [100].

3.3 Dopant's Effect

Numerous studies have looked at how adding a dopant to BFO affects its photocatalytic activity. An element as a dopant is added to change the band gap's structure. It has been reported that BFO was doped with rare-earth metals like lanthanum or yttrium to boost its photocatalytic activity. Zhang et al.'s [101] further research on lanthanum doping revealed that while it has no impact on BFO's optical characteristics, it increases photocatalytic activity [102]. Recently, employing strontium and lanthanum as A site dopants, Ghosh et al. investigated the doping impact on BFO in more detail. They reported that the system enters a metallic state due to increased magnetic, ferroelectric, and spontaneous polarization caused by the putting of Sr in place of the A site [103].

3.4 Impact of Dopants on BFO for Dye Degradation

The potential dopants for BFO's A and B sites were examined using data that has been released over the past four years, and Fig. 4 shows them in periodic table form. Bi ionic radii are associated with elements such as Pd, La, Sr, Y, Pr, Sm Ag, Gd, and Y, whereas Fe ionic radius is paired with Mn and Co ionic radius [104, 105] (Table 2).

3.5 Additional Factor

H_2O_2 can be used to boost photocatalytic activity in addition to the previously mentioned factors [112]. According to Wang et al., pH plays a major part in enhancing photocatalytic activity [104].

4 Important Properties of BFO

$BiFeO_3$ nanostructures' structural characteristics are important because they affect their optical, magnetic, and electrical capabilities. In contrast to its bulk counterpart, $BiFeO_3$ exhibits different characteristics at the nanoscale. Nanostructures—such as nanowires, thin films, and nanoparticles—have a major impact on their magnetic,

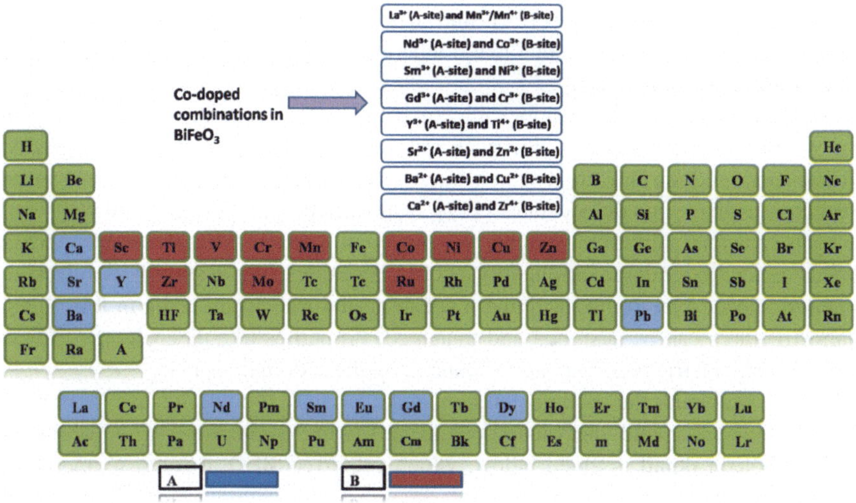

Fig. 4 Possible dopants in A site and B site of BFO

Table 2 Possible dopants in A site and B site of BFO band gap with/without dopant

Material	Dye	Band gap without dopant (eV)	Band gap with dopant (eV)	References
Mn-doped$BiFeO_3$	MB	2.26	1.9	[106]
Sm-Co codoped $BiFeO_3$	MB	2.32	2.25	[107]
Y-doped $BiFeO_3$	MO	1.97	1.83	[108]
Sm/Mn-doped $BiFeO_3$	CR	2.08	1.45	[109]
Bi0.97Sm0.03FeO3	RhB	2.17	2.14	[110]
Gd-doped$BiFeO_3$	MB	2.53	2.35	[111]
Pd-doped $BiFeO_3$	MB	2.23	2.03	[67]
Sr-Co codoped $BiFeO_3$	RhB	2.37	1.87	[112]
Yb-Co codoped $BiFeO_3$	Congo red dye	2.15	1.15	[113]

ferroelectric, and optical characteristics due to having a higher surface-to-volume ratio. The following is a list of these effects:

Strain in nanostructures can result from surface stresses in nanoparticles or from a lattice incongruity between the BFO and its thin-film substrate. Ferroelectric and magnetic characteristics can be adjusted by strain engineering. For example, shrinking the size can cause ferroelectric polarization to alter or long-range magnetic order to be suppressed [114, 115].

In the nanoscale range, size effects and structural confinement can alter ferroelectric Curie and antiferromagnetic Neel temperatures as well as the associated phase transition temperatures [116].

Surface effects are more noticeable in $BiFeO_3$ nanostructures, particularly in nanoparticles. The material's electrical and magnetic properties may be impacted by the increased surface area, which may result in surface reconstructions, more oxygen vacancies, and defect states [117]. Nanostructured BFO's domain walls and oxygen vacancies can affect fatigue resistance and leakage current, two properties that are essential for device applications [118]. By using external stimuli like electric/magnetic fields, patterning, or domain substrate-induced strain walls in thin films. Nanostructures can be designed and managed. Properties like piezoelectric effects and magnetoelectric coupling are influenced by these domains [119].

When combined, these characteristics make $BiFeO_3$ suitable for use in energy-related, magnetic, and electrical devices at room temperature. The following explaination in detail is given below, which preferred BFO at room temperature for applications.

4.1 *Electrical and Magnetic Characteristics and Uses of BFO*

One well-known room-temperature multiferroic material that concurrently has ferroelectric and antiferromagnetic characteristics is bismuth ferrite ($BiFeO_3$, BFO). Because of this special quality, BFO is a material that shows promise for several cutting-edge uses in energy harvesting, sensors, memory devices, spintronics, and catalysis. Electronics, sensors, memory devices, and energy storage applications depend heavily on these characteristics. Unique BFO has antiferromagnetism with a high Néel temperature (~ 640 K), ferroelectricity and high polarization (~ 90–100 $\mu C/cm^2$ at room temperature), magnetoelectric coupling (strong electric–magnetic order interaction, photoelectric, catalytic, and photovoltaic activity which make BFO unique or novel as compared to any other reported perovskite material [120].

An outline of the main electrical characteristics of $BiFeO_3$ nanoparticles is provided below.

4.1.1 Application and Ferroelectric Properties

Random polarization of BFO is created from the displacement of Bi^{3+} from the oxygen octahedron, which leads to the creation of an electric dipole moment. Because of the volatility of the Bi and the variability of the valence of the Fe elements, the manufacturing of BFO results in the production of large numbers of oxygen vacancies and the second phase. Yoshimura et al. reported a polarization value of 6.1 $\mu C\ cm^{-2}$ along the [67] at 77 K in the 1970s [121], and the ferroelectric hysteresis loop of BFO ceramic was first discovered. BFO thin film displayed a fully saturated polarization hysteresis loop with a Pr of 55 $\mu C\ cm^{-2}$. Ion displacement inside the crystal structure of $BiFeO_3$ results in spontaneous electric polarization. Ferroelectric materials are characterized by their capability to reverse this polarization by the use of an externally applied electric field. Additionally, it has a high Curie temperature (T_C) of about 830 °C and shows substantial ferroelectric polarization at ambient temperature. As a result, even at comparatively high temperatures, it can retain its ferroelectric characteristics [121]. The perovskite structure of $BiFeO_3$ is rhombohedrally deformed (space group R3c). The primary source of the ferroelectricity in $BiFeO_3$ is the deformation of the oxygen octahedra around the Fe^{3+} ions produced by the stereochemically active lone pair of Bi^{3+} ions [122].

The ferroelectric characteristics of $BiFeO_3$ might vary greatly from those of the bulk material when it is produced as nanoparticles. Their size may affect the switching behavior, domain structure, and polarization of the nanoparticles. In NPs, the surface-to-volume ratio rises dramatically, producing more pronounced surface effects. Because surface atoms are not fully coordinated and surface states are present, this decreases ferroelectric polarization [123, 124]. As the size of BFO nanoparticles decreases, the coercive field needed to change the polarization lean toward to rise. This is because switching the smaller ferroelectric domains requires more energy. $BiFeO_3$ transitions to a paraelectric phase at its high Curie temperature (T_C), which is around 830 °C. It is appropriate for high-temperature uses due to its high T_C.

For example, Zhang et al. reported ferroelectric polarization versus electric field hysteresis loops for Sm-doped $BiFeO_3$ (BFO) sintered at 825 °C, which is illustrated in Fig. 5 as P–E loop, with measurements taken at 100 Hz frequency. These P-E loops are not completely saturated due to leakage issues at higher electric fields. The residual polarization (Pr) increases from 10.5 μC/cm^2 in the Sm1 to 16.1 μC/cm^2 in the Sm3, reaching a maximum of 28.0 μC/cm^2 in the Sm5 before slightly decreasing to 26.3 μC/cm^2 as the doping level rises to 8 mol%. The coercive fields (Ec) show a similar pattern, changing from 69.0 kV/cm in Sm1 to 80.0 kV/cm in Sm3, 97.3 kV/cm in Sm5, and then 95.6 kV/cm in Sm8. The inset of Fig. 5 summarizes the Pr and Ec values. Furthermore, the Pr increases to 35.0 μC/cm^2 (Ec = 98.0 kV/cm) in the Sm5 when the sintering temperature is raised to 835 °C [125].

Using a ferroelectric analyzer, wang et al., reported (P–E) loops at ambient temperature for Nd/Nb co-doped $BiFeO_3$ and 10 Hz to examine the ferroelectric characteristics of the prepared ceramics. Because of a high current density J, BF ceramics showed an unsaturated and significantly inflated P–E loop area [126]. It is also important to note that the ferroelectric polarization was significantly influenced by the Nd–Nb concentration, with the 10NdNb ceramics showing a notable boost (see Fig. 5Ba2). Interestingly, coercive field (Ec) and polarizatio (Pr) showed nonlinear behavior, as Fig. 5Bb2 shows. Pr and Ec values for BF were around 0.41 μC cm^{-2} and 4.45 kV cm^{-1}, respectively. The form of P-E loops was significantly enhanced by Nd–Nb doping. As the doping content varied, the Pr showed an ideal value of around 3.12 μC cm − 2 for 10NdNb, which is significantly greater than that of BF (0.41 μC cm^{-2}). Nd–Nb co-doping is therefore a suitable tactic for improving the ferroelectric properties of BF ceramics. Nd^{3+} and Nb^{5+} ions replaced the Bi and Fe sites of BF, respectively, changing the crystals' grain size and structural properties, which in turn affected the materials Ec and Pr. Because of their structural and electrical

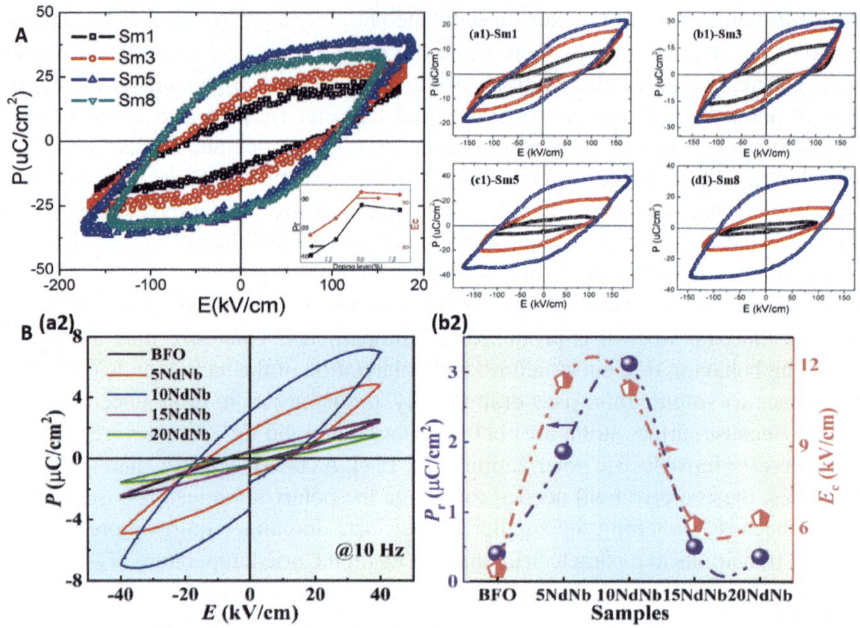

Fig. 5 As shown in prior investigations of BFO materials, the P–E loops of the Sm1–Sm5 samples at low electric fields (< 120 kV/cm) resemble antiferroelectric (AFE) double-hysteresis. Figure 5 **A** a1, b1 demonstrates that this feature is most noticeable in the Sm1 and Sm3 samples, whereas Fig. 5c1, d1 indicates that it progressively diminishes with increasing Sm-doping **B (a2)** Ferroelectric hysteresis loops at room temperature (P – E) and **(b2)** ferroelectric characteristics for undoped and Nd–Nb co-doped BFO measured at 10 Hz (Pr: residual polarization and Ec: coercive field) [125, 126], Copyright, Open access, MDPI publishing

changes, Nb–Nd co-doped $BiFeO_3$ (BFO) and Sm-doped BFO have different ferroelectric properties. By decreasing oxygen vacancies, enhancing polarization (Pr), and lowering leakage current, Nb–Nd co-doping improves ferroelectricity and makes it appropriate for memory applications. Additionally, it produces a P-E loop that is well-saturated and has a smaller coercive field (Ec), which makes domain switching simple (Fig. 5B). Sm-doping, on the other hand, creates a morphotropic phase boundary (MPB), which helps stabilize domains for energy storage applications by boosting coercivity and lattice distortion while simultaneously improving polarization at ideal values. Nevertheless, the leakage current of Sm-doped BFO is somewhat larger than that of Nb–Nd co-doped BFO [125, 126]. In general, Sm-doping is more appropriate for energy storage and electrocaloric applications, whereas Nb–Nd co-doping is better for ferroelectric memory and piezoelectric applications.

Ferroelectric Energy Harvester

BFO's ferroelectric polarization allows it to store and release electric energy, which is used in ferroelectric energy harvesters. The BFO nanostructure can produce an electric current when exposed to a fluctuating electric field. Applications like as power electronics and energy retrieval systems are a good fit for ferroelectric energy harvesters. In low-power electronics, where energy productivity is crucial, they can also be employed [114, 115]. It may not be as great for BFO nanostructures as it is for some other ferroelectric materials, including BFO, but it is nonetheless noteworthy [116].

Ferroelectric (FeRAM)

Because of their strong room-temperature ferroelectricity and potential for incorporation into semiconductor technology, BFO nanostructures are popular choices for FeRAM applications [117]. BFO's magnetoelectric interaction may enable magnetic field-assisted polarization switching, opening the door to new memory device designs. BFO-based nanostructures FeRAM, which offer non-volatile packing with quick write/read speeds, may find application in implanted memory for microcontrollers or other integrated circuits. Additionally, BFO-based FeRAM provides the benefit of high-speed operation and nonvolatility systems used mainly to make it perfect for systems that need real-time data processing and frequent data logging [118, 119].

4.1.2 Applications and Characteristics of Pyroelectric

BFO can produce electrical energy in reaction to temperature variations thanks to its pyroelectric characteristics. This can be used to create infrared detectors or temperature sensors, which are essential for non-contact temperature measurement, thermal imaging, and environmental monitoring. BFO shows temperature-dependent spontaneous electric polarization. Temperature variations can be inferred by measuring this temperature-dependent polarization [127]. Additionally, BFO has a T_c of about 830 °C, which enables it to function well throughout a broad temperature range and makes it appropriate for both room-temperature and high-temperature applications. Furthermore, BFO's ferroelectric characteristics can be combined with its temperature-dependent antiferromagnetic ordering to provide sensitive temperature sensors. Further assisting in temperature measurement, BFO's piezoelectric characteristics enable it to transform mechanical stress brought on by thermal expansion into an electrical energy [128].

Pyroelectric Energy Harvester

The pyroelectric action in BFO, where temperature changes alter the polarization of material and produce an electric charge, is used by pyroelectric energy harvesters. This makes it possible to capture waste heat or changes in the surrounding temperature. Pyroelectric energy harvesters can be utilized to power lower-energy devices from raising body temperature or in contexts like factories where thermal energy is easily accessible [129].

Temperature Sensor

Chemical reactors, manufacturing equipment, and furnaces are just a few examples of industrial processes where BFO sensors can be utilized to monitor and regulate temperature. Because BFO can function at high temperatures, it can be used in automotive and aerospace applications where precise temperature measurement is essential for both performance and safety [130]. Temperature variations in urban and natural settings can be measured using BFO temperature sensors in environmental monitoring systems. BFO-based sensors are appropriate for temperature monitoring in biomedical devices and also analytical equipment due to their sensitivity and stability [131]. Because the lattice characteristics and the orientation of the electric dipoles vary with temperature, BFO's spontaneous polarization also varies. The electric field or voltage produced across the material can be measured to identify this temperature-dependent polarization. When measured using the proper electrodes, the electrical signal (voltage/capacitance) varies as a result of the change in polarization. Accurate temperature monitoring is made possible by the correlation between this signal and temperature changes [132]. The temperature-dependent nature of the antiferromagnetic ordering in BFO is noteworthy [133]. Temperature variations can be inferred by tracking changes in magnetic characteristics, such as magnetization. Enhanced sensing capabilities can be achieved by combining this with the ferroelectric response. BFO experiences mechanical strain as a result of temperature-induced thermal expansion or contraction. By changing the material's piezoelectric sensitivity, this strain produces an electrical signal that is directly dependent on the temperature [134]. Temperature sensors can be made by depositing BFO thin films onto surfaces. These films may give precise temperature readings over a range of temperatures and are appropriate for incorporation into electronic circuits. When small and accurate temperature sensing is needed, thin-film sensors are employed in industrial processes, electrical devices, and environmental monitoring [135]. BFO in bulk forms can be utilized in temperature sensors for applications requiring high temperatures. Usually, these sensors are employed in settings requiring strong, long-lasting sensors. Because bulk ceramic sensors can tolerate high temperatures and challenging environments, this can be used in industrial, automotive, and aerospace applications [136, 137]. Small, extremely sensitive temperature sensors can be made using $BiFeO_3$ nanostructures like nanowires or nanoparticles. High-resolution temperature readings are possible with these tiny sensors. For uses in advanced research, biomedical devices, and microelectronics have high sensitivity and size are crucial, nanoscale sensors are perfect [138, 139].

4.1.3 Application and Magnetic Properties

Bismuth ferrite (BFO) is an antiferromagnetic material, which has a high Néel temperature (T_N) approx 370 °C, transitioning to a paramagnetic state above this temperature. This elevated T_N makes $BiFeO_3$ appropriate for high-temperature magnetic uses. The material exhibits a unique cycloidal spin structure, which may be disturbed under specific conditions, leading to various magnetic phenomena. In its bulk form, $BiFeO_3$ demonstrates G-type antiferromagnetic, where the Fe^{3+} ions' magnetic moments are ranged in alternating opposite directions, resulting in an overall zero magnetic moment typical of antiferromagnetic behavior. When reduced to the nanoscale, $BiFeO_3$'s antiferromagnetic characteristics undergo significant modifications. Due to the increased surface-to-volume ratio in nanoparticles, surface atoms exert a more significant influence, leading to effects such as uncompensated surface spins. This change can yield a weak ferromagnetic component/alter the antiferromagnetic ordering, a singularity known as "spin canting." This spin canting is often observed as weak ferromagnetism in $BiFeO_3$ nanoparticles, attributed to the enhanced surface effects and limited size. To optimize these magnetic properties, researchers frequently enhance $BiFeO_3$ by doping with other elements like rare-earth metals or by creating composite materials. Doping can adjust the antiferromagnetic ordering and improve magnetoelectric coupling, essential for applications in spintronics, magnetoelectric devices, and data storage technologies. Studies indicate that Sm-doping $BiFeO_3$ enhances its magnetoelectric properties, while research by Zhang et al. showed that Sm-doping significantly strengthens $BiFeO_3$'s magnetic characteristics as compared to some reported articles [140–142]. Because of the structural distortions brought about by the substitution of Sm^{3+} at the Bi-site, Sm-doped $BiFeO_3$ (BFO) ceramics have improved magnetic characteristics. The G-type antiferromagnetic ordering of pure BFO leads to zero remnant magnetization (Mr); nevertheless, as the Sm content increases from 1 to 8%, Sm-doping breaks the magnetic cycloidal structure, raising Mr from 0.002 to 0.007 emu/g (Fig. 6A). With Sm8, the coercive field (Hc) approaches 0.8 kOe. Increased magnetism results from the deformed Fe^{3+}–O–Fe^{3+} links' weakening of antiferromagnetic interactions. Furthermore, Sm^{3+} adds to magnetization since it is magnetically active. Sm-doped BFO is a viable option for magnetoelectric (ME) applications because of its enhanced ferroic characteristics [125].

For another example, Bia et al. performed the VSM analysis on BFO and co-doped BFO thin films in the applied magnetic field range 0–10 kOe to investigate the effect of co-doping on the magnetic characteristics at ambient temperature. Figure 6B a–f displays the saturated magnetic hysteresis loops of BFO and co-doped thin films. Both the saturation magnetization (M_s) and remanent magnetization (M_r) values dramatically increased as the co-doping concentration rose. M_s and M_r have the following values: 25.57 and 2.40, 21.38 and 2.06, 34.48 and 3.00, 42.38 and 3.60, 34.17 and

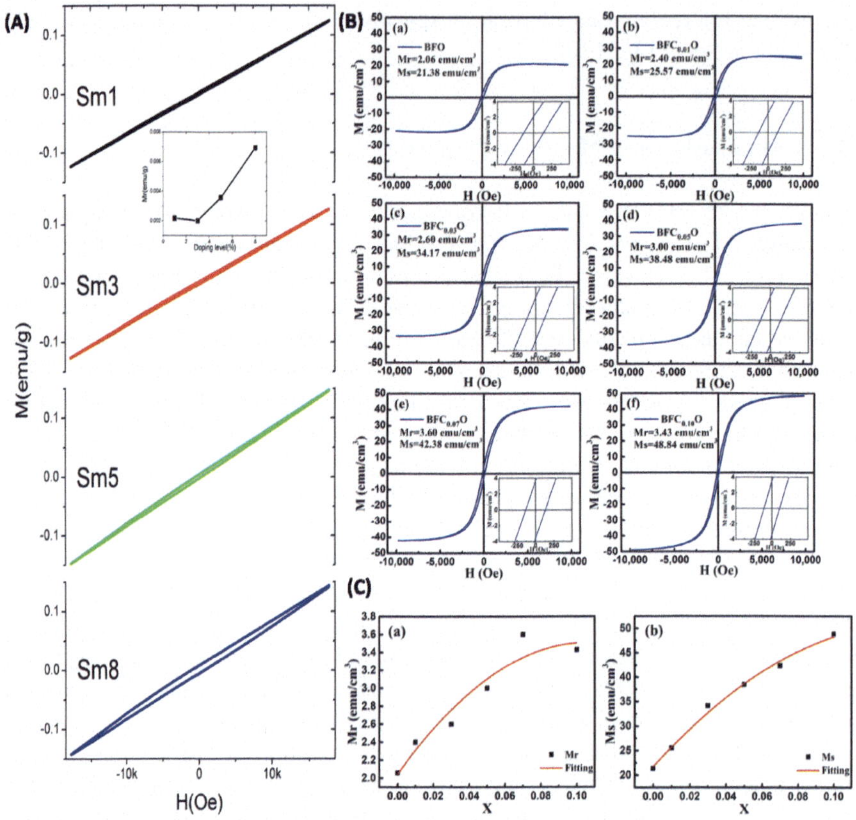

Fig. 6 **A** The samples sintered at 825 °C and their magnetic M-H loops. The figure's insets indicate the remnant magnetization (Mr) composition dependency [125]. MDPI publication, open access, and copyright. **B** The magnetic hysteresis (M–H) loops for room-temperature thin films of **a** BFO, **b** BFC_XO ($x = 0.01$), **c** BFC_XO ($x = 0.03$), **d** BFC_XO ($x = 0.05$), **e** BFC_XO ($x = 0.07$), and **f** BFC_XO ($x = 0.10$) [143], licensed under the MDPI publishing license. (C) The variations in saturation magnetization (Ms) and residual magnetization (Mr) values with the concentration of co-doping. The simulated outcomes based on the binomial polynomial equation are represented by red lines [143] Copyright, Open Access, MDPI publishing

2.60, and 48.84 and 3.43 emu/cm_3, for BFO and co-doped BFO. Figure 6C illustrates how the co-doping concentration affects M_s and M_r. As the concentration of Co^{3+} rises over $x = 0.10$, M_s always grows linearly, whereas M_r initially rises and then falls. At a concentration of 0.07 doped Co, the highest value of M_r is 3.60 emu/cm^3 [143–153]. Additionally, when the doping concentration reaches 0.10, the highest value of Ms is 48.84 emu/cm^3 [154–157]. The Nd–Nb co-doped $BiFeO_3$ (BFO) ceramics' magnetic hysteresis loops show a change in behavior from antiferromagnetic to weak ferromagnetic. Because pure BF is G-type antiferromagnetic, it showed very little magnetism, with a minimal remanent magnetization (Mr) of about 5.67×10^{-4} emu/g. But structural distortions brought about by co-doping with Nd and Nb broke the spin

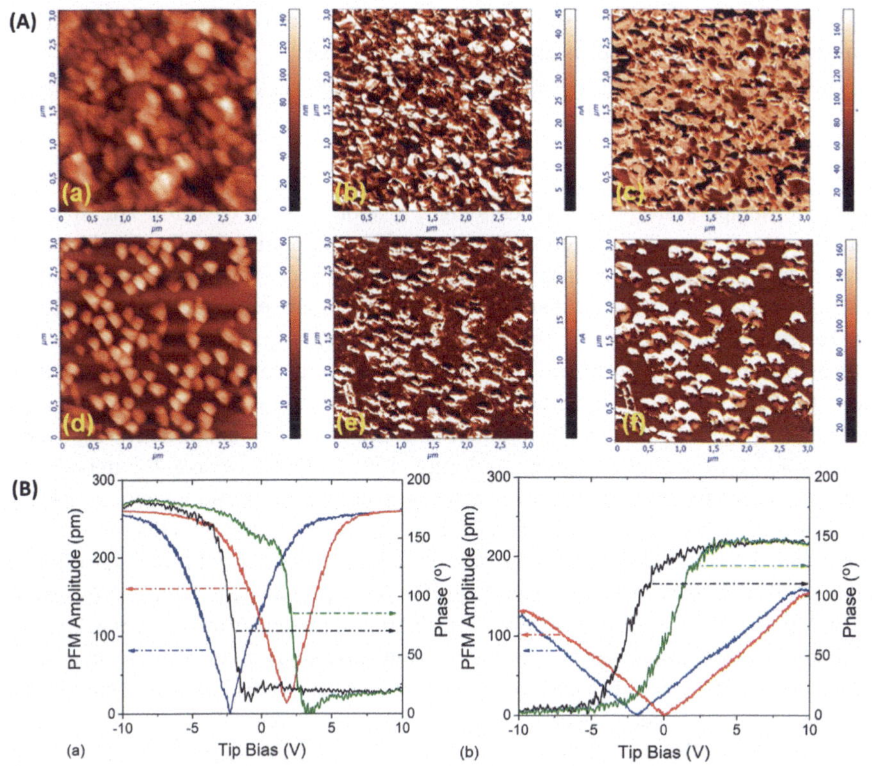

Fig. 7 **A** Thin-film PFM images (**a**, **d**), out-of-plane piezoresponse amplitude (**b**, **e**), and plane phase (**c**, **f**) of BFO/FTO (first row) and BFCO/STO (second row) [200], Copyright, Open Access, MDPI publishing. **B** Piezoresponse amplitude butterfly loops and phase hysteresis of BFO on FTO (**a**) and BFCO on STO (**b**) thin films [200], Copyright, Open Access, MDPI publishing

helical shape and increased net magnetization. Phase coexistence and spin tilt effects brought on by ionic substitution were responsible for the notable increase in Mr values to around 0.07 to 0.15 emu/g. Magnetic anisotropy produced high coercivity, which further supported the enhanced magnetic characteristics. These results are consistent with earlier research, emphasizing grain engineering and compositional design as useful techniques for improving BFO's multiferroic behavior at ambient temperature. Both Sm-doped and Nd–Nb co-doped BFO improve the magnetic characteristics of pure BFO, although the mechanism and strength of their effects vary. Because of structural phase coexistence, greater spin tilting, and disruption of the antiferromagnetic order, Nd–Nb co-doping [126] produces a stronger ferromagnetic response, which significantly increases remnant magnetization. Sm-doping, on the other hand, mainly alters the crystallographic structure by collapsing the spin cycloid and bending Fe–O–Fe links, leading to a slight increase in magnetism. Furthermore, Sm^{3+} contributes to the observed magnetic response due to its magnetic activity. Nd–Nb co-doping is a more successful method of improving the magnetic characteristics of BFO because it generally shows a more noticeable ferromagnetic character than Sm-doped BFO, which shows a more gradual increase.

Antiferromagnetism

The Neel temperature (T_N) of BFO is high, at about 370 °C (640 K). The material changes into a paramagnetic condition above this temperature [158]. BFO is appropriate for high-temperature magnetic uses due to its high T_N. Interesting magnetic events might result from the disruption of the cycloidal spin structure that characterizes the magnetic order [159]. Its antiferromagnetic qualities, especially when present in nanoparticle form, are highly desirable because of its possible uses in data storage, spintronic, and magnetoelectric devices. The magnetic moments of Fe with +3 ions are regularly arranged in opposite orientations in BFO nanostructures, which display a G-type antiferromagnetic ordering [158, 160]. This causes the material to have a net zero magnetic moment, which is a feature of antiferromagnetism. BFO's antiferromagnetic characteristics can be considerably changed by reducing it to the nanoscale. Because of the increased effect of surface atoms which limited size effects, changes in magnetic ordering result from the reduction in particle size. Compared to bulk materials, nanoparticles have a substantially greater surface-to-volume ratio. Uncompensated surface spins may result from this, altering the antiferromagnetic ordering or adding to a weak ferromagnetic component. This behavior, also known as spin canting, has been observed in diluted magnetic materials like as Fe, Ni, and Co-doped PZT NPs as well as magnetic alloys like Fe/Ni nanoparticles [161–163]. Since spin-based information storage and transfer are essential features of spintronic devices, BFO nanoparticles' antiferromagnetic characteristics make them promising candidates.

Due to spin canting, BFO nanostructures frequently show modest ferromagnetism in practice. Because of their restricted size and enhanced surface effects, weak ferromagnetic behavior becomes more pronounced in nanoparticles. Researchers frequently dope BFO with additional elements (rare-earth metals) or produce composite to improve magnetic characteristics. Doping can enhance the material's magnetoelectric interaction and alter the antiferromagnetic ordering. Co-doping Gd and Ni has been found to improve this BFO characteristic [160]. Bai et al. showed in another study that co-doping greatly enhances BFO's magnetic characteristics. Both the saturation magnetization Ms and the residual magnetization Mr exhibit a notable increase with increasing co-doping concentration. The main causes of these improvements in Ms and Mr are the disturbance of the spiral spin structure, denser particle collection, greater super-exchange contacts, and a decrease in the concentration of oxygen defects, as seen in Fig. 8 [164].

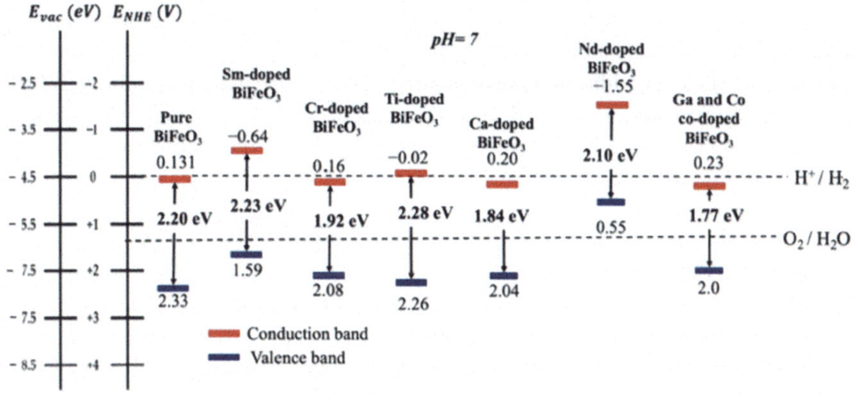

Fig. 8 Band gap positions concerning valence and conduction band at pH = 7 of BFO, Sm–BFO, Cr–BFO, Ti–BFO, Ca–BFO, Nd–BFO, Ga–, and Co–co–BFO [239] Copyright, Open access, MDPI publishing

Multiferroic Magnetoelectric Coupling

Materials that simultaneously display multiple basic ferroic orders, such as ferroelectricity, ferroelasticity, and ferromagnetism, are known as multiferroics. Because of their capacity to combine electric and magnetic properties, these materials are of great interest for next-generation devices [165]. Because of its room-temperature multiferroic characteristics, $BiFeO_3$ is one of the most researched multiferroic materials [166]. Because of the displacement of Bi with 3+ ions about the oxygen octahedra in the crystal, BFO displays spontaneous electric polarization. With polarization along the [67] direction, this ferroelectricity is present even at normal temperatures [167]. Because the Fe^{3+} spins are aligned so that adjacent spins are antiparallel, $BiFeO_3$ displays G-type antiferromagnetism. However, a small canting of spins indicates to weak ferromagnetism because of the Dzyaloshinskii–Moriya (DM) interaction [168].

The magnetoelectric coupling of BiFeO3 nanostructures, in which the magnetic characteristics of the material can affect its electric polarization and vice versa, is one of its most remarkable features. Additionally, $BiFeO_3$ nanoparticles may show an exchange bias effect, in which the hysteresis loop shifts as a result of the interaction between the ferromagnetic and antiferromagnetic phases [169]. In spintronic devices, this phenomenon is especially helpful. This linkage makes it possible to regulate electrical qualities with a magnetic field and magnetic order with an electric field, which is very promising for devices with many uses. The antiferromagnetic structure of BFO nanoparticles can be affected by the uses of an external magnetic field, which can alter the ferroelectric characteristics of the particles. One important area of study is this field-induced magnetoelectric coupling [170, 171]. Additionally, this characteristic can be utilized to create penetrating magnetic field sensors, which are crucial for a variety of applications, from industrial automation to medical diagnostics (such as identifying magnetic fields in the brain). $BiFeO_3$-based sensors are extremely energy-efficient because they can detect magnetic fields by electric polarization with no requirement for an additional power source. $BiFeO_3$ has a high T_N (about 370 °C) and Curie temperature (around 830 °C), as was previously reported. This ensures steady presentation over a broad temperature range, which is beneficial for real-world sensor uses [172–174].

Magnetic Field Sensors

Wearable technology, drones, and autonomous cars are just a few examples of advanced navigation systems that can make use of nanostructured BFO-based magnetic field sensors. These sensors can be used in imaging the magnetic particle for biomedical purposes or magnetic resonance imaging (MRI) systems. BFO magnetic sensors ensure the efficiency and safety of equipment by monitoring magnetic fields in industrial operations. High-sensitivity magnetic sensors are useful for identifying magnetic abnormalities, which can be used for underwater mine detection or security screening [175, 176]. Because of the magnetoelectric interaction, an external magnetic field can cause changes in the electric polarization of a BFO_3-based sensor. The sensor can determine the strength or direction of the magnetic field by detecting this change in polarization as an electrical output signal. On the other hand, BFO's magnetic state can be altered by an electric field, which makes it suitable for sensor setups that call for active control/modification of the magnetic sensitivity of the sensor [177, 178]. BFO is frequently combined with a magnetostrictive substance in nanocomposite sensors to improve the magnetoelectric response. When exposed to a magnetic field, the magnetostrictive material deforms, causing strain in the $BiFeO_3$ layer. This changes the polarization of the layer and produces a detectable electrical signal. Weak magnetic fields can be detected thanks to this strain-mediated coupling, which can greatly increase the sensor's sensitivity [179, 180]. $BiFeO_3$ and magnetostrictive materials are combined in these sensors to form a composite structure with improved magnetoelectric coupling. The $BiFeO_3$ layer transforms the mechanical reaction of the magnetostrictive layer to magnetic fields into an electrical signal. Applications needing high sensitivity to magnetic fields, like industrial monitoring, defense technologies, and scientific instrumentation, benefit greatly from composite sensors [181, 182]. Planar magnetic sensors are created by depositing $BiFeO_3$ thin films onto electrode-equipped substrates. Compact sensor devices can be made and integrated with other electronic components thanks to the thin-film structure. These sensors can observe magnetic fields in handheld devices like navigation systems, cellphones, and biomedical equipment, making them appropriate for application in microelectromechanical systems (MEMS) [183]. To construct tiny, very sensitive magnetic sensors, $BiFeO_3$ can be manufactured into nanoscales like nanowires, nanodots, or nanopillars. Data storage devices, high-resolution magnetic imaging, and the recognition of magnetic NPs for medical purposes all make use of nanoscale magnetic sensors [184].

Without the requirement for cooling systems, nanostructured BFO magnetic field sensors can be used in various real-world applications since they function well at ambient temperatures. Additionally, very low magnetic field strengths can be detected thanks to $BiFeO_3$'s high magnetoelectric interaction, particularly when paired with magnetostrictive materials. Thin-film fabrication of $BiFeO_3$ makes it simple to integrate with traditional semiconductor technologies, resulting in small and adaptable sensor systems. Furthermore, $BiFeO_3$'s combination of ferroelectric and magnetic characteristics opens up new opportunities for sensor technologies by enabling the creation of multipurpose devices that can analyze electric and magnetic fields [185, 186]. Superior BFO films are necessary for dependable sensor operation. Improvements in manufacturing procedures are the focus of ongoing study because crystal structure defects can lower the sensitivity and efficiency of magnetoelectric coupling. It is difficult to maintain the material's characteristics at tiny sizes, which makes it difficult to develop nanoscale $BiFeO_3$-based sensors with reliable performance. The goal of research is to comprehend and overcome these obstacles. Another ongoing research field aiming at improving the sensitivity and robustness of these sensors is improving the interface manufacturing between magnetostrictive layers and $BiFeO_3$ to enhance the magnetoelectric coupling.

4.1.4 Properties of Piezoelectrics

It is possible to detect and quantify strain or pressure since the strength of this electric field directly depends on its applied mechanical stress. Electrodes affixed to the $BiFeO_3$ material may gather the produced electric charge, transforming the mechanical stress into electrical energy suitable for measurement and analysis [187, 188]. Nanostructures of $BiFeO_3$ show piezoelectric characteristics, especially when it is nanostructured [189, 190].

Despite generally is less powerful than those of specialized piezoelectric materials. Nonetheless, $BiFeO_3$ is an intriguing material for multipurpose devices due to its piezoelectricity combined with ferroelectric and magnetic characteristics [191, 192]. Furthermore, the ferroelectric characteristics of $BiFeO_3$ are intimately linked to its piezoelectric action [193]. The spontaneous polarization brought on by the ion's internal movement inside the crystal.

The multiferroic material bismuth ferrite ($BiFeO_3$) has piezoelectric characteristics, especially when it is in the form of nanoparticles [194, 195]. Despite this, they usually lack the strength of certain piezoelectric materials. Nonetheless, $BiFeO_3$ is an intriguing material for multipurpose devices due to its ferroelectric, magnetic, and piezoelectric characteristics [196, 197]. An outline of $BiFeO_3$ nanoparticles piezoelectric characteristics is shown below.

Intrinsic Piezoelectricity: $BiFeO_3$ exhibits piezoelectric activity due to its non-centrosymmetric crystal structure (rhombohedral with space group R3c). Accordingly, it produces an electric charge when mechanical force is applied and experiences mechanical deformation when an electric field is provided. Additionally, $BiFeO_3$ ferroelectric characteristics are intimately linked to its piezoelectric action

[143]. The piezoelectric response is influenced by spontaneous polarization that results from ion displacement inside the crystal structure [198]. There is a technique to know the response of the piezoelectric effect so many researchers reported their results by using the piezoelectric force microscopy (PFM) technique. The ferroelectric and piezoelectric characteristics of BFO and Cr-doped BFO (BFCO) films were investigated using piezoresponse force microscopy (PFM), which in addition to surface topography gave important information about piezoelectric and ferroelectric behavior through recording as piezoresponse amplitude butterfly curve and phase hysteresis curves within a particular grid. Figure 7A a, d shows the morphology of the BFO-FTO and BFCO-STO thin films, respectively, within a $3 \times 3\ \mu m^2$ region. Both topographic images show that the films' surfaces have uniformly distributed small grains, whereas the BFO has a compact grain structure. As more active sites take part in the degradation process, the presence of evenly distributed particles may often guarantee a larger surface area and boost the photocatalytic reaction. Furthermore, as shown in the next section, the intrinsic electric field generated at the $Bi_2Fe_4O_9$ and $BiFeO_3$ heterojunction [199] further improves the photocatalytic activity. The BFO-FTO and BFCO-STO thin films' respective out-of-plane piezoresponse amplitude maps are shown in Fig. 7. The presence of spontaneous and separate piezoelectric domains with an impressive piezoelectric response is shown by the BFO thin film's (Fig. 7b) discernible amplitude fluctuation in various places. Figure shows the equivalent out-of-plane phase images. Domains with varying polarization and orientation inside the films are shown by the phase angle, which shows the link with the applied tip bias voltage and the piezoresponse. The existence of well-defined and consistently formed domain barriers is seen in Fig. 7.

Figure 7B presents the PFM data detailing the polarization switching curves of BFO and BFCO, showing observable variations in the piezoresponse amplitude (butterfly curves) and phase angle shifts within the local hysteresis loops. These differences verify that both BFO and BFCO mean thin films exhibit ferroelectricity and piezoelectricity. Changes in strain by an external electric field applied by the AFM tip are responsible for the observed variations in the PFM amplitude butterfly curves. Both BFO and BFCO piezoresponse curves have the distinctive butterfly form; Fig. 7a shows that BFO's maximum amplitude is approximately 250 pm under $\pm$ 5 V DC tip bias voltage, while Fig. 7b shows that BFCO's maximum amplitude is about 130 pm under $\pm$ 10 V DC tip bias voltage [200].

BFO experiences deformation in response to mechanical forces like strain or pressure, which results from changes in atomic locations in the crystal lattice. This deformation causes a rearrangement of charges inside the material, producing an electric field, much like in perovskite materials like lead-free (KNN) and lead-based (PZT) piezoelectric materials [201]. Strain or pressure can be detected and measured since the strength of this electric field directly depends on the applied mechanical pressure. Electrodes affixed to the $BiFeO_3$ material can gather the produced electric charge, transforming the mechanical energy into electricity suitable for measurement and analysis [201]. Piezoelectric characteristics are demonstrated by $BiFeO_3$ nanostructures, especially when they are nanostructured [201, 202]. Despite this, they usually lack the strength of specific piezoelectric materials. However, $BiFeO_3$

is an intriguing material for multipurpose devices due to its mix of ferroelectric, magnetic, and piezoelectric capabilities [203, 204]. Additionally, $BiFeO_3$'s ferroelectric characteristics are intimately linked to its piezoelectric action. The piezoelectric response is influenced by the random polarization that results from the shift of ions within the crystal, [205]. When $BiFeO_3$ is in nanoparticle form, its piezoelectric qualities can be improved. The enhanced surface area and surface states, which can affect the mechanical response and polarization, are the causes of this improvement. The piezoelectric response may alter as the particle size reduces [206]. For example, compared to bulk materials, smaller nanoparticles may have a greater piezoelectric response because of increased surface effects and fewer flaws [207]. Additionally, temperature is a major factor; in the phase diagram and its temperature dependency. To improve safety and provide early warnings of possible structural failures, $BiFeO_3$-based strain sensors can be integrated into bridges, buildings, and other structures to monitor strain/pressure changes over time [208]. $BiFeO_3$ sensors can be utilized to make tactile sensors in robotics that can sense force, pressure, and touch, giving robots more precise and controllable control over how they interact with their surroundings [209, 210]. By keeping an eye on mechanical stresses in machinery and equipment, these sensors can guarantee peak performance and guard against damage. By measuring forces at the cellular level, nanoscale BFO sensors can shed light on cellular mechanics and facilitate the creation of cutting-edge biomedical instruments for study and diagnosis [211, 212]. Because $BiFeO_3$ thin films are easily integrated with other electronic components and can be produced on flexible substrates, they are frequently utilized in strain and pressure sensors. Small mechanical deformations can be detected by these thin-film sensors in a variety of applications. These sensors are perfect for applications requiring small and light sensors, such as microelectromechanical systems (MEMS), wearable technology, and flexible electronics [213]. To improve strain and pressure sensing capabilities, BFO is mixed with other piezoelectric or magnetostrictive materials in composite systems. BFO can be used with polymers, for instance, to make elastic and flexible sensors. In sophisticated robotics, industrial automation, and structure health monitoring, where accurate mechanical force measurement is crucial, composite sensors are employed [214, 215]. Highly sensitive strain and pressure sensors with a compact size can be made using BFO nanostructures, such as nanowires/nanoparticles. These tiny sensors have a great resolution for detecting even the smallest mechanical deformations. Tactile sensors for robotic systems and biomedical applications like the measurement of cellular forces both use nanoscale sensors [216]. BFO's piezoelectric quality makes it appropriate for several uses, including:

Piezoelectric Energy

Piezoelectric energy harvester transforms mechanical energy from pressure or strain, vibrations, or movements into electrical energy. $BiFeO_3$'s piezoelectric characteristics provide an electrical charge when it experiences mechanical stress. These gadgets can be employed in settings with a lot of mechanical energy, like autonomous sensors, wearable technology, and structural health monitoring systems. To generate power from walking, for example, a piezoelectric harvester based on $BiFeO_3$ might be implanted in the sole of a shoe [217, 218].

Pressure and Strain Sensor

BFO may transform mechanical stress, like strain or pressure, into electrical energy thanks to its piezoelectric characteristics. This makes it appropriate for use in touch sensors, pressure sensors, and strain gauges. BiFeO3 is valuable in precision sensing applications, such as civil engineering and structural health monitoring in aerospace, and, because of its strong piezoelectric response, which allows it to detect even the smallest changes in pressure or deformation [219]. BFO can produce an electric charge in reaction to mechanical pressure because it is intrinsically piezoelectric. This characteristic is essential for its use in pressure and strain sensors. By introducing an external mechanical/electric field, BFO's spontaneous electric polarization can be reoriented. As a result, mechanical energy can be directly transformed into electrical impulses. BFO is appropriate for tough conditions because of its high T_C ~ 830 °C, which guarantees that its ferroelectric and piezoelectric capabilities stay stable across a wide temperature range. Furthermore, the material has a link between its magnetic and electric characteristics, which makes it possible to detect mechanical deformations by observing variations in both magnetization and electric polarization [220].

MEMS Microactuators

By using the piezoelectric characteristics of $BiFeO_3$, precise actuators that transform electrical signals into mechanical motion can be produced. These actuators are essential for applications that need precise control, like robotics, optoelectronic devices, and microelectromechanical systems (MEMS). $BiFeO_3$'s potent piezoelectric effect allows for extremely accurate positioning and actuation, which makes it appropriate for applications such as industrial micro-part movement, valve control, and optical component alignment [221]. These micro-actuators are essential for use in inkjet printers, biomedical devices, and micro-robotics because they can power small mechanical parts in equipment like microscale pumps, switches, and sensors [222]. $BiFeO_3$-based actuators provide flexible actuation capabilities for intricate MEMS systems by integrating piezoelectric, ferroelectric, and magnetoelectric effects to function in different modes [223]. BFO's robust piezoelectric and magnetoelectric responses make these sensors extremely sensitive since they can detect even minute mechanical deformations. Additionally, BFO-based sensors offer multifunctional sensing capability in a device by simultaneously detecting strain, pressure, and magnetic fields. Furthermore, BFO's high Curie temperature guarantees consistent sensor performance in high-temperature settings, which makes it appropriate for industrial and aerospace applications. Lastly, the expansion of scalable sensor technologies that can be incorporated into a range of stages is made possible by the fabrication of BFO in thin-film/nanoscale forms, as shown in Fig. 6 [224, 225].

Leakage currents are a problem when using $BiFeO_3$ for piezoelectric applications since they can reduce the material's piezoelectric performance. Defects like oxygen vacancies are frequently to blame for this to lower leakage currents and enhance piezoelectric characteristics [226, 227]. $BiFeO_3$'s ferroelectric qualities may be hampered by its high leakage currents. Grain boundaries, oxygen vacancies, and other flaws are frequently blamed for this. This leakage is a serious problem for memory device applications since it can conceal the actual ferroelectric behavior. It is difficult and demands exact control over the synthesis conditions to produce high-quality BFO nanoparticles with regulated shape and size [228]. To lower leakage currents in $BiFeO_3$ nanoparticles, a variety of techniques are used, including doping and the development of composite architectures. The electrical performance of the material can be enhanced and oxygen vacancies reduced by doping with elements such as manganese (Mn) or lanthanum (La) [229–232].

4.1.5 Applications and Characteristics of Pyroelectric

$BiFeO_3$ can produce an electrical signal in reaction to temperature variations thanks to its pyroelectric characteristics. This can be used to create infrared detectors or temperature sensors, which are essential for non-contact temperature measurement, thermal imaging, and environmental monitoring. $BiFeO_3$ shows temperature-dependent spontaneous electric polarization. Temperature variations can be inferred by measuring this temperature-dependent polarization [233]. Additionally, $BiFeO_3$ has a Curie temperature of about 830 °C, which enables it to function well throughout a broad temperature range and makes it appropriate for both room-temperature and high-temperature applications. Furthermore, $BiFeO_3$'s ferroelectric characteristics can be combined with its temperature-dependent antiferromagnetic ordering to provide sensitive temperature sensors. Further assisting in temperature measurement, $BiFeO_3$'s piezoelectric characteristics enable it to transform mechanical strain brought on by thermal expansion into an electrical signal [234].

Pyroelectric Energy Harvester

The pyroelectric action in $BiFeO_3$, where temperature changes alter the polarization of material and produce an electric charge, is used by pyroelectric energy harvesters. This makes it possible to capture waste heat or changes in the surrounding temperature. Pyroelectric energy harvesters can be used to power low-energy instruments from body heat or in contexts like factories where thermal energy is easily accessible [235, 236].

Temperature Sensor

Manufacturing equipment, chemical reactors, and furnaces are just a few examples of industrial processes where $BiFeO_3$ sensors can be utilized to monitor and regulate temperature. Because $BiFeO_3$ may function at high temperatures, it can be used in automotive and aerospace applications where precise temperature measurement is essential for both performance and safety [47]. Temperature variations in urban and natural settings can be measured using $BiFeO_3$ temperature sensors in environmental monitoring systems. $BiFeO_3$-based sensors are appropriate for temperature monitoring in medical devices and diagnostic sensors/device due to their sensitivity and stability, as shown in Fig. 7 [48]. Because the lattice characteristics and the orientation of the electric dipoles vary with temperature, BFO's spontaneous polarization also varies. The electric field or voltage produced across the material can be measured to identify this temperature-dependent polarization. When measured using the proper electrodes, the electrical signal (voltage or capacitance) varies as a result of the change in polarization. Accurate temperature monitoring is made possible by the correlation between this signal and temperature changes [49]. The temperature-dependent nature of the antiferromagnetic ordering in $BiFeO_3$ is noteworthy [50]. Temperature variations can be inferred by tracking changes in magnetic characteristics, such as magnetization. Enhanced sensing capabilities can be achieved by combining this with the ferroelectric response. $BiFeO_3$ experiences mechanical strain as a result of temperature-induced thermal expansion or contraction. By changing the material's piezoelectric sensitivity, this strain produces an electrical signal that is proportional to the temperature [51]. Temperature sensors can be made by depositing $BiFeO_3$ thin films onto surfaces. These films may give precise temperature readings over various temperatures and are appropriate for incorporation into electronic circuits. When small and accurate temperature sensing is needed, thin-film sensors are employed in industrial processes, electrical devices, and environmental monitoring [52]. $BiFeO_3$ in bulk ceramic forms can be utilized in temperature sensors for applications requiring high temperatures. Usually, these sensors are employed in settings requiring strong, long-lasting sensors. Because bulk ceramic sensors can tolerate high temperatures and challenging environments, they can be used in industrial, automotive, and aerospace applications [53, 54]. Small, extremely sensitive temperature sensors can be made using $BiFeO_3$ nanostructures like nanowires or nanoparticles. High-resolution temperature readings are possible with these tiny sensors. For applications in advanced research, microelectronics, and biomedical devices, where high sensitivity and small size are crucial, nanoscale sensors are perfect [55, 56].

4.1.6 Optical Properties

Bismuth ferrite ($BiFeO_3$) is particularly useful when produced as nanoparticles. Because of these characteristics, $BiFeO_3$ is useful for photovoltaic, optoelectronic, and photocatalytic citation applications. $BiFeO_3$ nanoparticles are distinguished by their straight band gap, which normally falls between 2.2 and 2.8 eV. $BiFeO_3$ is ideal for applications powered by visible light because of its band gap, which is located within the visible light spectrum. Particle size, synthesis technique, and doping are among the variables affecting the precise band gap value. Notably, $BiFeO_3$ nanoparticles' band gap tends to widen as their particle size decreases, further tuning their optical properties for certain application. Yassine Nassereddine et al. reported a comparative study of novel perovskite material BFO, which has been reported Gd in place of Bi and Co in place of Fe to improve the photocatalytic dissociation of water splitting and produce H_2 by photoelectrochemical splitting of water [237]. Figure 8 illustrates how the doping levels cause the band gap engineering to shift from 2.23 eV to 1.77 eV. The resultant materials' photocatalytic response is enhanced by this band gap reduction. With high photo-conversion capability of 2.29%, Gd/Co co-doped BFO has the lowest band gap (1.77 eV) and the largest H_2 generation rate (74.57 mmol h^{-1} cm^{-2}). Therefore, the improved solar spectrum response of Co–co-doped BFO is the reason for its increased rate of H_2 generation and improved photo-conversion efficiency [239]. $BiFeO_3$ (BFO) exhibits stable optical properties that support its multifunctional photocatalytic performance. In piezo-photocatalysis, BFO combines light absorption and mechanical stress to enhance charge separation. Upon visible or UV light exposure, electrons are excited from the valence band to the conduction band, forming electron–hole pairs. Simultaneously, mechanical stress induces a piezoelectric field that drives these charges apart, reducing recombination and boosting photocatalytic efficiency. This synergy highlights BFO's potential in light-driven energy and environmental applications.

5 Conclusion

At ambient temperature, bismuth ferrite ($BiFeO_3$) exhibits a unique combination of ferroelectric, antiferromagnetic, and piezoelectric properties, making it one of the most promising multiferroic materials. The substantial study on $BiFeO_3$-based materials is highlighted in this chapter, with particular attention paid to their important characteristics, including Neel temperatures and high Curie, photovoltaic effects, notable piezoelectric response, and magnetoelectric coupling. These characteristics make $BiFeO_3$ suitable for a broad range of applications, including optoelectronics, sensors, actuators, memory devices, and energy harvesting devices like solar cells. $BiFeO_3$'s capacity to grip visible light and produce photocurrent makes it a promising material for energy harvesting for plasmonic photovoltaic devices and next-generation solar cells. It is utilized in optoelectronic devices including waveguides and light-emitting diodes (LEDs), and its piezoelectric and magnetoelectric properties make it appropriate for good-performance sensors and actuators. Notwithstanding the material's potential, commercial $BiFeO_3$-based device development has been hampered by issues such as limited leakage currents, carrier mobility, and fabrication complexity. Ongoing developments in interface engineering and thin-film deposition, however, are enhancing these materials' scalability, stability, and performance. This chapter highlights how $BiFeO_3$-based materials are anticipated to revolutionize future technologies by providing potential answers in areas such as sophisticated electronics, sensing systems, and renewable energy. $BiFeO_3$ is positioned as a crucial material for innovation across a range of industries due to its special qualities and ongoing research into overcoming present constraints. $BiFeO_3$ nanostructures are extremely valuable for next-generation electronic, memory, sensor, energy, and environmental applications because they provide a special mix of ferroelectric, piezoelectric, dielectric, photovoltaic, and magnetic properties. Improvements to BFO-based heterostructures, magnetoelectric coupling, and leakage current reduction are the main areas of future research.

References

1. Schmid H (1994) Multi-ferroic magnetoelectrics. Ferroelectrics 162(1):317–338. https://doi.org/10.1080/00150199408245120
2. Spaldin NA, Fiebig M (2005) The renaissance of magnetoelectric multiferroics. Science 309(5733):391–392. https://doi.org/10.1126/science.1113357
3. Eerenstein W, Mathur ND, Scott JF (2006) Multiferroic and magnetoelectric materials. Nature 442(7104):759–765. https://doi.org/10.1038/nature05023
4. Wang J et al (2003) Epitaxial $BiFeO_3$ multiferroic thin film heterostructures. Science 299(5613):1719–1722. https://doi.org/10.1126/science.1080615
5. Hill NA (2000) Why are there so few magnetic ferroelectrics? J Phys Chem B 104(29):6694–6709. https://doi.org/10.1021/jp000114x
6. Spaldin NA, Cheong SW, Ramesh R (2010) Multiferroics: past, present, and future. Phys Today 63(10):38–43. https://doi.org/10.1063/1.3502547
7. Fiebig M (2005) Revival of the magnetoelectric effect. J Phys D Appl Phys 38(8):R123–R152. https://doi.org/10.1088/0022-3727/38/8/R01
8. Chappert C, Fert A, Van Dau FN (2007) The emergence of spin electronics in data storage. Nat Mater 6(11):813–823. https://doi.org/10.1038/nmat2024
9. Ramesh R, Spaldin NA (2007) Multiferroics: progress and prospects in thin films. Nat Mater 6(1):21–29. https://doi.org/10.1038/nmat1805
10. Martin LW, Crane SP, Chu YH, Holcomb MB, Gajek M, Huijben M, Ramesh R (2008) Multiferroics and magnetoelectrics: thin films and nanostructures. J Phys Condens Matter 20(43):434220. https://doi.org/10.1088/0953-8984/20/43/434220
11. Chauhan S, Kumar M, Chhoker S, Katyal SC, Singh H, Jewariya M, Yadav KL (2012) Multiferroic, magnetoelectric and optical properties of Mn doped $BiFeO_3$ nanoparticles. Solid State Commun 152(6):525–529. https://doi.org/10.1016/j.ssc.2011.12.037
12. Chauhan S, Kumar M, Chhoker S, Katyal SC, Singh M (2016) Substitution driven structural and magnetic transformation in Ca-doped $BiFeO_3$ nanoparticles. RSC Adv 6(49):43080–43090. https://doi.org/10.1039/C6RA02316A
13. Heron JT, Schlom DG, Ramesh R (2014) Electric field control of magnetism using $BiFeO_3$-based heterostructures. Appl Phys Rev 1(2). https://doi.org/10.1063/1.4870957
14. Ramesh R (2014) Electric field control of ferromagnetism using multi-ferroics: the bismuth ferrite story. Philos Trans R Soc A Math Phys Eng Sci 372(2009):20120437. https://doi.org/10.1098/rsta.2012.0437
15. Yang JC, He Q, Yu P, Chu YH (2015) $BiFeO_3$ thin films: a playground for exploring electric-field control of multifunctionalities. Annu Rev Mater Res 45(1):249–275. https://doi.org/10.1146/annurev-matsci-070214-020837

16. Catalan G, Scott JF (2009) Physics and applications of bismuth ferrite. Adv Mater 21(24):2463–2485. https://doi.org/10.1002/adma.200802849
17. Silva J, Reyes A, Esparza H, Camacho H, Fuentes L (2011) $BiFeO_3$: a review on synthesis, doping and crystal structure. Integr Ferroelectr 126(1):47–59. https://doi.org/10.1080/10584587.2011.574986
18. Hussain A, Xu X, Yuan G, Wang Y, Yang Y, Yin J, Liu Z (2014) The development of $BiFeO_3$-based ceramics. Chin Sci Bull 59:5161–5169. https://doi.org/10.1007/s11434-014-0648-0
19. Rojac T, Bencan A, Malic B, Tutuncu G, Jones JL, Daniels JE, Damjanovic D (2014) $BiFeO_3$ ceramics: processing, electrical, and electromechanical properties. J Am Ceram Soc 97(7):1993–2011. https://doi.org/10.1111/jace.12982
20. Chu YH, Martin LW, Zhan Q, Yang PL, Cruz MP, Lee K, Ramesh R (2007) Epitaxial multiferroic $BiFeO_3$ thin films: progress and future directions. Ferroelectrics 354(1):167–177. https://doi.org/10.1080/00150190701454867
21. Sando D, Barthélémy A, Bibes M (2014) $BiFeO_3$ epitaxial thin films and devices: past, present and future. J Phys Condens Matter 26(47):473201. https://doi.org/10.1088/0953-8984/26/47/473201
22. Hill NA, Filippetti A (2002) Why are there any magnetic ferroelectrics? J Magn Magn Mater 242:976–979. https://doi.org/10.1016/S0304-8853(01)01078-2
23. Spaldin NA, Ramesh R (2019) Advances in magnetoelectric multiferroics. Nat Mater 18(3):203–212. https://doi.org/10.1038/s41563-018-0275-2
24. Nan CW, Bichurin MI, Dong S, Viehland D, Srinivasan G (2008) Multiferroic magnetoelectric composites: historical perspective, status, and future directions. J Appl Phys 103(3). https://doi.org/10.1063/1.2836410
25. Martin LW, Chu YH, Ramesh RJMS (2010) Advances in the growth and characterization of magnetic, ferroelectric, and multiferroic oxide thin films. Mater Sci Eng R Rep 68(4–6):89–133. https://doi.org/10.1016/j.mser.2010.03.001
26. Ma J, Hu J, Li Z, Nan CW (2011) Recent progress in multiferroic magnetoelectric composites: from bulk to thin films. Adv Mater 23(9):1062–1087. https://doi.org/10.1002/adma.201003636
27. Khomchenko VA, Kiselev DA, Kopcewicz M, Maglione M, Shvartsman VV, Borisov P, Kholkin AL (2009) Doping strategies for increased performance in BiFeO3. J Magn Magn Mater 321(11):1692–1698. https://doi.org/10.1016/j.jmmm.2009.02.008
28. Martin LW, Ramesh R (2012) Multiferroic and magnetoelectric heterostructures. Acta Mater 60(6–7):2449–2470. https://doi.org/10.1016/j.actamat.2011.12.024
29. Kim DH, Lee HN, Biegalski MD, Christen HM (2008) Effect of epitaxial strain on ferroelectric polarization in multiferroic $BiFeO_3$ films. Appl Phys Lett 92(1). https://doi.org/10.1063/1.2830799
30. Reddy VA, Pathak NP, Nath R (2013) Enhanced magnetoelectric coupling in transition-metal-doped $BiFeO_3$ thin films. Solid State Commun 171:40–45. https://doi.org/10.1016/j.ssc.2013.07.032
31. Lu Z, Fan Z, Li P, Fan H, Tian G, Song X, Liu J (2016) Ferroelectric resistive switching in high-density nanocapacitor arrays based on $BiFeO_3$ ultrathin films and ordered Pt nanoelectrodes. ACS Appl Mater Interfaces 8(36):23963–23968. https://doi.org/10.1021/acsami.6b07792
32. Guo Y, Xiao P, Wen R, Wan Y, Zheng Q, Shi D, Lin D (2015) Critical roles of Mn-ions in enhancing the insulation, piezoelectricity and multiferroicity of $BiFeO_3$-based lead-free high temperature ceramics. J Mater Chem C 3(22):5811–5824. https://doi.org/10.1039/C5TC00507H
33. Zhang W, Fan M, Li L, Chen A, Su Q, Jia Q, Wang H (2015) Heterointerface design and strain tuning in epitaxial $BiFeO_3$: $CoFe_2O_4$ nanocomposite films. Appl Phys Lett 107(21). https://doi.org/10.1063/1.4936157
34. Zhao Y, Yin Z, Zhang X, Fu Z, Wu J (2015) Formation and local conduction of nanopits in $BiFeO_3$ epitaxial films. J Mater Chem C 3(42):11250–11256. https://doi.org/10.1039/C5TC02704G

35. Xie X, Yang S, Zhang F, Fan S, Che Q, Wang C, Zhang L (2015) Effects of excess Bi on structure and electrical properties of $BiFeO_3$ thin films deposited on indium tin oxide substrate using sol–gel method. J Mater Sci Mater Electron 26:10095–10101. https://doi.org/10.1007/s10854-015-3693-9
36. Li Y, Cao MS, Wang DW, Yuan J (2015) High-efficiency and dynamic stable electromagnetic wave attenuation for La doped bismuth ferrite at elevated temperature and gigahertz frequency. RSC Adv 5(94):77184–77191. https://doi.org/10.1039/C5RA15458H
37. Li S, Zhang G, Zheng H, Wang N, Zheng Y, Wang P (2016) Microwave-assisted synthesis of $BiFeO_3$ nanoparticles with high catalytic performance in microwave-enhanced Fenton-like process. RSC Adv 6(85):82439–82446. https://doi.org/10.1039/C6RA12728B
38. Zhou W, Deng H, Cao H, He J, Liu J, Yang P, Chu J (2015) Effects of Sm and Mn co-doping on structural, optical and magnetic properties of $BiFeO_3$ films prepared by a sol–gel technique. Mater Lett 144:93–96. https://doi.org/10.1016/j.matlet.2015.01.038
39. Tang P, Kuang D, Yang S, Zhang Y (2015) The structural, optical and enhanced magnetic properties of $Bi_{1-x}Gd_xFe_{1-y}Mn_yO_3$ nanoparticles synthesized by sol–gel. J Alloy Compd 622:194–199. https://doi.org/10.1016/j.jallcom.2014.10.035
40. Zhang X, Wang B, Wang X, Xiao X, Dai Z, Wu W, Jiang C (2015) Preparation of M@ $BiFeO_3$ nanocomposites (M = Ag, Au) bowl arrays with enhanced visible light photocatalytic activity. J Am Ceram Soc 98(7):2255–2263. https://doi.org/10.1111/jace.13543
41. Ye W, Tan G, Yan X, Ren H, Xia A (2015) Influence of Mn dopants on the structure and multiferroic properties of a $Bi_{0.90}$ $Ho_{0.10}$ FeO_3 thin film. RSC Advances 5(54):43594–43600. https://doi.org/10.1039/C5RA00609K
42. Min K, Huang F, Jin Y, Lu X, Wu H, Zhu J (2015) Control of oxygen vacancies and their kinetic behaviours via reversible oxygen loss in $BiFeO_3$ ceramics. J Phys D Appl Phys 48(44):445301. https://doi.org/10.1088/0022-3727/48/44/445301
43. Jha PK, Jha PA, Kumar P, Asokan K, Dwivedi RK (2016) Gradient core-shell microstructure in mixed valence multiferroic: TM (Ti, Nb, W) substituted bismuth ferrite. J Alloy Compd 667:178–183. https://doi.org/10.1016/j.jallcom.2016.01.170
44. Priya AS, Banu IS, Anwar S (2016) Influence of Dy and Cu doping on the room temperature multiferroic properties of BiFeO3. J Magn Magn Mater 401:333–338. https://doi.org/10.1016/j.jmmm.2015.10.059
45. Li ZJ, Hou ZL, Song WL, Liu XD, Cao WQ, Shao XH, Cao MS (2016) Unusual continuous dual absorption peaks in Ca-doped $BiFeO_3$ nanostructures for broadened microwave absorption. Nanoscale 8(19):10415–10424. https://doi.org/10.1039/C6NR00223D
46. Agarwal R, Sharma Y, Katiyar RS (2015) Switchable photovoltaic and polarization modulated rectification in Si-integrated Pt/$(Bi_{0.9}Sm_{0.1})(Fe_{0.97}Hf_{0.03})$ O_3/$LaNiO_3$ heterostructures. Appl Phys Lett 107(16). https://doi.org/10.1063/1.4934665
47. Li L, Lu L, Zhang D, Su R, Yang G, Zhai J, Yang Y (2016) Direct observation of magnetic field induced ferroelectric domain evolution in self-assembled Quasi (0–3) $BiFeO_3$–$CoFe_2O_4$ thin films. ACS Appl Mater Interfaces 8(1):442–448. https://doi.org/10.1021/acsami.5b09265
48. Xu X, Liu W, Wu P, Zhang H, Guo M, Han Y, Wang S (2015) Abnormal variation of band gap in Zn doped $Bi_{0.9}La_{0.1}FeO_3$ nanoparticles: role of Fe–O–Fe bond angle and Fe–O bond anisotropy. Appl Phys Lett 107(4). https://doi.org/10.1063/1.4927644
49. Hu JM, Chen LQ, Nan CW (2016) Multiferroic heterostructures integrating ferroelectric and magnetic materials. Adv Mater 28(1):15–39. https://doi.org/10.1002/adma.201502824
50. Wang Y, Hu J, Lin Y, Nan CW (2010) Multiferroic magnetoelectric composite nanostructures. NPG Asia Mater 2(2):61–68. https://doi.org/10.1038/asiamat.2010.32
51. Wang KF, Liu JM, Ren ZF (2009) Multiferroicity: the coupling between magnetic and polarization orders. Adv Phys 58(4):321–448. https://doi.org/10.1080/00018730902920554
52. Dong S, Liu JM, Cheong SW, Ren Z (2015) Multiferroic materials and magnetoelectric physics: symmetry, entanglement, excitation, and topology. Adv Phys 64(5–6):519–626. https://doi.org/10.1080/00018732.2015.1114338
53. Safi R, Shokrollahi H (2012) Physics, chemistry and synthesis methods of nanostructured bismuth ferrite ($BiFeO_3$) as a ferroelectro-magnetic material. Prog Solid State Chem 40(1–2):6–15. https://doi.org/10.1016/j.progsolidstchem.2012.03.001

54. Kreisel J, Noheda B, Dkhil B (2009) Phase transitions and ferroelectrics: revival and the future in the field. https://doi.org/10.1080/01411590903335249
55. Ederer C (2009) First principles studies of multiferroic materials
56. Kan D, Long CJ, Steinmetz C, Lofland SE, Takeuchi I (2012) Combinatorial search of structural transitions: Systematic investigation of morphotropic phase boundaries in chemically substituted $BiFeO_3$. J Mater Res 27(21):2691–2704. https://doi.org/10.1557/jmr.2012.314
57. Gao T, Chen Z, Huang Q, Niu F, Huang X, Qin L, Huang Y (2015) A review: preparation of bismuth ferrite nanoparticles and its applications in visible-light induced photocatalyses. Rev Adv Mater Sci 40(2):97–109
58. Yan L, Yang Y, Wang Z, Xing Z, Li J, Viehland D (2009) Review of magnetoelectric perovskite–spinel self-assembled nano-composite thin films. J Mater Sci 44:5080–5094. https://doi.org/10.1007/s10853-009-3679-1
59. Schlom DG, Ramesh R (2014) Electric field control of magnetism using $BiFeO_3$ based heterostructures. Appl Phys Rev 1(2). https://doi.org/10.1063/1.4870957
60. Baek SH, Eom CB (2012) Reliable polarization switching of $BiFeO_3$. Philos Trans R Soc A Math Phys Eng Sci 370(1977):4872–4889. https://doi.org/10.1098/rsta.2012.0197
61. Chu YH, Martin LW, Holcomb MB, Gajek M, Han SJ, He Q, Ramesh R (2008) Electric-field control of local ferromagnetism using a magnetoelectric multiferroic. Nature Mater 7(6):478–482. https://doi.org/10.1038/nmat2184
62. Ge M, Li Y, Liu L, Zhou Z, Chen W (2011) Bi_2O_3- Bi_2WO_6 composite microspheres: hydrothermal synthesis and photocatalytic performances. J Phys Chem C 115(13):5220–5225. https://doi.org/10.1021/jp108414e
63. Zeches RJ, Rossell MD, Zhang JX, Hatt AJ, He Q, Yang CH, Ramesh R (2009) A strain-driven morphotropic phase boundary in $BiFeO_3$. Science 326(5955):977–980. https://doi.org/10.1126/science.1177046
64. Michel C, Moreau JM, Achenbach GD, Gerson R, James WJ (1969) The atomic structure of $BiFeO_3$. Solid State Commun 7(9):701–704. https://doi.org/10.1016/0038-1098(69)90597-3
65. Kubel F, Schmid H (1990) Structure of a ferroelectric and ferroelastic monodomain crystal of the perovskite $BiFeO_3$. Struct Sci 46(6):698–702. https://doi.org/10.1107/S0108768190006887
66. Moreau JM, Michel C, Gerson R, James WJ (1971) Ferroelectric $BiFeO_3$ X-ray and neutron diffraction study. J Phys Chem Solids 32(6):1315–1320. https://doi.org/10.1016/S0022-3697(71)80189-0
67. Lam SM, Jaffari ZH, Sin JC (2018) Hydrothermal synthesis of coral-like palladium-doped $BiFeO_3$ nanocomposites with enhanced photocatalytic and magnetic properties. Mater Lett 224:1–4. https://doi.org/10.1016/j.matlet.2018.04.058
68. Thomas NW, Beitollahi A (1994) Inter-relationship of octahedral geometry, polyhedral volume ratio and ferroelectric properties in rhombohedral perovskites. Struct Sci 50(5):549–560. https://doi.org/10.1107/S0108768194002764
69. Castillo ME, Shvartsman VV, Gobeljic D, Gao Y, Landers J, Wende H, Lupascu DC (2013) Effect of particle size on ferroelectric and magnetic properties of $BiFeO_3$ nanopowders. Nanotechnology 24(35):355701. https://doi.org/10.1088/0957-4484/24/35/355701
70. Park TJ, Papaefthymiou GC, Viescas AJ, Moodenbaugh AR, Wong SS (2007) Size-dependent magnetic properties of single-crystalline multiferroic $BiFeO_3$ nanoparticles. Nano Lett 7(3):766–772. https://doi.org/10.1021/nl063039w
71. Yang CH, Kan D, Takeuchi I, Nagarajan V, Seidel J (2012) Doping $BiFeO_3$: approaches and enhanced functionality. Phys Chem Chem Phys 14(46):15953–15962. https://doi.org/10.1039/C2CP43082G
72. Bhushan B, Basumallick A, Bandopadhyay SK, Vasanthacharya NY, Das D (2009) Effect of alkaline earth metal doping on thermal, optical, magnetic and dielectric properties of $BiFeO_3$ nanoparticles. J Phys D Appl Phys 42(6):065004. https://doi.org/10.1088/0022-3727/42/6/065004
73. Cheng ZX, Wang XL, Du Y, Dou SX (2010) A way to enhance the magnetic moment of multiferroic bismuth ferrite. J Phys D Appl Phys 43(24):242001. https://doi.org/10.1088/0022-3727/43/24/242001

74. Karthik T, Rao TD, Srinivas A, Asthana S (2012) A-site cation disorder and size variance effects on the physical properties of multiferroic $Bi_{0.9}RE_{0.1}FeO_3$ Ceramics (RE = Gd^{3+}, Tb^{3+}, Dy^{3+}). arXiv preprint arXiv:1206.5606. https://doi.org/10.48550/arXiv.1206.5606
75. Jortner J, Rao CNR (2002) Nanostructured advanced materials. Perspectives and directions. Pure Appl Chem 74(9):1491–1506. https://doi.org/10.1351/pac200274091491
76. Bawendi MG, Steigerwald ML, Brus LE (1990) The quantum mechanics of larger semiconductor clusters ("quantum dots"). Annu Rev Phys Chem 41(1):477–496.0066-426X/90/1101-0477502.00
77. Bucci JD, Robertson BK, James WJ (1972) The precision determination of the lattice parameters and the coefficients of thermal expansion of $BiFeO_3$. Appl Crystallogr 5(3):187–191. https://doi.org/10.1107/S0021889872009173
78. Kumar MM, Palkar VR, Srinivas K, Suryanarayana SV (2000) Ferroelectricity in a pure $BiFeO_3$ ceramic. Appl Phys Lett 76(19):2764–2766. https://doi.org/10.1063/1.126468
79. Lebeugle D, Colson D, Forget A, Viret M, Bonville P, Marucco JF, Fusil S (2007) Room-temperature coexistence of large electric polarization and magnetic order in $BiFeO_3$ single crystals. Phys Rev B Condensed Matter Mater Phys 76(2):024116. https://doi.org/10.1103/PhysRevB.76.024116
80. Neaton JB, Ederer C, Waghmare UV, Spaldin NA, Rabe KM (2005) First-principles study of spontaneous polarization in multiferroic BiFeO3. Phys Rev B Condensed Matter Mater Phys 71(1):014113. https://doi.org/10.1103/PhysRevB.71.014113
81. Ravindran P, Vidya R, Kjekshus A, Fjellvåg H, Eriksson O (2006) Theoretical investigation of magnetoelectric behavior in $BiFeO_3$. Phys Rev B Condensed Matter Mater Phys 74(22):224412. https://doi.org/10.1103/PhysRevB.74.224412
82. Shvartsman VV, Kleemann W, Haumont R, Kreisel J (2007) Large bulk polarization and regular domain structure in ceramic $BiFeO_3$. Appl Phys Lett 90(17). https://doi.org/10.1063/1.2731312
83. Gradauskaite E, Meisenheimer P, Müller M, Heron J, Trassin M (2021) Multiferroic heterostructures for spintronics. Phys Sci Rev 6(2):20190072. https://doi.org/10.1515/psr-2019-0072
84. Kamba S, Nuzhnyy D, Savinov M, Šebek J, Petzelt J, Prokleška J, Kreisel J (2007) Infrared and terahertz studies of polar phonons and magnetodielectric effect in multiferroic $BiFeO_3$ ceramics. Phys Rev B Condensed Matter Mater Phys 75(2):024403. https://doi.org/10.1103/PhysRevB.75.024403
85. Campos LM, Tontcheva A, Günes S, Sonmez G, Neugebauer H, Sariciftci NS, Wudl F (2005) Extended photocurrent spectrum of a low band gap polymer in a bulk heterojunction solar cell. Chem Mater 17(16):4031–4033. https://doi.org/10.1021/cm050463+
86. Deng J, Banerjee S, Mohapatra SK, Smith YR, Misra M (2011) Bismuth iron oxide nanoparticles as photocatalyst for solar hydrogen generation from water. J Fundam Renew Energy Appl 1. https://doi.org/10.4303/jfrea/R101204
87. Soltani T, Entezari MH (2013) Sono-synthesis of bismuth ferrite nanoparticles with high photocatalytic activity in degradation of Rhodamine B under solar light irradiation. Chem Eng J 223:145–154. https://doi.org/10.1016/j.cej.2013.02.124
88. Mohapatra SK, John SE, Banerjee S, Misra M (2009) Water photooxidation by smooth and ultrathin α-Fe_2O_3 nanotube arrays. Chem Mater 21(14):3048–3055. https://doi.org/10.1021/cm8030208
89. Banerjee S, Mohapatra SK, Misra M (2009) Synthesis of TaON nanotube arrays by sonoelectrochemical anodization followed by nitridation: a novel catalyst for photoelectrochemical hydrogen generation from water. Chem Commun 46:7137–7139. https://doi.org/10.1039/b912549c
90. Rajeshwar K, McConnell R, Harrison K, Licht S (2008) Renewable energy and the hydrogen economy
91. Zou Z, Ye J, Sayama K, Arakawa H (2001) Direct splitting of water under visible light irradiation with an oxide semiconductor photocatalyst. Nature 414(6864):625–627. https://doi.org/10.1038/414625a

92. Fei L, Yuan J, Hu Y, Wu C, Wang J, Wang Y (2011) Visible light responsive perovskite BiFeO3 pills and rods with dominant 111 c facets. Cryst Growth Des 11(4):1049–1053. https://doi.org/10.1021/cg101144s
93. Yi HT, Choi T, Choi SG, Oh YS, Cheong SW (2011) Mechanism of the switchable photovoltaic effect in ferroelectric $BiFeO_3$. Adv Mater 23(30):3403–3407. https://doi.org/10.1002/adma.201100805
94. Wang X, Lin Y, Zhang ZC, Bian JY (2011) Photocatalytic activities of multiferroic bismuth ferrite nanoparticles prepared by glycol-based sol–gel process. J Sol-Gel Sci Technol 60:1–5. https://doi.org/10.1007/s10971-011-2542-4
95. Cai D, Du D, Yu S, Cheng J (2012) Oriented growth of $Bi_2Fe_4O_9$ crystal and its photocatalytic activity. Procedia Eng 27:577–582. https://doi.org/10.1016/j.proeng.2011.12.490
96. Zhu X, Hang Q, Xing Z, Yang Y, Zhu J, Liu Z, Zou Z (2011) Microwave hydrothermal synthesis, structural characterization, and visible-light photocatalytic activities of single-crystalline bismuth ferric nanocrystals. J Am Ceramic Soc 94(8):2688–2693. https://doi.org/10.1111/j.1551-2916.2011.04430.x
97. Su J, Guo L, Bao N, Grimes CA (2011) Nanostructured $WO_3/BiVO_4$ heterojunction films for efficient photoelectrochemical water splitting. Nano Lett 11(5):1928–1933. https://doi.org/10.1021/nl2000743
98. Ge M, Li y, Liu L, Zhou Z, Chen W (2011) Bi_2O_3–Bi_2WO_6 composite microspheres: hydrothermal synthesis and photocatalytic performances. J Phys Chem C 115(13):5220–5225. https://doi.org/10.1021/jp108414e
99. Mukherjee A, Hossain SM, Pal M, Basu S (2012) Effect of Y-doping on optical properties of multiferroics $BiFeO_3$ nanoparticles. Appl Nanosci 2:305–310. https://doi.org/10.1007/s13204-012-0114-8
100. Sivula K, Formal FL, Gratzel M (2009) $WO_3 - Fe_2O_3$ photoanodes for water splitting: a host scaffold, guest absorber approach. Chem Mater 21(13):2862–2867. https://doi.org/10.1021/cm900565a
101. Zhang X, Lv J, Bourgeois L, Cui J, Wu Y, Wang H, Webley PA (2011) Formation and photocatalytic properties of bismuth ferrite submicrocrystals with tunable morphologies. New J Chem 35(4):937–941. https://doi.org/10.1039/C1NJ00008J
102. Bellakki MB, Manivannan V (2010) Citrate-gel synthesis and characterization of yttrium-doped multiferroic $BiFeO_3$. J Sol-Gel Sci Technol 53:184–192. https://doi.org/10.1007/s10971-009-2076-1
103. Ghosh A, Trujillo DP, Choi H, Nakhmanson SM, Alpay SP, Zhu JX (2019) Electronic and magnetic properties of lanthanum and strontium doped bismuth ferrite: a first-principles study. Sci Rep 9(1). 194. https://doi.org/10.1038/s41598-018-37339-3
104. Wang X, Lin Y, Ding X, Jiang J (2011) Enhanced visible-light-response photocatalytic activity of bismuth ferrite nanoparticles. J Alloy Compd 509(23):6585–6588. https://doi.org/10.1016/j.jallcom.2011.03.074
105. Li T, Shen J, Li N, Ye M (2013) Hydrothermal preparation, characterization and enhanced properties of reduced graphene- $BiFeO_3$ nanocomposite. Mater Lett 91:42–44. https://doi.org/10.1016/j.matlet.2012.09.045
106. Si YH, Xia Y, Li YY, Shang SK, Xiong XB, Zeng XR, Zhou J (2018) Enhanced visible-active photocatalytic behaviors observed in Mn-doped $BiFeO_3$. Mod Phys Lett B 32(17):185018. https://doi.org/10.1142/S0217984918501853
107. Arya G, Yogiraj J, Negi NS, Shah J, Kotnala RK (2017) Observation of enhanced multiferroic, magnetoelectric and photocatalytic properties in Sm-Co codoped $BiFeO_3$ nanoparticles. J Alloy Compd 723:983–994. https://doi.org/10.1016/j.jallcom.2017.06.325
108. Maleki H (2019) Characterization and photocatalytic activity of Y-doped BiFeO3 ceramics prepared by solid-state reaction method. Adv Powder Technol 30(11):2832–2840. https://doi.org/10.1016/j.apt.2019.08.031
109. Hu Z, Chen D, Wang S, Zhang N, Qin L, Huang Y (2017) Facile synthesis of Sm-doped $BiFeO_3$ nanoparticles for enhanced visible light photocatalytic performance. Mater Sci Eng B 220:1–12. https://doi.org/10.1016/j.mseb.2017.03.005

110. Irfan S, Shen Y, Rizwan S, Wang HC, Khan SB, Nan CW (2017) Band-gap engineering and enhanced photocatalytic activity of Sm and Mn doped $BiFeO_3$ nanoparticles. J Am Ceram Soc 100(1):31–40. https://doi.org/10.1111/jace.14487
111. Reddy P, Cui H, Vattikuti SP, Suh Y, Park SH (2019) Influence of Gd doping on the visible-light photocatalytic activity and magnetic properties of $BiFeO_3$ particles. Mater Res Express 6(11):115044. https://doi.org/10.1088/2053-1591/ab463d
112. Puhan A, Bhushan B, Satpathy S, Meena SS, Nayak AK, Rout D (2019) Facile single phase synthesis of Sr, Co co-doped $BiFeO_3$ nanoparticles for boosting photocatalytic and magnetic properties. Appl Surf Sci 493:593–604. https://doi.org/10.1016/j.apsusc.2019.07.002
113. Ahmad S, Khan MA, Sarfraz M, ur Rehman, A., Warsi, M. F., & Shakir, I. (2017) The impact of Yb and Co on structural, magnetic, electrical and photocatalytic behavior of nanocrystalline multiferroic $BiFeO_3$ particles. Ceram Int 43(18):16880–16887. https://doi.org/10.1016/j.ceramint.2017.09.088
114. Polking MJ, Han MG, Yourdkhani A, Petkov V, Kisielowski CF, Volkov VV, Ramesh R (2012) Ferroelectric order in individual nanometre-scale crystals. Nature Mater 11(8):700–709. https://doi.org/10.1038/nmat3371
115. Yadlovker D, Berger S (2005) Uniform orientation and size of ferroelectric domains. Phys Rev B Condensed Matter Mater Phys 71(18):184112. https://doi.org/10.1103/PhysRevB.71.184112
116. Rosalin M, Telang P, Singh S, Muthu DVS, Sood AK (2023) Raman signatures of quadratic band touching state and strong spin-phonon coupling in pyrochlore iridates $(Sm_{1-x}Bi_x)_2Ir_2O_7$. Phys Rev B 108(3):035133. https://doi.org/10.1103/PhysRevB.108.035133
117. Saxena A, Aeppli G (2009) Phase transitions at the nanoscale in functional materials. MRS Bull 34(11):804–813. https://doi.org/10.1557/mrs2009.230
118. Bai X (2016) Size and doping effect on the structure, transitions and optical properties of multiferroic $BiFeO_3$ particles for photocatalytic applications (Doctoral dissertation, Université Paris Saclay (COmUE)). https://theses.hal.science/tel-01315558
119. Srihari NV, Bayari S, Kumar S, Padaki M, Nagaraja KK (2024) Comprehensive investigation into the influence of oxygen vacancies on the ferroelectric properties of spin coated bismuth ferrite thin films. Phys Scr 99(7):075917. https://doi.org/10.1088/1402-4896/ad4fec
120. Fernandez A, Acharya M, Lee HG, Schimpf J, Jiang Y, Lou D, Martin LW (2022) Thin-film ferroelectrics. Adv Mater 34(30):2108841. https://doi.org/10.1002/adma.202108841
121. Yoshimura T, Murakami S, Wakazono K, Kariya K, Fujimura N (2013) Piezoelectric vibrational energy harvester using lead-free ferroelectric $BiFeO_3$ films. Appl Phys Express 6(5):051501. https://doi.org/10.7567/APEX.6.051501
122. Sasmal A, Patra A, Devi PS, Sen S (2021) Hydroxylated $BiFeO_3$ as efficient fillers in poly (vinylidene fluoride) for flexible dielectric, ferroelectric, energy storage and mechanical energy harvesting application. Dalton Trans 50(5):1824–1837. https://doi.org/10.1039/D0DT04017G
123. Bowen CR, Kim HA, Weaver PM, Dunn S (2014) Piezoelectric and ferroelectric materials and structures for energy harvesting applications. Energy Environ Sci 7(1):25–44. https://doi.org/10.1039/C3EE42454E
124. Aghayan M, Zak AK, Behdani M, Hashim AM (2014) Sol–gel combustion synthesis of Zr-doped $BiFeO_3$ nanopowders and ceramics: Dielectric and ferroelectric studies. Ceram Int 40(10):16141–16146. https://doi.org/10.1016/j.ceramint.2014.07.045
125. Zhang F, Zeng X, Bi D, Guo K, Yao Y, Lu S (2018) Dielectric, ferroelectric, and magnetic properties of Sm-doped $BiFeO_3$ ceramics prepared by a modified solid-state-reaction method. Materials 11(11):2208. https://doi.org/10.3390/ma11112208
126. Wang T, Ye H, Wang X, Cui Y, Mei H, Song S, Ma Q (2025) Improved ferroelectric and magnetic properties of bismuth ferrite-based ceramics by introduction of non-isovalent ions and grain engineering. Nanomaterials 15(3):215. https://doi.org/10.3390/nano15030215
127. Yao Y, Tao T, Liang B, Mak CL, Lu SG (2019) Pyroelectric properties and AC impedance study of bismuth ferrite ($BiFeO_3$) ceramics. Ceram Int 45(1):1308–1313. https://doi.org/10.1016/j.ceramint.2018.10.017

128. Yao Y, Ploss B, Mak CL, Wong KH (2010) Pyroelectric properties of $BiFeO_3$ ceramics prepared byáaámodified solid-state-reaction method. Appl Phys A 99:211–216. https://doi.org/10.1007/s00339-009-5499-1
129. Sharma S, Singh V, Kotnala RK, Dwivedi RK (2014) Comparative studies of pure $BiFeO_3$prepared by sol–gel versus conventional solid-state-reaction method. J Mater Sci: Mater Electron 25:1915–1921. https://doi.org/10.1007/s10854-014-1820-7
130. Halder S, Bhuyana S, Tripathy A, Al Zaabi O, Swain B, Muduli UR (2023) Development of a capacitive temperature sensor using a lead-free ferroelectric Bi (Fe $_{2/3}Ta_{1/3}$) O_3 ceramic. IEEE Sens J 23(14):15382–15390. https://doi.org/10.1109/JSEN.2023.3277795
131. Ray J, Hing P (2000) Sol-gel derived Ba (Fe, Ti) O_3 ferroelectric materials for infrared sensors. J Appl Phys 88(2):1008–1014. https://doi.org/10.1063/1.373770
132. Lin Y, Zhang L, Yu J (2015) Stable piezoelectric property of modified $BiFeO_3$–$BaTiO_3$ lead-free piezoceramics. J Mater Sci: Mater Electron 26:8432–8441. https://doi.org/10.1007/s10854-015-3512-3
133. Sosnowska I, Przeniosło R (2011) Low-temperature evolution of the modulated magnetic structure in the ferroelectric antiferromagnet $BiFeO_3$. Phys Rev B Condensed Matter Mater Phys 84(14):144404. https://doi.org/10.1103/PhysRevB.84.144404
134. Zhu J, Zhang J, Xu H, Vogel SC, Jin C, Frantti J, Zhao Y (2014) Pressure-induced reversal between thermal contraction and expansion in ferroelectric $PbTiO_3$. Sci Rep 4(1):3700. https://doi.org/10.1038/srep03700
135. Fourmont P, Bai Y, Fortier FX, Cloutier SG (2022) Graphene-enhanced screen-printed $BiFeO_3$-based thermistors. ACS Appl Electron Mater 4(12):5905–5913. https://doi.org/10.1021/acsaelm.2c01093
136. Narayan B, Li Z, Wang B, Haugen AB, Hall D, Khanbareh H, Roscow J (2024) Temperature-dependent ferroelectric properties and aging behavior of freeze-cast bismuth ferrite-barium titanate ceramics. ACS Appl Mater Interfaces 16(15):19283–19297. https://doi.org/10.1021/acsami.4c03002
137. Jia J, Qian J, Shi Y, Zhang L, Xu L, Shen B, Zhai J (2024) Synergistic design strategy for enhancing the piezoelectric performance of $BiFeO_3$-based high-temperature piezoelectric ceramics. Ceram Int 50(17):31107–31113. https://doi.org/10.1016/j.ceramint.2024.05.416
138. Prashanthi K, Antić Ž, Thakur G, Dramićanin MD, Thundat T (2018) Surface state-induced anomalous negative thermal quenching of multiferroic $BiFeO_3$ nanowires. Phys Status Solidi (RRL) Rapid Res Lett 12(1):1700352. https://doi.org/10.1002/pssr.201700352
139. Ibrahim EMM, Farghal G, Khalaf MM, El-Lateef HMA (2017) Effect of calcination temperature on magnetic and electrical properties of $BiFeO_3$ nanoparticles prepared by sol-gel method. J Nano Adv Mat 5(1):33–39. https://doi.org/10.18576/jnam/050105
140. Yu L, Deng H, Zhou W, Zhang Q, Yang P, Chu J (2016) Effects of (Sm, Mn and Ni) co-doping on structural, optical and magnetic properties of $BiFeO_3$ thin films fabricated by a sol-gel technique. Mater Lett 170:85–88. https://doi.org/10.1016/j.matlet.2016.02.004
141. Gu Y, Zhao J, Zhang W, Zheng H, Liu L, Chen W (2017) Structural transformation and multiferroic properties of Sm and Ti co-doped $BiFeO_3$ ceramics with Fe vacancies. Ceram Int 43(17):14666–14671. https://doi.org/10.1016/j.ceramint.2017.07.187
142. Mukherjee A, Basu S, Green LAW, Thanh NTK, Pal M (2015) Enhanced multiferroic properties of Y and Mn codoped multiferroic $BiFeO_3$ nanoparticles. J Mater Sci 50:1891–1900. https://doi.org/10.1007/s10853-014-8752-8
143. Bai L, Sun M, Ma W, Yang J, Zhang J, Liu Y (2020) Enhanced magnetic properties of co-doped $BiFeO_3$ thin films via structural progression. Nanomaterials 10(9):1798. https://doi.org/10.3390/nano10091798
144. Yu Z, Zeng J, Zheng L, Liu W, Li G, Kassiba A (2020) Large piezoelectricity and high Curie temperature in novel bismuth ferrite-based ferroelectric ceramics. J Am Ceram Soc 103(11):6435–6444. https://doi.org/10.1111/jace.17382
145. Jiang AQ, Wang C, Jin KJ, Liu XB, Scott JF, Hwang CS, Yang GZ (2011) A resistive memory in semiconducting $BiFeO_3$ thin-film capacitors. Adv Mater 23(10):1277–1281. https://doi.org/10.1002/adma.201004317

146. Si YH, Xia Y, Shang SK, Xiong XB, Zeng XR, Zhou J, Li YY (2018) Enhanced visible light driven photocatalytic behavior of $BiFeO_3$/reduced graphene oxide composites. Nanomaterials 8(7):526. https://doi.org/10.3390/nano8070526
147. Marzouki A, Harzali H, Loyau V, Gemeiner P, Zehani K, Dkhil B, Megriche A (2018) Large magnetoelectric response and its origin in bulk Co-doped BiFeO3 synthesized by a stirred hydrothermal process. Acta Materialia 145:316–321. https://doi.org/10.1016/j.actamat.2017.11.055
148. Zhang Y, Qi J, Wang Y, Tian Y, Zhang J, Hu T, Yang J (2018) Tuning magnetic properties of $BiFeO_3$ thin films by controlling Mn doping concentration. Ceram Int 44(6):6054–6061. https://doi.org/10.1016/j.ceramint.2017.12.230
149. Zhang Y, Wang Y, Qi J, Tian Y, Sun M, Zhang J, Yang J (2018) Enhanced magnetic properties of $BiFeO_3$ thin films by doping: analysis of structure and morphology. Nanomaterials 8(9):711. https://doi.org/10.3390/nano8090711
150. Liu Y, Qi J, Zhang Y, Wang Y, Feng M, Zhang J, Yang J (2018) Surface agglomeration is beneficial for release of magnetic property via research of rare earth (RE) element-substitution. Appl Surface Sci 427:745–752. https://doi.org/10.1016/j.apsusc.2017.08.083
151. Qi J, Zhang Y, Wang Y, Liu Y, Wei M, Zhang J, Yang J (2017) Effect of Cr doping on the phase structure, surface appearance and magnetic property of $BiFeO_3$ thin films prepared via sol–gel technology. J Mater Sci Mater Electron 28:17490–17498. https://doi.org/10.1007/s10854-017-7684-x
152. Yamashita T, Hayes P (2008) Analysis of XPS spectra of Fe^{2+} and Fe^{3+} ions in oxide materials. Appl Surf Sci 254(8):2441–2449. https://doi.org/10.1016/j.apsusc.2007.09.063
153. Sinha AK, Bhushan B, Gupta N, Sen S, Prajapat CL, Nuwad J, Priyam A (2020) Effect of cobalt-doping on dielectric, magnetic and optical properties of BiFeO3 nanocrystals synthesized by sol–gel technique. Solid State Sci 102:106168. https://doi.org/10.1016/j.solidstatesciences.2020.106168
154. Lahmar A, Habouti S, Dietze M, Solterbeck CH, Es-Souni M (2009) Effects of rare earth manganites on structural, ferroelectric, and magnetic properties of $BiFeO_3$ thin films. Appl Phys Lett 94(1). https://doi.org/10.1063/1.3064954
155. Hou P, Liu B, Guo Z, Zhou P, Wang B, Zhao L (2019) Effect of Ho doping on the crystal structure, surface morphology and magnetic property of $BiFeO_3$ thin films prepared via the sol-gel technology. J Alloy Compd 775:59–62. https://doi.org/10.1016/j.jallcom.2018.10.112
156. Kuang D, Tang P, Ding X, Yang S, Zhang Y (2015) Effects of Y doping on multiferroic properties of sol–gel deposited $BiFeO_3$ thin films. J Mater Sci: Mater Electron 26:3001–3007. https://doi.org/10.1007/s10854-015-2789-6
157. Yamashita T, Hayes P (2006) Effect of curve fitting parameters on quantitative analysis of $Fe_{0.94}O$ and Fe_2O_3 using XPS. J Electron Spectrosc Relat Phenom 152(1–2):6–11. https://doi.org/10.1016/j.elspec.2006.02.002
158. Kaczkowski J, Pugaczowa-Michalska M, Jezierski A (2015) Electronic structure of $BiFeO_3$ in different crystal phases. Acta Phys Pol, A 127(2):266–268. https://doi.org/10.12693/APhysPolA.127.266
159. Ramazanov S, Sobola D, Orudzhev F, Knápek A, Polčák J, Potoček M, Dallaev R (2020) Surface modification and enhancement of ferromagnetism in $BiFeO_3$ nanofilms deposited on HOPG. Nanomaterials 10(10):1990. https://doi.org/10.3390/nano10101990
160. Patel SKS, Lee JH, Kim MK, Kim SK (2019) Structural and magnetic properties of Gd-Ni-co-doped $BiFeO_3$ nanoparticles. J Magn 24(3):371–378. https://doi.org/10.4283/JMAG.2019.24.3.371
161. Krupinski K, Marks A, Chernikov A, Fischer V (2022) Poznan, Poland July 19–24, 2021. Bull Symb Log 28(2). https://doi.org/10.1017/bsl.2022.17
162. Hosseinzadeh L, Baedi J, Zak AK (2014) X-ray peak broadening analysis of Fe_{50} Ni_{50} nanocrystalline alloys prepared under different milling times and BPR using size strain plot (SSP) method. Bull Mater Sci 37:1147–1152. https://doi.org/10.1007/s12034-014-0055-9
163. Khorrami GH, Zak AK, Banihashemian SM (2014) Magnetic and dielectric properties on sol–gel combustion synthesis of Pb ($Zr_{0.52}$, $Ti_{0.43}X_{0.05}$) O_3 (X= Fe, Ni, and Co) nanoparticles. Adv Powder Technol 25(4):1319–1324. https://doi.org/10.1016/j.apt.2014.03.011

164. Hong NH, Thu Huong N, Kim TY, Goumri-Said S, Kanoun MB (2015) Tuning magnetic properties of $BiFeO_3$ thin films by controlling rare-earth doping: experimental and first-principles studies. J Phys Chem C 119(25):14351–14357. https://doi.org/10.1021/acs.jpcc.5b03834
165. Achary SN, Jayakumar OD, Tyagi AK (2012) Multiferroic materials. Funct Mater Prep Process Appl 155–191. https://doi.org/10.1016/B978-012-385142-0.00004-0
166. Khikhlovskyi SVV, Blake G (2010) The renaissance of multiferroics: bismuth ferrite ($BiFeO_3$)–a candidate multiferroic material in nanoscience. University of Groningen 1(3)
167. Ding HC, Li YW, Zhu W, Gao YC, Gong SJ, Duan CG (2013) Improved multiferroic behavior in [111]-oriented $BiFeO_3/BiAlO_3$ superlattice. J Appl Phys 113(12). https://doi.org/10.1063/1.4795847
168. Meyer S, Xu B, Verstraete MJ, Bellaiche L, Dupé B (2023) Spin-current driven Dzyaloshinskii-Moriya interaction in multiferroic $BiFeO_3$ from first principles. Phys Rev B 108(2):024403. https://doi.org/10.1103/PhysRevB.108.024403
169. Tokunaga M, Azuma M, Shimakawa Y (2010) High-field study of strong magnetoelectric coupling in single-domain crystals of $BiFeO_3$. J Phys Soc Jpn 79(6):064713. https://doi.org/10.1143/JPSJ.79.064713
170. Baek SH, Jang HW, Folkman CM, Li YL, Winchester B, Zhang JX, Eom CB (2010) Ferroelastic switching for nanoscale non-volatile magnetoelectric devices. Nature Mater 9(4):309–314. https://doi.org/10.1038/nmat2703
171. Bai F, Wang J, Wuttig M, Li J, Wang N, Pyatakov AP, Viehland D (2005) Destruction of spin cycloid in (111) c-oriented $BiFeO_3$ thin films by epitiaxial constraint: enhanced polarization and release of latent magnetization. Appl Phys Lett 86(3). https://doi.org/10.1063/1.1851612
172. Khan AH, Atiq S, Mahmood A, Ramay SM, Abbas SK, Naseem S (2018) Optimisation of giant magnetoresistance in Mn-substituted $BiFeO_3$ for low field sensors. Ceram Int 44(12):14677–14685. https://doi.org/10.1016/j.ceramint.2018.05.094
173. Paliwal A, Tomar M, Gupta V (2018) Study of optical properties of Ce and Mn doped $BiFeO_3$ thin films using SPR technique for magnetic field sensing. Vacuum 158:48–51. https://doi.org/10.1016/j.vacuum.2018.09.018
174. Montanher DZ, Freitas VF, Pereira JRD, Cótica LF, Garcia D, Eiras JA, Santos IA (2013) Overlap of the intrinsic and extrinsic magnetoelectric effects in $BiFeO_3$–$PbTiO_3$ compounds: potentialities for magnetic-sensing applications. J Appl Phys 113(3). https://doi.org/10.1063/1.4775796
175. Ratha S, Kuppan M, Egawa G, Yoshimura S (2023) Excellent magnetic properties in multiferroic $BiFeO_3$ based thin films for magnetic devices application. Nano Struct Nano Objects 35:101007. https://doi.org/10.1016/j.nanoso.2023.101007
176. Amrillah T, Hermawan A, Wulandari CP, Muthi'Ah AD, Simanjuntak FM (2021) Crafting the multiferroic $BiFeO_3$–$CoFe_2O_4$ nanocomposite for next-generation devices: a review. Mater Manuf Process 36(14):1579–1596. https://doi.org/10.1080/10426914.2021.1945096
177. Ji H, Zhang L, Zhang R (2023) Gas sensitive performance and mechanism of multiferroic $BiFeO_3$ under thermal-magnetic synergetic excitation. Inorg Chem Commun 150:110491. https://doi.org/10.1016/j.inoche.2023.110491
178. Sharma S, Paliwal A, Tomar M, Gupta V (2019) Multiferroic BFO/BTO multilayer structures based magnetic field sensor. Physica B 571:1–4. https://doi.org/10.1016/j.physb.2019.06.056
179. Kumari S, Pradhan DK, Pradhan NR, Rack PD (2021) Recent developments on 2D magnetic materials: challenges and opportunities. Emergent Mater 4(4):827–846. https://doi.org/10.1007/s42247-021-00214-5
180. Trivedi H (2015) Mapping local manifestations of the strain mediated magnetoelectric effect in composites. Universitaet Duisburg-Essen (Germany)
181. Zhang S, Luo W, Wang L, Wang D, Ma Y (2010) Simultaneously improved magnetization and polarization in $BiFeO_3$ based multiferroic composites. J Appl Phys 107(5). https://doi.org/10.1063/1.3332040
182. Pan Q, Fang C, Luo H, Chu B (2019) Magnetoelectric response from the enhanced ferromagnetism and flexoelectric response in reduced $BiFeO_3$–based ceramics. J Eur Ceram Soc 39(4):1057–1064. https://doi.org/10.1016/j.jeurceramsoc.2018.12.045

183. Ratha S, Egawa G, Yoshimura S (2024) Optimizing the reactive ion etching conditions with minimal damage for high functional magnetic nano device application in $BiFeO_3$-based thin film by Eu/Co substitution. Appl Surf Sci 643:158704. https://doi.org/10.1016/j.apsusc.2023.158704
184. Cheng Y, Zhao S, Zhou Z, Liu M (2023) Recent development of E-field control of interfacial magnetism in multiferroic heterostructures. Nano Res 16(4):5983–6000. https://doi.org/10.1007/s12274-022-5125-5
185. Peng S, Ma M, Yang W, Wang Z, Wang Z, Bi J, Wu J (2020) Acetone sensing with parts-per-billion limit of detection using a $BiFeO_3$-based solid solution sensor at the morphotropic phase boundary. Sens Actuators B Chem 313:128060. https://doi.org/10.1016/j.snb.2020.128060
186. Puhan A, Bhushan B, Nayak AK, Rout D (2021) $BiFeO_3$-based multiferroic materials and their properties. In: Fundamentals and properties of multifunctional nanomaterials. Elsevier, pp 275–293. https://doi.org/10.1016/B978-0-12-822352-9.00008-0
187. Ahmed T, Khan SA, Bae J, Habib M, Akram F, Choi SY, Lee S (2021) Role of Bi chemical pressure on electrical properties of $BiFeO_3$–$BaTiO_3$–based ceramics. Solid State Sci 114:106562. https://doi.org/10.1016/j.solidstatesciences.2021.106562
188. Ganesh RS, Sharma SK, Sankar S, Divyapriya B, Durgadevi E, Raji P, Kim DY (2017) Microstructure, structural, optical and piezoelectric properties of $BiFeO_3$ nanopowder synthesized from sol-gel. Curr Appl Phys 17(3):409–416. https://doi.org/10.1016/j.cap.2016.12.008
189. Amdouni W, Fricaudet M, Otoničar M, Garcia V, Fusil S, Kreisel J, Dkhil B (2023) $BiFeO_3$ nanoparticles: the "Holy-Grail" of piezo-photocatalysts? Adv Mater 35(31):2301841. https://doi.org/10.1002/adma.202301841
190. Khadtare S, Kumar A, Jung U, Sang C, Park J (2022) Piezoelectric nanogenerator based on lead-free $BiFeO_3$: Sr perovskite. J Alloy Compd 926:16673. https://doi.org/10.1016/j.jallcom.2022.166738
191. Yang L, Chen C, Jiang X, Huang X, Nie X, Chang S (2022) Enhanced ferroelectric and piezoelectric properties of $BiFeO_3$–$BaTiO_3$ lead-free ceramics by simultaneous optimization of Bi compensation and sintering conditions. Ceram Int 48(9):12866–12874. https://doi.org/10.1016/j.ceramint.2022.01.158
192. Sun Y, Li X, Vijayakumar A, Liu H, Wang C, Zhang S, Cheng Z (2021) Hydrogen generation and degradation of organic dyes by new piezocatalytic 0.7 $BiFeO_3$–0.3 $BaTiO_3$ nanoparticles with proper band alignment. ACS Appl Mater Interfaces 13(9):11050–11057. https://doi.org/10.1021/acsami.1c01407
193. Wang N, Luo X, Han L, Zhang Z, Zhang R, Olin H, Yang Y (2020) Structure, performance, and application of $BiFeO_3$ nanomaterials. Nano-Micro Lett 12:1–23. https://doi.org/10.1007/s40820-020-00420-6
194. Lee S, Pirogov A, Kang M, Jang KH, Yonemura M, Kamiyama T, Park JG (2008) Giant magneto-elastic coupling in multiferroic hexagonal manganites. Nature 451(7180):805–808. https://doi.org/10.1038/nature06507
195. Sun H, Liu Y, Zhang Y, Lv L, Zhou J, Chen W (2014) Synthesis of Bi_2 Fe_4 O_9/reduced graphene oxide composite by one-step hydrothermal method and its high photocatalytic performance. J Mater Sci: Mater Electron 25:4212–4218. https://doi.org/10.1007/s10854-014-2151-4
196. Shi XX, Liu XQ, Chen XM (2017) Readdressing of magnetoelectric effect in bulk $BiFeO_3$. Adv Func Mater 27(12):1604037. https://doi.org/10.1002/adfm.201604037
197. Du Y, Cheng ZX, Dou SX, Shahbazi M, Wang XL (2010) Enhancement of magnetization and dielectric properties of chromium-doped $BiFeO_3$ with tunable morphologies. Thin Solid Films 518(24):e5–e8. https://doi.org/10.1016/j.tsf.2010.03.118
198. Liu YQ, Zhang J, Wu YH, Zhang YJ, Wei MB, Yang JH (2013) Enhancement of magnetization in Er doped BiFeO 3 thin Film. J Sol-Gel Sci Technol 67:1–7. https://doi.org/10.1007/s10971-013-3044-3
199. Mushtaq F, Chen X, Hoop M, Torlakcik H, Pellicer E, Sort J, Pane S (2018) Piezoelectrically enhanced photocatalysis with BiFeO3 nanostructures for efficient water remediation. Iscience 4:236–246. https://doi.org/10.1016/j.isci.2018.06.003

200. Ren X, Fan H, Zhao Y, Liu Z (2016) Flexible lead-free $BiFeO_3$/PDMS-based nanogenerator as piezoelectric energy harvester. ACS Appl Mater Interfaces 8(39):26190–26197. https://doi.org/10.1021/acsami.6b04497
201. Qin H, Zhao J, Chen X, Li H, Wang S, Du Y, Wang D (2023) Investigation of $BiFeO_3$-$BaTiO_3$ lead-free piezoelectric ceramics with nonstoichiometric bismuth. Microstructures 3(4):N-A. https://doi.org/10.20517/microstructures.2023.34
202. Wang X, Lu X, Zhao X, Chen WJ, Liu Y, Pan X, Liang S (2024) Crystal-facet-dependent piezocatalytic activity of $BiFeO_3$ nanosheets for H_2 evolution and environmental remediation. ACS Appl Nano Mater 7(10):11794–11802. https://doi.org/10.1021/acsanm.4c01407
203. Queraltó A, Frohnhoven R, Mathur S, Gómez A (2020) Intrinsic piezoelectric characterization of $BiFeO_3$ nanofibers and its implications for energy harvesting. Appl Surf Sci 509:144760. https://doi.org/10.1016/j.apsusc.2019.144760
204. Yoshiyama K, Mori M, Hagiwara M, Fujihara S (2020) Effect of particle size and morphology on the performance of $BiFeO_3$–PDMS piezoelectric generators. CrystEngComm 22(17):2919–2925. https://doi.org/10.1039/D0CE00067A
205. Yue Y, Xu X, Zhang M, Yan Z, Koval V, Whiteley RM, Yan H (2021) Grain size effects in Mn-modified 0.67 $BiFeO_3$–0.33 $BaTiO_3$ ceramics. ACS Appl Mater Interfaces 13(48):57548–57559. https://doi.org/10.1021/acsami.1c16083
206. Holcomb MB, Martin LW, Scholl A, He Q, Yu P, Yang CH, Ramesh R (2010) Probing the evolution of antiferromagnetism in multiferroics. Phys Rev B Condensed Matter Mater Phys 81(13):134406. https://doi.org/10.1103/PhysRevB.81.134406
207. Schlom DG, Chen LQ, Fennie CJ, Gopalan V, Muller DA, Pan X, Uecker R (2014) Elastic strain engineering of ferroic oxides. Mrs Bulletin 39(2):118–130. https://doi.org/10.1557/mrs.2014.1
208. Sando D, Yang Y, Bousquet E, Carrétéro C, Garcia V, Fusil S, Bibes M (2016) Large elasto-optic effect and reversible electrochromism in multiferroic BiFeO3. Nature Commun 7(1):10718. https://doi.org/10.1038/ncomms10718
209. Kan D, Pálová L, Anbusathaiah V, Cheng CJ, Fujino S, Nagarajan V, Takeuchi I (2010) Universal behavior and electric-field-induced structural transition in rare-earth-substituted $BiFeO_3$. Adv Functional Mater 20(7):1108–1115. https://doi.org/10.1002/adfm.200902017
210. Yang CH, Seidel J, Kim SY, Rossen PB, Yu P, Gajek M, Ramesh R (2009) Electric modulation of conduction in multiferroic Ca-doped BiFeO3 films. Nature Mater 8(6):485–493. https://doi.org/10.1038/nmat2432
211. Peng B, Peng RC, Zhang YQ, Dong G, Zhou Z, Zhou Y, Liu M (2020) Phase transition enhanced superior elasticity in freestanding single-crystalline multiferroic $BiFeO_3$ membranes. Sci Adv 6(34):eaba5847. https://doi.org/10.1126/sciadv.aba5847
212. Panda PK, Sahoo B, Thejas TS (2023) High strain lead-free piezo ceramics for sensor and actuator applications: a review. Sensors Int 4:100226. https://doi.org/10.1016/j.sintl.2022.100226
213. Zou Q, Xie Y, Yin Y, Liu B, Yu Y (2023) Flexible pressure sensors based on microcrack structure and composite conductive mechanism for medical robotic applications. Micromachines 14(6):1110. https://doi.org/10.3390/mi14061110
214. Scarisoreanu ND, Craciun F, Birjega R, Ion V, Teodorescu VS, Ghica C, Dinescu M (2016) Joining chemical pressure and epitaxial strain to yield Y-doped $BiFeO_3$ thin films with high dielectric response. Sci Rep 6(1):25535. https://doi.org/10.1038/srep25535
215. Graham SA, Manchi P, Paranjape MV, Yu JS (2024) Mechanical and acoustic-driven $BiFeO_3$ composite films-based hybrid nanogenerator for energy harvesting and sensing applications. Small 20(20):2308428. https://doi.org/10.1002/smll.202308428
216. Kohno T, Mihara M, Tanabe A, Abe T, Okuyama M, Sohgawa M (2017) Deposition and characterization of Al_2O_3 and BiFeO3 thin films on titanium substrates for tough MEMS devices. IEEJ Trans Sens Micromachines 137(1):46–47. https://doi.org/10.1541/ieejsmas.137.46
217. Lin GL et al (2022) Barium titanate–bismuth ferrite/polyvinylidene fluoride nanocomposites as flexible piezoelectric sensors with excellent thermal stability. Sensors Actuators A Phys 346:113885. https://doi.org/10.1016/j.sna.2022.113885

218. Jabeen N, Hussain A, Nawaz S, Qaiser MA, Zafar R, Ishtiaq S, Muntaha ST (2022) Thermally stable piezoelectric performance of MnO_2 inserted pseudo-tetragonal phase existent $CaBi_2Nb_2O_9$-based ceramics. Mater Technol 37(13):2702–2710. https://doi.org/10.1080/10667857.2022.2059046
219. Liu J, Prashanthi K, Li Z, McGee RT, Ahadi K, Thundat T (2016) Strain-induced electrostatic enhancements of $BiFeO_3$ nanowire loops. Phys Chem Chem Phys 18(33):22772–22777. https://doi.org/10.1039/C6CP03068H
220. Aramaki M, Yoshimura T, Murakami S, Satoh K, Fujimura N (2019) Demonstration of high-performance piezoelectric MEMS vibration energy harvester using $BiFeO_3$ film with improved electromechanical coupling factor. Sens Actuators A 291:167–173. https://doi.org/10.1016/j.sna.2019.03.050
221. Murakami S, Yoshimura T, Satoh K, Wakazono K, Kariya K, Fujimura N (2013) Development of piezoelectric MEMS vibration energy harvester using (100) oriented $BiFeO_3$ ferroelectric film. J Phys Conf Ser 476(1):012007. IOP Publishing. https://doi.org/10.1088/1742-6596/476/1/012007
222. Li Y, Dong M, Zou X, Zhang J, Zhang J, Huang X (2023) Freeze-drying assisted liquid exfoliation of $BiFeO_3$ for pressure sensing. Front Phys 18(6):63303. https://doi.org/10.1007/s11467-023-1301-7
223. Lim SH, Murakami M, Sarney WL, Ren SQ, Varatharajan A, Nagarajan V, Salamanca-Riba LG (2007) The effects of multiphase formation on strain relaxation and magnetization in multiferroic $BiFeO_3$ thin films. Adv Functional Mater 17(14):2594–2599. https://doi.org/10.1002/adfm.200700055
224. Yuan H, Li L, Hong H, Ying Z, Zheng X, Zhang L, Wang G (2021) Low sintering temperature, large strain and reduced strain hysteresis of $BiFeO_3$–$BaTiO_3$ ceramics for piezoelectric multilayer actuator applications. Ceram Int 47(22):31349–31356. https://doi.org/10.1016/j.ceramint.2021.08.008
225. Li L, Chew ZJ (2018) Microactuators: design and technology. In: Smart sensors and MEMS. Woodhead Publishing, pp 313–354. https://doi.org/10.1016/B978-0-08-102055-5.00012-7
226. Kim SH, Leung A, Koo CY, Kuhn L, Jiang W, Kim DJ, Kingon AI (2012) Lead-free $(Na_{0.5}K_{0.5})$(Nb0.95Ta0.05) O_3–$BiFeO_3$ thin films for MEMS piezoelectric vibration energy harvesting devices. Mater Lett 69:24–26. https://doi.org/10.1016/j.matlet.2011.11.069
227. Tang X, Cheng D, Ran J, Li D, He C, Bi S, Wang X (2021) Recent advances on the fabrication methods of nanocomposite yarn-based strain sensor. Nanotechnol Rev 10(1):221–236. https://doi.org/10.1515/ntrev-2021-0021
228. Zavaliche F, Yang SY, Zhao T, Chu YH, Cruz MP, Eom CB, Ramesh R (2006) Multiferroic $BiFeO_3$ films: domain structure and polarization dynamics. Phase Transitions 79(12):991–1017. https://doi.org/10.1080/01411590601067144
229. Hu YQ, Liu NT, Lao J, Liang RH, Deng X, Guan Z, Duan CG (2022) Ultrahigh ferroelectric and piezoelectric properties in $BiFeO_3$–$BaTiO_3$ epitaxial films near morphotropic phase boundary. ACS Appl Mater Interfaces 14(32):36825–36833. https://doi.org/10.1021/acsami.2c09062
230. Xue M, Tang Y, Shan Z, Hao Y, Zhou X, Gao X, Zhang B (2023) Deciphering the leakage conduction mechanism of $BiFeO_3$–$BaTiO_3$ lead-free piezoelectric ceramics. J Adv Ceram 12(10):1844–1856. https://doi.org/10.26599/JAC.2023.9220792
231. Iqbal S, Rafique HM, Mustafa GM, Younas B, Imran M, Alwadai N, Iqbal M (2023) Tuning of relaxation phenomenon by substitution of La in $BiFeO_3$. Mater Today Commun 34:105337. https://doi.org/10.1016/j.mtcomm.2023.105337
232. Singh SK, Maruyama K, Ishiwara H (2007) Reduced leakage current in La and Ni codoped $BiFeO_3$ thin films. Appl Phys Lett 91(11). https://doi.org/10.1063/1.2784968
233. Yoneda Y, Kitanaka Y, Noguchi Y, Miyayama M (2012) Electronic and local structures of Mn-doped $BiFeO_3$ crystals. Phys Rev B Condensed Matter Mater Phys 86(18):184112. https://doi.org/10.1103/PhysRevB.86.184112

234. Choi JY, Lee SY (2017) Comprehensive review on the development of high mobility in oxide thin film transistors. J Korean Phys Soc 71:516–527. https://doi.org/10.3938/jkps.71.516
235. Teague JR, Gerson R, James WJ (1970) Dielectric hysteresis in single crystal $BiFeO_3$. Solid State Commun 8(13):1073–1074. https://doi.org/10.1016/0038-1098(70)90262-0
236. Vishwakarma AK, Tripathi P, Srivastava A, Sinha ASK, Srivastava ON (2017) Band gap engineering of Gd and Co doped $BiFeO_3$ and their application in hydrogen production through photoelectrochemical route. Int J Hydrogen Energy 42(36):22677–22686. https://doi.org/10.1016/j.ijhydene.2017.07.153
237. Nassereddine Y, Benyoussef M, Asbani B, El Marssi M, Jouiad M (2023) Recent advances toward enhanced photocatalytic proprieties of $BiFeO_3$-based materials. Nanomaterials 14(1):51. https://doi.org/10.3390/nano14010051

Chapter 5
Quantum Criticality in the Sustainable Perovskite Multiferroic Materials

Vinay Pratap Verma, Upkar Kumar Verma, and Sandeep Kumar Pundir

1 Introduction

The rapid advancement in technologies and deep understanding of new concepts setup new dimensions in the field of material science. To focus on materials that can exhibit multiple functional properties within a single phase is important in the field of material science [1, 2]. Multiferroic materials show multiple ferroic orders simultaneously [3–5], and rapidly grown as advantageous material for next-generation technologies such as in the field of energy-efficient electronic devices, memory storage, spintronics, sensors, and solar cells [6–8]. Figure 5.1 shows the application of quantum criticality in perovskite multiferroic materials in different fields. Perovskite materials having highly flexible crystal structure with adjustable characteristics and tunable bandgap are well-suitable for multiferroic applications [9–13].

To design new advanced sustainable multiferroic perovskite materials of high efficiency and better performance requires a deep understanding of quantum criticality phenomena at various conditions, such as at low temperatures and high pressure. [14]. It is a milestone concept that stimulates many exciting phenomena which occurs when a system approaches to QCP [15]. QCP is a point in its phase diagram where the system undergoes a phase transition at absolute zero temperature dominated by quantum fluctuations rather than thermal variations [16, 17]. Near a QCP material's properties can show exciting changes such as they may set to new phases that are not seen under normal conditions and not explained by classical theory [18–21]. Deep

V. P. Verma · S. K. Pundir (✉)
Department of Physics, University of Lucknow, Lucknow, Uttar Pradesh, India
e-mail: pundir_s@lkouniv.ac.in

V. P. Verma
e-mail: vinaypratapverma@gmail.com

U. K. Verma
Department of Physics, B S N V PG College Lucknow, Lucknow, Uttar Pradesh, India
e-mail: upkarthegreat@gmail.com

M. Kumar (ed.), *Sustainable Perovskite Multiferroic Materials*, Materials Horizons: From Nature to Nanomaterials, https://doi.org/10.1007/978-981-96-7419-0_5

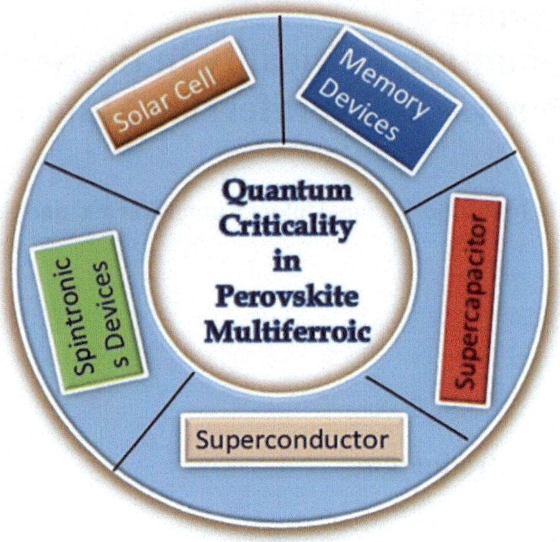

Fig. 5.1 Illustrative figure of application of quantum criticality in perovskite multiferroic materials in different fields

understanding of quantum criticality in multiferroic perovskites ensures unlocking their potential and can provide explanations how quantum fluctuations affect their coupling between magnetism, electricity, and lattice dynamics [22–24].

Perovskite multiferroics are emerging materials for new technological applications and behavior of these materials at the quantum critical point is not fully explained by traditional theories [6, 25–27]. This creates interest to think about the new boundaries of material design and development of new theory to explain the quantum fluctuations influence on ferroic ordering [28]. The advantage of perovskite multiferroics has also potential beyond their functional capabilities, like many of these materials can be synthesized using earth-abundant nontoxic elements which are environmental friendly alternative to materials which traditionally used in fabrication of electronic devices [29, 30]. Therefore, it becomes important to investigate the role of quantum criticality in the generation of advanced sustainable perovskite multiferroic materials.

Before going forward, we must know the basics of quantum criticality and related phenomena.

2 Quantum Criticality

Quantum criticality is a phenomenon that are seen in many types of quantum systems, especially those showing complex nature at low temperatures [16]. Quantum criticality arises when a system approaches to quantum critical point, a condition where quantum fluctuation plays vital role in the behavior of materials due to dominant quantum interaction such as those between spin, charge, or lattice degrees of freedom which converge at critical temperature of $T = 0$ K [31, 32]. Classical phase transitions are mainly governed by thermal effects and occur at finite temperatures, whereas quantum criticality governs with the delicate balance between quantum mechanical effects at low temperatures and quantum interactions at the microscopic level [33]. At or near QCP material undergoes a phase transition where the material's ground state exhibits a drastic change due to quantum fluctuations [34–36]. The systems exhibiting quantum criticality, the critical point, is mainly governed by divergent physical quantities such as the correlation length, susceptibility, or specific heat [35, 36]. These are driven by breakdown of conventional ordering, the appearance of non-Fermi liquid behavior, and the formation of novel quantum phases like spin liquids or quantum critical metals [37]. These anomalies create interest to investigate nature of quantum phase transitions and help to understand hidden mechanism of the phenomenon like non-Fermi liquid behavior, the breakdown of conventional ordering, and the emergence of exotic phases like spin liquids.

2.1 The Quantum Critical Point (QCP)

Quantum critical point is a point in phase space at absolute zero temperature at which continuous phase transition takes place driven by nonthermal transition such as pressure, field, or doping [21]. This phase transition does not govern by the thermal excitations of materials particles such as atoms or electrons. Instead of this it is governed by quantum fluctuations of the material's microscopic degrees of freedom which is inherent quantum properties of the particles associated with such transition like wave-particle duality nature, nature of uncertainty, and superposition of quantum states [38]. Figure 5.2 shows the schematic of quantum critical point with different ordered and disordered region in phase diagram.

Phase transition mainly occurs in two ways:

1. First-order phase transition: It is generally governed with thermal fluctuation, and it is discrete in nature [39].
2. Second-order phase transition: It is a continuous phenomenon occurs at QCP due to quantum fluctuations governed by quantum mechanical effect [40].

2.2 Quantum Phase Transitions

When system approaches to QCP, the material characteristics are strongly governed with quantum fluctuations. These fluctuations influence the ordering of particles, like the magnetic or electric ordering which generates new and exotic phases of matter [41]. Classical transitions generally take place by discrete jumps between different phases like solid to liquid, whereas quantum transitions are continuous process which continuously changes material properties as the system approaches to QCP [42].

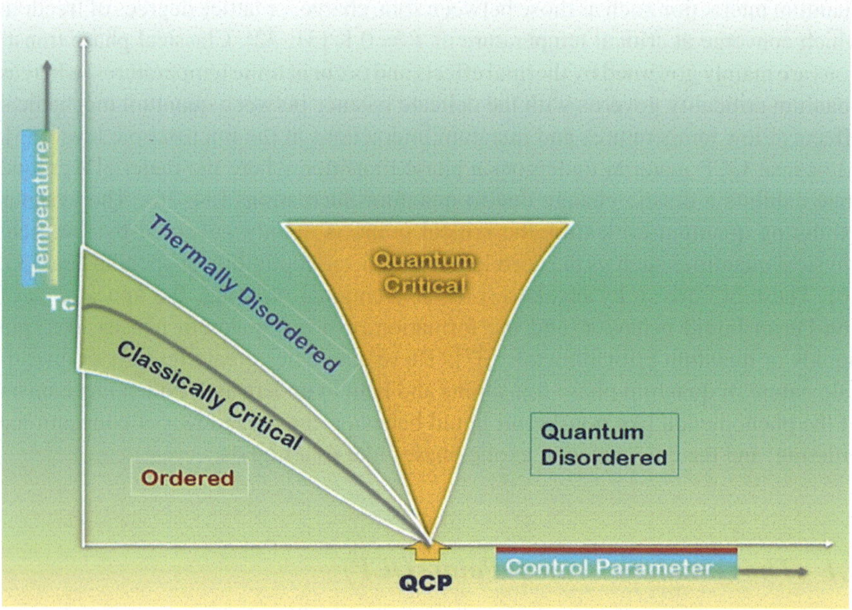

Fig. 5.2 Schematic figure showing Quantum Critical Point and different region in phase diagram

2.3 Non-fermi Liquid Behavior

One of the most important phenomena observed near QCP is non-Fermi liquid behavior [43]. According to classical theory, the electronic states at low temperature are described by the Fermi liquid theory. In Fermi liquid theory, electrons are weakly interacting quasi-particles and behavior of electron is predictable [44]. However, near the quantum critical point, these quasi-particles break down and the material behavior cannot be explained based on traditional Fermi liquid theory [45, 46]. This phenomena of Fermi and non-Fermi liquid phases in quantum critical region is shown in Fig. 5.3.

In a non-Fermi liquid, the electron–electron interactions become strong and the material show unusual properties, for example, power-law dependence in the electrical resistivity or anomalous temperature dependencies in thermodynamic properties [47]. Undefined quasi-particles and the strong correlation between the particles which cannot be understood by conventional theory of electronic behavior [48]. This phenomenon creates interest to study the high-temperature superconductivity, heavy-fermion systems, and other quantum materials properties where we observed non-Fermi liquid behavior [49].

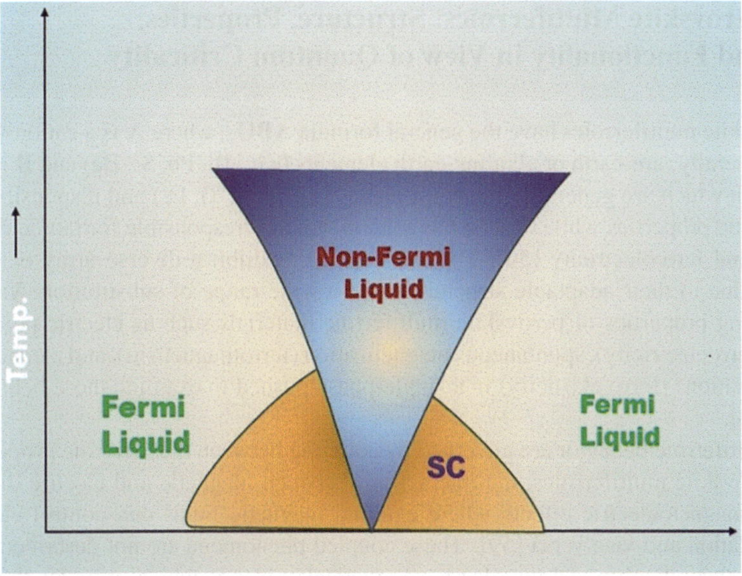

Fig. 5.3 Schematics figure showing fermi and non-fermi liquid phases

2.4 Quantum Fluctuations and Their Role in Quantum Criticality

Quantum fluctuations refer to the temporary changes in the energy of a point in space and governed by the uncertainty principle [50]. These fluctuations are existed at absolute zero temperature, whereas classical fluctuations are absent at absolute zero temperature [51]. At quantum critical point, the material becomes highly sensitive to these fluctuations which can strongly affect the material's properties [52].

Quantum fluctuations are the key factor of continuous nature of quantum phase transitions [53]. In classical domain, phase transitions take place at the system reaching to a certain temperature or energy threshold, whereas in quantum domain at the QCP phase, transitions are continuously fluctuating due to the interactions between different quantum states [54]. These fluctuations are key factor of critical phenomenon where material's properties show exciting changes as it is move toward the QCP.

For example, in quantum magnets, the quantum fluctuations of electron spins at low temperatures show a disordered quantum spin liquid phase instead of material traditional ordered phase transition (like ferromagnetism) [55]. In such systems, the quantum fluctuations lead to exotic behaviors that cannot be described by classical models of magnetism.

3 Perovskite Multiferroics: Structure, Properties, and Functionality in View of Quantum Criticality

Perovskite multiferroics have the general formula $\mathbf{ABO_3}$, where **A** is a cation which are generally rare-earth or alkaline-earth elements (e.g., Bi, Pb, Sr, Ba) and **B** is also a cation which are generally transition metals (e.g., Mn, Ti, Fe) and responsible for magnetic properties while O is an oxygen ion which is responsible for lattice distortions and ferroelectricity [56]. This material can exhibit a diverse array of properties due to their adaptable structures with a wide range of substitution. Various changing properties of perovskite multiferroic materials such as electric polarization (ferroelectricity), spontaneous magnetization (ferromagnetism), and mechanical deformations (ferro elasticity) in a single-phase pushed to examine these behaviors [57, 58].

Multiferroic behavior are governed by coupling between these ferroic properties. In perovskite multiferroics, the interactions between magnetic and electric dipoles show magnetoelectric effects where external magnetic fields can control electric polarization and vice versa [59]. These coupled phenomena are not described only from a theoretical model but also have practical evidence which provides the path for practical applications in various technologies, including memory devices, energy harvesters, and quantum information systems [60, 61].

Study of perovskite multiferroics in prospective of quantum criticality can lead to understand how quantum phase transitions are beneficial to design improved multiferroic materials which enhance remarkable properties with optimal performance. Figure 5.4 shows the schematic diagram of material exhibit multiferroic properties.

The interaction between electric and magnetic orders plays crucial role near the QCP, in the perovskite multiferroics materials [62]. Phase transition like ferromagnetic to antiferromagnetic or the simultaneous existence of electric polarization in multiferroic materials causes to change material behavior continuously [63, 64]. Furthermore, the exotic states like spin liquids or quantum spin glasses can govern the magnetoelectric coupling by quantum fluctuations near the QCP [65].

In perovskite multiferroics, the continuously changing ferroelectric and magnetic ordering can create exciting phenomena at the QCP, such as an electric field cause reorientation of the magnetic moment or vice versa [66]. For example, phenomena like this have been observed for bismuth ferrite ($BiFeO_3$) in which the magnetic and electric ordering can be controlled by external conditions such as temperature, pressure, or chemical doping [67].

As discussed above, quantum criticality in perovskite multiferroics becomes dominant by the suppression of ferroelectric or magnetic order by external control parameters [20] as discussed below in subsection. The suppression of ferroelectric can be achieved by following ways.

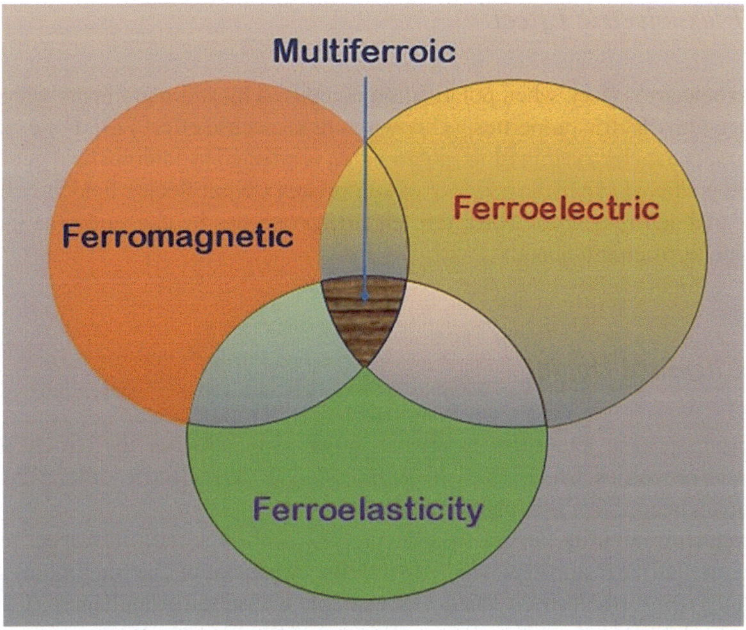

Fig. 5.4 Schematic of materials exhibiting multiferroic properties

3.1 *Using External Electric Field*

In the presence of external electric field, the ferroelectric materials reorient their polarization domains [68]. In some cases, strong electric field cause suppression the ferroelectric order in the materials, in which ferroelectricity and magnetism are coupled. For example, multiferroic materials like $Ni_3V_2O_8$ having different stoichiometry in which electric fields can control magnetic phase transitions, which modulate ferroelectric properties of this materials [69]. This suppression takes place by various factors such as depolarizing fields, phase transitions, and flexoelectric effects. In ferroelectric capacitors, incomplete charge compensation at the interfaces can create internal electric fields which oppose the spontaneous polarization, known as depolarizing fields [70]. Since it is observed that a small compensation in the polarization charge can influence ferroelectric properties, complete compensating charge nearly eliminate polarization. It is crucial in the development of thin films and nanoscale devices in which surface-to-volume ratios are high. Also, an application of external electric field can induce phase transitions that disturb long-range polar order [71]. For example, in certain polycrystalline oxides, an ordered ferroelectric phase is converted into a disordered relaxer phase in the presence of an electric field [72]. This transition suppresses the original ferroelectric order.

3.2 Flexoelectric Effect

The flexoelectric effect, when polarization is induced by strain gradients which can influence ferroelectric properties, is known as flexoelectric effect [73]. For example, in epitaxial ferroelectric thin films, polarization is affected by internal field generated by stress gradient [74]. In the persence of apropriate external electric field this effect is disaapeared and restore the basic ferroelectric properties by canceling the internal fields due to strain gradients.

3.3 Chemical Doping

Implanting specific impurities in crystal lattice also influence the structural and electronic properties which leads to suppression of ferroelectric order [75]. The mechanism behind this is explained as follows.

Introducing impurity dopant ions having different ionic radii in respect to host ions creates lattice distortions which disturb the alignment of electric dipoles and is responsible for ferroelectricity [76]. For example, substituting lanthanum (La^{3+}) in place of lead (Pb^{2+}) in lead-based perovskite materials affects the tolerance factor and suppresses the coupling between off-centered Pb and B cations [77]. This substitution enhances antiphase tilting of BO_6, which results to unordered antiferroelectric order at low temperatures.

3.3.1 Charge Carriers

Doping introduced charge carriers electrons or holes in the material, which conceal the internal electric fields that is responsible for ferroelectric polarization [78]. In barium titanate ($BaTiO_3$), both electron and hole introduced by doping can decrease ferroelectricity properties. The strength of suppression varies between electrons and holes, depending on the structures of conduction and valence bands [79]. Carrier-induced screening reduces the material strength to sustain their spontaneous polarization [80].

3.3.2 Phase Stabilization

Impurity doping ions can regularize nonferroelectric phases instead of ferroelectric [81]. For example, when hafnium dioxide (HfO_2) is doped with certain elements can form nonpolar monoclinic phase in place of the polar orthorhombic phase, hence suppressing ferroelectricity [82]. The effect of doping on phase stability is complex and depends on various factors like dopant size and concentration [83]. Hence, understanding the effect of chemical doping on ferroelectric materials gives clue for tuning the properties for specific applications. By selecting suitable dopants and concentrations, researchers can design materials with required ferroelectric characteristics and suppress unwanted ferroelectric behavior to enhance their functionalities.

3.4 *Strain Engineering*

With application of mechanical strain, the lattice parameters are disturbed and hence affect the ferroelectric phase [84]. Due to mechanical strain, ferroelectricity can be either enhanced or suppressed, depending on the material properties and the nature of the applied strain [85]. This approach is useful in the development of thin films and layered structures in which strain can be controlled during fabrication process. It is a powerful tool which modulates the ferroelectric properties of materials using mechanical deforming forces, hence disturb their crystal lattice and affect polarization states. It suppresses ferroelectric order, and the mechanisms behind this is discussed below in subsections.

3.4.1 Lattice Distortion

Strain disturbed the interatomic distances and bond angles in ferroelectric materials impacting polarization [86]. For example, in barium titanate ($BaTiO_3$), dislocations through high-temperature plastic deformation can cause internal strain fields [87]. These fields affect domain configurations, resulting to the suppression of ferroelectric order by nonpolar domain orientations.

3.4.2 Phase Transitions

Phase transitions between ferroelectric and nonferroelectric phases can take place due to external factors such as temperature, pressure, or strain [88]. For example, in strontium titanate ($SrTiO_3$), a typical nonferroelectric at room temperature converted to ferroelectric phase, due to biaxial strain [89]. Conversely, changing the strain material returns to its paraelectric state, suppressing ferroelectricity. This transition is occurred due to the structural changes in the crystal lattice. Thermally induced transition takes place when the temperature exceeds the Curie temperature leading to the disappearance of spontaneous polarization [90–92]. Thermal vibrations increase at high temperatures resulting in the reduction in the displacement of ions responsible for polarization [93–95]. Further, high pressure can suppress ferroelectricity by altering the crystal structure and preventing ion displacement [96, 97]. A sufficiently strong external electric field can also affect the ferroelectric order, leading to a transition to a nonpolar phase [98, 99].

3.4.3 Domain Reorientation

The orientation and distribution of ferroelectric domains are also affected by strain [100]. For example, strain engineering used to control domain configurations in potassium sodium niobate thin films [101]. With specific strain patterns, researchers can modulate domain structures that regulate polarization, hence suppressing the overall ferroelectric response.

3.5 Suppression of Magnetic Order

Suppressing magnetic ordering in perovskite materials plays a vital role to tuning the properties for various applications. General strategies used to achieve suppression are explained below in subsections.

3.5.1 Chemical Doping

Inducing nonmagnetic ions in the perovskite lattice disturb magnetic interactions, which results suppression of magnetic order [102]. This is achieved by the suppression of spin disorder using magnetic field, which is enhanced by nonmagnetic ions. Chemical doping is an effective way to suppress magnetic order by modifying lattice distortions, charge carrier concentrations, or exchange interactions [103–105]. This suppression can lead to enhanced ferroelectricity, emergent functionalities, or even exotic quantum phases [106, 107]. Charge carrier doping with ions alters the oxidation state of magnetic cations, weakening magnetic interactions. Further, replacing magnetic ions with nonmagnetic or weakly magnetic ions also suppresses long-range magnetic order [108, 109].

3.5.2 Strain Engineering

Strain engineering is used for controlling and suppressing magnetic order in multiferroics by modifying lattice symmetry, bond angles, and exchange interactions [110–112]. This approach is widely deployed in thin films and heterostructures, where substrate-induced strain can significantly alter the magnetic and ferroelectric properties. Epitaxial strain changes lattice parameters, modifying magnetic interactions. Strain can introduce competing interactions, leading to spin-glass-like states or suppression of long-range magnetic order [113–115]. Application of mechanical strain to perovskite materials disturb their crystal lattice parameters; hence, magnetic interactions are modulated [116]. This is achieved by varying crystal structure using controlled disorder, resulting in magnetic ordering.

3.5.3 Light-Induced Effects

When perovskite materials are exposed to light creates skin defects, causes suppression of magnetic order [117]. Recent researches shown photo-excitation results to suppression of magnetic order. The magnetic order can be completely suppressed at higher photo-excitation densities. This nonthermal breaking of magnetic order by photo-generated spin defects allow to control magnetic properties in perovskites.

3.5.4 Electric Field Control

In multiferroic and magnetoelectric materials, magnetic ordering is controlled by electric fields based on magnetoelectric coupling [118]. This cross-control are utilized to vary magnetization using electric fields, which becomes more advantageous for low-power spintronic devices.

3.5.5 Chemical Substitution

Replacing atoms in the lattice of material with different elements (isovalent substitution) can affect magnetic interactions and suppress magnetic order [119]. In CeFeAsO, substituting arsenic with phosphorus changes the electronic environment, resulting to the suppression of spin-density wave order.

3.5.6 Magnetic Field Application

An external magnetic field can affect magnetic ordering by aligning spins or inducing phase transitions [120]. A strong magnetic field can suppress antiferromagnetic order, yielding to a transition to a paramagnetic state in some materials.

3.6 Quantum Criticality in Ferroelectric Perovskites

Some perovskites like $SrTiO_3$ and $KTaO_3$ are known as quantum paraelectric [121], i.e., they are close to a ferroelectric instability but remain paraelectric state at absolute zero temperature due to quantum fluctuations. Since quantum fluctuations suppress the ferroelectric transition to prevent long-range dipole ordering, it results to a continuous change of permittivity with a temperature [122].

3.7 *Magnetic Quantum Criticality*

A phenomenon that occurs at or near QCP in magnetic systems are called magnetic quantum criticality (MQC), due to dominate quantum fluctuations and resulting a phase transition at absolute zero temperature ($T = 0$) [37]. It followed the uncertainty principle and the wave-like nature of particles.

A material goes to phase transition between different magnetic phases like from a magnetically ordered phase ferromagnetic or antiferromagnetic to a disordered phase paramagnetic state at magnetic quantum critical point [123]. This transition take place at $T = 0$ K by tuning external parameters such as pressure, magnetic field, or doping. The study of magnetic quantum criticality new dimensions in the quantum states of matter. Phase transition takes place at a magnetic QCP with changing the magnetic order.

Ferromagnetic to Paramagnetic Transition: Ferromagnetic materials have the spins of atoms aligned in the same direction, resulting to a net magnetic moment. At QCP, these spins become disordered and phase transition takes place to a paramagnetic state in which the spins are randomly arranged [124].

Antiferromagnetic to Paramagnetic Transition: Since antiferromagnetic materials have the spins of atoms are aligned in opposite directions but near the QCP, this order is suppressed by quantum fluctuations, resulting to a loss of magnetic order [40].

3.8 *Multicritical Points and Coupled Quantum Criticality*

Some materials having coexisting or competing quantum critical points in which both ferroelectric and magnetic instabilities lie simultaneously known as multicritical points and coupled quantum criticality due to the intersection and interaction of multiple quantum phase transitions, resulting to rich and complex behaviors of materials [125].

3.8.1 Multicritical Points

A multicritical point (MCP) is a junction point in a phase diagram where multiple phases phase transitions coincide due to quantum fluctuations. These points are explained based on different order parameters and fluctuations, resulting to novel universal classes and critical behaviors of materials [126]. For example, the two- and three-dimensional random quantum Ising model shows QMCPs arise due to the intersection of different quantum phase transitions [127]. The complex interaction of various fluctuations at these points results in unique critical phenomena which is different from single quantum critical points.

3.8.2 Coupled Quantum Criticality

Coupled quantum criticality occurs when numerus quantum systems or order parameters are interlinked, and their collective critical behavior cannot be explained based on each system in isolation [128]. The coupling among different quantum critical modes can emerge new phenomena and modulate the nature of the phase transitions in the material. For example, hybrid metal–semiconductor systems, in which interaction between dynamical Coulomb blockade and the integer quantum Hall effect give birth of resonant tunneling in Luttinger liquids [129].

4 Theoretical Models of Quantum Criticality in Multiferroics

Several models used to explain quantum criticality in perovskite multiferroics are discussed below.

4.1 Soft Phonon Mode Theory

Soft phonon mode theory is a basic concept in material science that explains the mechanisms behind structural phase transitions in crystalline materials. It proposes that certain vibrational modes, known as soft modes, show decrease in frequency as the system is under phase transition, resulting to a transformation in the crystal structure [130]. In a crystalline solid, atoms oscillate at lattice positions, and all these oscillations results to phonons. A soft phonon mode theory is governed by specific vibrational pattern in which frequency decreases as the temperature or another external parameter close to critical value [131]. This softening results to instability in the lattice, shift to a new lower-symmetry phase. For example, in ferroelectric materials, the softening of a transverse optical phonon mode results to phase transition from paraelectric to ferroelectric phase, due to spontaneous electric polarization [132].

Macroscopic explanation of soft mode behavior is explained by Landau theory of phase transitions. According to Landau theory of phase transitions, an order parameter quantifies the degree of structural distortion. The free energy of the system is written as a power series in order parameter, and the coefficients of this series are temperature dependent [133]. As the system goes to critical temperature, coefficient associated with the quadratic term tends to zero, resulting to the softening of the phonon mode and phase transition take place.

On a microscopic level, the soft mode theory is explained by dynamics of specific phonon modes using models like the transverse Ising model and Green's function techniques [134].

For example, materials like barium titanate ($BaTiO_3$) have soft phonon modes associated with transitions from paraelectric to ferroelectric phases, with temperature-dependent frequency decrease [135].

4.2 Landau Theory with Quantum Corrections

Landau theory, proposed by Lev Landau, is based on classical phase transition model. According to this theory, free energy is expressed as a power series in terms of an order parameter as given in Eq. 5.1 [136]:

$$F(\rho) = a(T)\rho^2 + b\rho^4 + c\rho^6 + \ldots \tag{5.1}$$

where ρ represents the order parameter, $a(T)$ is temperature-dependent coefficient as expressed in Eq. 5.2, and b & c are higher-order coefficients.

$$a(T) \propto \frac{1}{T + T_\mathrm{q}} \tag{5.2}$$

where T_q is a characteristic temperature and as system approaches QCP, T_q tends to zero.

This classical approach explains many thermodynamic phenomena, but it does not explain quantum mechanical effects, near absolute zero temperature or where quantum fluctuations are dominant. Quantum corrections in Landau theory enhance its applicability and made it applicable to quantum mechanical effects, near absolute zero temperature or where quantum fluctuations are dominant.

4.2.1 Quantum Corrections in Landau Theory

Quantum corrections in Landau theory is done by modifying the free energy expansion considering quantum fluctuations, resulting to a temperature-dependent renormalization of the expansion coefficients with taking in account fluctuations arise from quantum effects [136]. Now Quantum Corrections in $a(T)$ at finite temperatures is expressed as [137],

$$a(T) = a_0(T - T_c) \tag{5.3}$$

where T_c is the transition temperature and a_0 is a material-dependent constant.

Now integrating high-energy quantum fluctuations in renormalized form [138], the general equation of free energy becomes:

$$F(\rho) = a_0(T - T_c)\rho^2 + b\rho^4 + c\rho^6 + \ldots \tag{5.4}$$

$F(\rho)$ determines the phase of the system such as:

Case. I—$T > T_c$; system is in disordered phase paraelectric or paramagnetic phase.

Case. II—$T < T_c$; system is in ordered phase ferroelectric or ferromagnetic state.

Case. II—$T = T_c$; phase transition occurs, and small fluctuations can determine transition is first order or second order.

Note that if **b > 0**, the transition is continuous. Also order parameter increases smoothly from zero when T decreases below Tc.

4.3 Quantum Heisenberg and Using Models

It is very useful theory to examine magnetic quantum criticality in perovskites with competing interactions which are discussed below:

4.3.1 Quantum Ising Model Hamiltonian

The transverse-field Ising model (TFIM) is the simplest model explain a quantum phase transition [139]. The Hamiltonian is given by [140]:

$$H = -J \sum_{\langle i,j \rangle} \sigma_i^z . \sigma_j^z - h_x \sum_i \sigma_i^x \tag{5.5}$$

where σ_i^z and σ_i^x are Pauli matrices with quantum spin-1/2 operators. J is the exchange interaction, h_x is the transverse field. Quantum phase transition occurs at a critical field h_x^c, between an ordered (ferromagnetic) and a disordered (paramagnetic) phase.

4.3.2 Quantum Phase Transition in the Ising Model

Case I—Low Transverse Field ($h_x < h_x^c$):

The interaction term J becomes dominates, resulting to long-range order in the z-direction. This is a ferromagnetic phase.

Case II—High Transverse Field ($h_x > h_x^c$):

The transverse field h_x becomes dominates, and spins are aligned along the x-direction, disturbing the long-range order. This is a paramagnetic phase.

Case III—At the Critical Point ($h_x = h_x^c$):

Quantum superposition of ordered and disordered states take place. The correlation length ξ diverges and system obeys quantum critical scaling laws.

4.3.3 Critical Exponents and Universality

The correlation length ξ diverges, at QCP given as,

$$\xi \propto \left|h_x - h_x^c\right|^{-\nu} \tag{5.6}$$

where ν is the critical exponent.

4.3.4 Heisenberg Model Hamiltonian

Quantum Heisenberg model generalizes the Ising model by allowing spin components to interact in all directions [141, 142]:

$$H = -J \sum_{i,j} \left(s_i^x s_j^x + s_i^y s_j^y + s_i^z s_j^z \right) \tag{5.7}$$

where $s_i^\alpha (\alpha = x, y, z)$ are spin-1/2 operators.

4.4 *Multicritical Scaling Theories*

In multiferroic perovskites, the interaction between ferroelectric (electric polarization) and magnetic (spin ordering) degrees of freedom results to multiple phase states, at multicritical point [143] as discuss below:

The effective free energy considering from both order parameters (P for polarization and M for magnetization) and their coupling terms.

Landau–Ginzburg free energy expansion given as [144, 145]:

$$F = a_P(T)P^2 + b_P T^4 + a_M(T)M^2 + b_M M^4 + \lambda P^2 M^2 + \ldots\ldots \tag{5.8}$$

where λ is coupling between the ferroelectric and magnetic orders.

4.4.1 Multicritical Scaling Framework: Scaling Hypotheses

At multicritical point, physical observables follow scaling laws that depend on more than one variable (e.g., temperature, pressure, magnetic field, and electric field) [146]. For example, the correlation length ξ given as [147],

$$\xi \sim |g - g_c|^{-\nu} \tag{5.9}$$

where g is a generalized tuning parameter (which could be a combination of pressure, doping, or strain), and v is the critical exponents.

5 Application and Future Directions

Quantum criticality in sustainable perovskite multiferroic materials establishes new dimensions in field of material science which open the divergent path to design advanced materials with variable properties according to requirement of new generation. As the interaction between magnetic and electric dipoles can be finely tuned for enhanced multiferroic behaviors in these materials. For example, quantum fluctuations near the QCP, the energy barriers between different magnetic configurations become low, making it more sensitive to external fields, which can result to improved magnetoelectric coupling, due to more efficient control over both the magnetic and electric properties simultaneously.

Furthermore, the vicinity to the QCP may open the path for the emergence of novel phases, like quantum spin liquids or topologically protected states, which is emerging as entirely new types of multiferroic materials. These phases may exhibit unique properties, like higher strength to external disturbance or greater stability under varying temperature and pressure conditions. By manipulating quantum criticality in sustainable perovskite materials, it may be possible to develop new multiferroic systems that are more efficient, versatile, and durable. For example, the use of strain engineering or doping can push a material go into a quantum critical region, resulting to enhanced response to external electric or magnetic fields. Such materials will be ideal for use in applications where energy efficiency and speedy response are crucial, like in low-power memory devices, sensors, and actuators. Now we explore the practical applications of these materials with unique properties, along with the future directions which can make it more advantageous.

5.1 Applications

Quantum criticality in perovskite multiferroic materials have most attention due to its unique interaction between electric, magnetic, and structural order parameters near quantum critical points (QCPs) make it important for various high-performance applications. Some applications are discussed below.

Quantum criticality provides enhanced dielectric permittivity and magnetoelectric coupling, which can be used to in develop new generation electronic devices or components as discussed below.

Using sustainable perovskite multiferroics with quantum critical properties, we can enhance overall performance of FeRAM by reducing power consumption. Enhancing data retention fast polarization switching facilitates high-speed computing applications. Use of quantum criticality reducing energy dissipation since quantum fluctuations suppress resistive losses resulting to ultralow-power operation. Minimizing structures results to enhanced material properties which allow to minimize transistor sizes for next-generation processors. Boosting performance of sustainable multiferroic materials provides improved charge carrier mobility and adjustable electronic properties.

The strong coupling between electric and magnetic order helps to development of multistate logic gates which increase computational efficiency of device. Also, reconfigurable electronics materials with adjustable quantum phase transitions opens the path of adaptive circuit design for energy-efficient devices and new generation processors are designed tuning with electric and magnetic degrees of freedom, resulting in lower heat generation and power requirements. Perovskite multiferroics with quantum critical properties are good candidate for flexible or wearable devices due to their high sensitivity and flexibility and enable it to integrate in smart textiles processing and bioelectronics field. Using this material, we can design smart energy-efficient flexible displays.

Quantum fluctuations in multiferroic perovskites enable it for smart efficient energy applications like hybrid energy harvesters devices, which are designed combining piezoelectric and photovoltaic effects for sustainable power generation that are environment friendly. Also, low-loss supercapacitors using enhanced dielectric properties which provide superior charge storage capacity create new revolution in this field. High-efficiency batteries are also using Quantum-enhanced ionic conductivity setup a milestone in energy storage and conversion.

5.2 *Future Directions*

Quantum criticality in sustainable perovskite multiferroics is a becoming rapidly growing field with numerous exciting possibilities for development of new advanced technologies for both fundamental science and practical applications. These materials exhibit coupled magnetic, electric, and ferroelastic properties. Quantum critical phenomena where the material's behavior governed by quantum fluctuations at absolute zero temperature, setup new dimentions in field of material science. Some of the future directions for new research are discussed below.

5.2.1 Sustainable Materials Design

Quantum criticality in perovskite multiferroics opens new exciting possibilities in material science, when we talk about this it is essential to consider the sustainability of these materials for environmental considerations. Traditional multiferroics, like lead-based perovskites, create significant environmental problems due to the toxicity of lead and the limited availability. Now the development of materials need for environmentally friendly alternatives has force researches to think about lead-free and earth-abundant perovskite materials.

Sustainable perovskite materials may be developed from available numerous nontoxic and abundant elements. For example, halide perovskites, tin (Sn), calcium (Ca), or magnesium (Mg) perovskites are emerging as lead-free alternatives. These materials provide comparative enhanced multiferroic properties reducing the environmental concerns. Also, the synthesis of these materials generally requires less energy, giving contributions to sustainability.

Sustainable materials are designed using different methodologies as discussed below in subsections.

Future research will center to designing perovskite multiferroics using abundant, nontoxic, and sustainable elements. As we know rare earth metals are generally used in current materials, which will shift toward more environmentally friendly compositions are crucial point. Using elements such as magnesium, calcium, or manganese in perovskite materials and examining their effects in different condition will provide a better alternative to other traditional materials.

Use of hybrid materials, such as organic–inorganic perovskites, will opens new dimensions in multiferroic material with sustainability. These materials are reconstructed for optimal performance with tuning quantum criticality by in both the organic and inorganic components for advanced applications.

5.2.2 Quantum Criticality and Disorder

In multiferroics, disorder which appears due to structural imperfections or compositional variations play a crucial role in behavior of quantum critical points (QCPs). Investigating such disorder, can be controlled in sustainable perovskite multiferroics with tuning between quantum fluctuations and structural disorder at QCPs.

Perovskite multiferroics having electric polarization and magnetic order are coupled, which are excellent way to study quantum critical behavior among multiple transition order parameters. Understanding this, we able to explain how these transition orders can coexist or vary near the QCP which yield to generate novel high-performance materials for energy-efficient technologies.

5.2.3 Tailoring Critical Temperature and External Control

In perovskite multiferroic materials, we tailor the critical temperature and external control to develop advanced materials. Tuning of temperature at or near quantum critical point in multiferroic perovskites can results to novel phenomena, can be achieved by adjusting parameters such as pressure, doping, or strain. This flexibility will also enable the path of new development of devices that use quantum criticality for sensing or memory applications. Varying external fields, such as magnetic, electric, and strain, affects quantum criticality in which multiferroics will provide a wide area of future research. Quantum criticality can be adjusted using high-pressure techniques or through electrostatic control over the material's properties, and understanding nonequilibrium dynamics in these systems will result to new device functionalities for sustainable technologies.

5.2.4 Multifunctional Quantum Devices

Perovskite multiferroics, exhibiting quantum critical behavior, may be used for applications in quantum memory, where the nonvolatile state of quantum critical systems could be achieved. Additionally, these materials will serve as a basis for energy-efficient technologies, with their unique dielectric and piezoelectric properties. Exploring these concepts, we can design multifunctional devices. The extreme sensitivity of multiferroics makes them superior candidates for developing ultra-sensitive quantum sensors, and these are becoming more efficient at QCP, which are used in metrology, biomedical sensing, and environmental monitoring. With new advancement we can enhance the stability which results to sustainability of perovskite-based quantum sensors and provide green alternatives to existing technologies.

5.2.5 Modeling and Simulations

In this era, modeling and simulation are the center of research in each field, using this we can able to design various materials with varying different parameters and conditions. It accelerates the research without creating any environment hazard like in traditional processing of materials.

Theoretical and computational advances in first-principles modeling will play crucial role for predicting the behavior of sustainable perovskite multiferroics near quantum critical points. Using density functional theory (DFT) and other advanced techniques, researchers can predict how electronic, magnetic, and structural properties work in response to quantum criticality, open the new dimension to design new smart materials. The complex interventions among different orders (magnetic, electric, and ferroelastic), multiscale simulations which consider both quantum and classical effects will be better aspects of material design. These models will be helpful in designing perovskites that result to optimal performance at quantum critical behavior for specific applications.

5.2.6 Low-Dimensional and Heterostructure Systems

Using this, we can be able to design low dimensions such as zero dimension, one dimension, two dimension heterostructure material for optimal performance and efficiency. Exploring two-dimensional (2D) perovskite multiferroic materials or creating nano-sized perovskite structures will establish new avenues for optimal use of quantum criticality in confined geometries. These may enhance quantum effects due to quantum confinement and reduce dimensionality, probably generate advanced technology.

Combining perovskite multiferroics with other materials in heterostructures may provide novel quantum critical phenomena that cannot be observed in bulk systems. These heterostructures are observed by manipulating various parameters and different material to get enhanced multiferroic properties, and the addition of interfacial effects could result to new quantum states and critical phenomena.

5.2.7 Quantum Criticality in Nonequilibrium System

The quantum criticality of perovskite multiferroics in nonequilibrium states (such as under ultrafast laser pulses or time-dependent magnetic/electric fields) could show dynamic critical phenomena. This could provide a better understanding of how these materials behave under extreme conditions, which opens the way for novel quantum devices.

6 Conclusion

The quantum criticality in sustainable perovskite multiferroics encourages to establish a frontier in material science. By harvesting quantum critical points, it allows to alter the magnetoelectric coupling and ferroic characteristics of these perovskite multiferroics materials, which explore the new way to develop more efficient and user versatile devices. At the same time, these sustainable materials ensure that it is environmentally friendly and these advancements can be achieved without harming on ecological and economic parameters. With continuous improvement and innovation in material design, synthesis using quantum engineering sustainable perovskite multiferroics plays a crucial role in the development of next-generation electronic and quantum technologies.

References

1. Dahman Y (2017) Chapter 1—an Introduction to Nanotechnology. In: Dahman Y, Heng Lo H, Edney M (eds) Nanotechnology and functional materials for engineers, pp 1–17. Elsevier. https://doi.org/10.1016/B978-0-323-51256-5.00001-0
2. Lu C, Wu M, Lin L, Liu J-M (2019) Single-phase multiferroics: new materials, phenomena, and physics. Natl Sci Rev 6(4):653–668. https://doi.org/10.1093/nsr/nwz091
3. Zhao Y, Wang Y, Yang Y, Zhao J, Jiang X (2024) Realization of 2D multiferroic with strong magnetoelectric coupling by intercalation: a first-principles high-throughput prediction. NPJ Comput Mater 10(1):122. https://doi.org/10.1038/s41524-024-01301-x
4. Lu C, Liu J-M (2016) DyMnO3: a model system of type-II multiferroics. J Mater 2(3):213–224. https://doi.org/10.1016/j.jmat.2016.04.004
5. Khomskii D (2006) Multiferroics: different ways to combine magnetism and ferroelectricity. J Magn Magn Mater 306:1–8. https://doi.org/10.1016/j.jmmm.2006.01.238

6. Zhang L, Mei L, Wang K, Lv Y, Zhang S, Lian Y, Ding L et al (2023) Advances in the application of perovskite materials. Nano-Micro Lett 15(1):177. https://doi.org/10.1007/s40820-023-01140-3
7. Li X, Aftab S, Mukhtar M, Kabir F, Khan MF, Hegazy HH, Akman E (2024) Exploring nanoscale perovskite materials for next-generation photodetectors: a comprehensive review and future directions. Nano-Micro Lett 17(1):28. https://doi.org/10.1007/s40820-024-01501-6
8. Bui TH, Shin JH (2023) Perovskite materials for sensing applications: recent advances and challenges. Microchem J 191:108924. https://doi.org/10.1016/j.microc.2023.108924
9. Njema GG, Kibet JK (2025) A review of chalcogenide-based perovskites as the next novel materials: solar cell and optoelectronic applications, catalysis and future perspectives. Next Nanotechnol 7:100102. https://doi.org/10.1016/j.nxnano.2024.100102
10. Monama GR, Ramohlola KE, Iwuoha EI, Modibane KD (2022) Progress on perovskite materials for energy application. Results Chem 4:100321. https://doi.org/10.1016/j.rechem.2022.100321
11. Khirade P, Raut A (2022) Perovskite structured materials: synthesis, structure, physical properties and applications. https://doi.org/10.5772/intechopen.106252
12. Simenas M, Gagor A, Banys J, Maczka M (2024) Phase transitions and dynamics in mixed three- and low-dimensional lead halide perovskites. Chem Rev 124(5):2281–2326. https://doi.org/10.1021/acs.chemrev.3c00532
13. Hamukwaya SL, Hao H, Zhao Z, Dong J, Zhong T, Xing J, Mashingaidze MM et al (2022) A review of recent developments in preparation methods for large-area perovskite solar cells. Coatings 12(2):252. https://doi.org/10.3390/coatings12020252
14. Afre RA, Pugliese D (2024) Perovskite solar cells: a review of the latest advances in materials, fabrication techniques, and stability enhancement strategies. Micromachines 15(2):192. https://doi.org/10.3390/mi15020192
15. Shao C, Lu H, Zhang X, Yu C, Tohyama T, Lu R (2022) High-harmonic generation approaching the quantum critical point of strongly correlated systems. Phys Rev Lett 128(4):047401. https://doi.org/10.1103/PhysRevLett.128.047401
16. Grbić MS, O'Farrell ECT, Matsumoto Y, Kuga K, Brando M, Küchler R, Nakatsuji S et al (2022) Anisotropy-driven quantum criticality in an intermediate valence system. Nat Commun 13(1):2141. https://doi.org/10.1038/s41467-022-29757-9
17. Chandra P, Coleman P, Continentino MA, Lonzarich GG (2020) Quantum annealed criticality: a scaling description. Phys Rev Res 2(4):043440. https://doi.org/10.1103/PhysRevResearch.2.043440
18. Sharma S, Nag T, Rajak A, Bandyopadhyay S, Bhattacharjee S, Maity S, Bhattacharya U (2024) Unquenched—a memoir on non-equilibrium dynamics of quantum many-body systems: honoring amit dutta. Eur Phys J B 97(7):102. https://doi.org/10.1140/epjb/s10051-024-00740-1
19. Zhang G, Chung C-H, Ke C-T, Lin C-Y, Mebrahtu H, Smirnov AI, Baranger HU et al (2021) Nonequilibrium quantum critical steady state: Transport through a dissipative resonant level. Phys Rev Res 3(1):013136. https://doi.org/10.1103/PhysRevResearch.3.013136
20. Narayan A, Balatsky A, Spaldin N (2019) Multiferroic quantum criticality. Nat Mater 18. https://doi.org/10.1038/s41563-018-0255-6
21. Paschen S, Si Q (2020) Quantum phases driven by strong correlations. Nat Rev Phy 3. https://doi.org/10.1038/s42254-020-00262-6
22. Goyal RK, Maharaj S, Kumar P, Chandrasekhar M (2025) Exploring quantum materials and applications: a review. J Mater Sci: Mater Eng 20(1):4. https://doi.org/10.1186/s40712-024-00202-7
23. Lonzarich G (2005) Quantum criticality: magnetic quantum liquid enigma. Nat Phys 1:11–12. https://doi.org/10.1038/nphys139
24. Chandra P (2019) Multifunctionality goes quantum critical. Nat Mater 18:197–198. https://doi.org/10.1038/s41563-019-0302-y

25. Liu H, Yang X (2017) A brief review on perovskite multiferroics. Ferroelectrics 507:69–85. https://doi.org/10.1080/00150193.2017.1283171
26. Bati ASR, Zhong YL, Burn PL, Nazeeruddin MK, Shaw PE, Batmunkh M (2023) Next-generation applications for integrated perovskite solar cells. Commun Mater 4(1):2. https://doi.org/10.1038/s43246-022-00325-4
27. Huamán JLC, Rivera VAG, Pinto AH, Marega E (2023) 11—Multiferroic perovskite ceramics: properties and applications. In: Huamán JLC, Rivera VAG (eds) Perovskite ceramics, pp 339–381. Elsevier. https://doi.org/10.1016/B978-0-323-90586-2.00003-6
28. Sciences NAO, Engineering DO, Sciences P, Physics BO, Materials N, Board M, Survey AD (2019) Frontiers of materials research: a decadal survey. National Academies Press. https://doi.org/10.17226/25244
29. Alberi K, Nardelli MB, Zakutayev A, Mitas L, Curtarolo S, Jain A, Green ML et al (2018) The 2019 materials by design roadmap. J Phys D Appl Phys 52(1):013001. https://doi.org/10.1088/1361-6463/aad926
30. Hemminger JC, Sarrao J, Crabtree G, Flemming G, Ratner M (2015) Challenges at the frontiers of matter and energy: transformative opportunities for discovery science: USDOE Office of Science (SC)(United States). https://doi.org/10.2172/1283188
31. Sachdev S (2023) Quantum phases of matter. Cambridge University Press. https://doi.org/10.1017/9781009212717
32. Leonov I, Shorikov A, Anisimov V, Abrikosov I (2020) Emergence of quantum critical charge and spin-state fluctuations near the pressure-induced Mott transition in MnO, FeO, CoO, and NiO. Phys Rev B 101(24):245144. https://doi.org/10.1103/PhysRevB.101.245144
33. Täuber UC (2017) Phase transitions and scaling in systems far from equilibrium. Annu Rev Condens Matter Phys 8(1):185–210. https://doi.org/10.1146/annurev-conmatphys-031016-025444
34. Chibani S, Farina D, Massat P, Cazayous M, Sacuto A, Urata T, Canfield P (2021) Lattice-shifted nematic quantum critical point in FeSe1−x S x. NPJ Quantum Mater 6(1):37. https://doi.org/10.1038/s41535-021-00336-3
35. Munoz MA (2018) Colloquium: Criticality and dynamical scaling in living systems. Rev Mod Phys 90(3):031001. https://doi.org/10.1103/RevModPhys.90.031001
36. Licciardello S, Buhot J, Lu J, Ayres J, Kasahara S, Matsuda Y, Hussey N et al (2019) Electrical resistivity across a nematic quantum critical point. Nature 567(7747):213–217. https://doi.org/10.1038/s41586-019-0923-y
37. Haravifard S, Yamani Z, Gaulin BD (2015) Quantum phase transitions. In: Experimental methods in the physical sciences, vol 48. Elsevier, pp 43–144. https://doi.org/10.1016/B978-0-12-802049-4.00002-6
38. Rauch H, Werner SA (2015) Neutron interferometry: lessons in experimental quantum mechanics, wave-particle duality, and entanglement. OUP Oxford. https://doi.org/10.1093/acprof:oso/9780198712510.001.0001
39. Ma T, Wang S (2014) Phase transition dynamics. Springer. https://doi.org/10.1007/978-3-030-29260-7
40. Singh M (2023) Time-dependent phenomena in one-dimensional quantum systems: driven quantum spin chains. Indian Institute Of Science Education and Research Bhopal. https://doi.org/10.13140/RG.2.2.29985.79205
41. Rademaker L, Mydosh J (2016) Quantum critical matter and phase transitions in rare earths and actinides. In: Handbook on the physics and chemistry of rare earths, vol 49. Elsevier, pp 293–338. https://doi.org/10.1016/bs.hpcre.2016.03.002
42. Kremen A, Khan H, Loh Y, Baturina T, Trivedi N, Frydman A, Kalisky B (2018) Imaging quantum fluctuations near criticality. Nat Phys 14(12):1205–1210. https://doi.org/10.1038/s41567-018-0264-z
43. Khanenko P, Hafner D, Semeniuk K, Banda J, Lühmann T, Bärtl F, Geibel C et al (2025) Origin of the non-Fermi-liquid behavior in CeRh 2 As 2. Phys Rev B 111(4):045162. https://doi.org/10.1103/PhysRevB.111.045162

44. Shaginyan V, Stephanovich V, Msezane A, Schuck P, Clark J, Amusia MY, Kirichenko E et al (2017) New state of matter: heavy fermion systems, quantum spin liquids, quasicrystals, cold gases, and high-temperature superconductors. J Low Temp Phys 189:410–450. https://doi.org/10.1007/s10909-017-1801-3
45. Wölfle P (2018) Quasiparticles in condensed matter systems. Rep Prog Phys 81(3):032501. https://doi.org/10.1088/1361-6633/aa9bc4
46. Keimer B, Moore J (2017) The physics of quantum materials. Nat Phys 13(11):1045–1055
47. Schlottmann P (2015) Non-fermi liquid behavior in heavy fermion systems. In: Handbook of magnetic materials, vol 23. Elsevier, pp 85–160. https://doi.org/10.1038/nphys4302
48. Deslippe J, Samsonidze G, Strubbe DA, Jain M, Cohen ML, Louie SG (2012) BerkeleyGW: a massively parallel computer package for the calculation of the quasiparticle and optical properties of materials and nanostructures. Comput Phys Commun 183(6):1269–1289. https://doi.org/10.1016/j.cpc.2011.12.006
49. Lee S-S (2018) Recent developments in non-fermi liquid theory. Annu Rev Condens Matter Phys 9(1):227–244. https://doi.org/10.1146/annurev-conmatphys-031016-025531
50. Sen D (2014) The uncertainty relations in quantum mechanics. Curr Sci 203-218. https://doi.org/10.13140/2.1.5183.0406
51. Dunkel J, Hilbert S (2014) Consistent thermostatistics forbids negative absolute temperatures. Nat Phy 10(1):67–72. https://doi.org/10.13140/2.1.5183.0406
52. Basov D, Averitt R, Hsieh D (2017) Towards properties on demand in quantum materials. Nat Mater 16(11):1077–1088. https://doi.org/10.1038/nmat5017
53. Heyl M (2018) Dynamical quantum phase transitions: a review. Rep Prog Phys 81(5):054001. https://doi.org/10.1088/1361-6633/aaaf9a
54. Vasin M, Ryzhov V, Vinokur V (2015) Quantum-to-classical crossover near quantum critical point. Sci Rep 5(1):18600. https://doi.org/10.1038/srep18600
55. Broholm C, Cava R, Kivelson S, Nocera D, Norman M, Senthil T (2020) Quantum spin liquids. Science 367(6475):eaay0668. https://doi.org/10.1126/science.aay0668
56. Yadav AK, Rajput P, Alshammari O, Khan M, Kumar G, Kumar S, Sen S et al (2017) Structural distortion, ferroelectricity and ferromagnetism in Pb (Ti1−xFex) O3. J Alloy Compd 701:619–625. https://doi.org/10.1016/j.jallcom.2017.01.127
57. Mishra KR, Parida S, Behura SK Multiferroics Multifunctional material dielectric materials for energy storage and energy harvesting devices. River Publishers, pp 207–236. https://doi.org/10.1201/9781032630816-9
58. Singh M (2025) Multiferroics, ferroelectric, and piezoelectric materials. Contemp Adv Mater Technol: Nov Solut Complex Probl 41. https://doi.org/10.1201/9781003561859
59. Kumar M, Shankar S, Kumar A, Anshul A, Jayasimhadri M, Thakur O (2020) Progress in multiferroic and magnetoelectric materials: applications, opportunities and challenges. J Mater Sci: Mater Electron 31:19487–19510. https://doi.org/10.1007/s10854-020-04574-2
60. Ku M-L, Li W, Chen Y, Liu KR (2015) Advances in energy harvesting communications: past, present, and future challenges. IEEE Commun Surv Tutor 18(2):1384–1412. https://doi.org/10.1109/COMST.2015.2497324
61. Schleich WP, Ranade KS, Anton C, Arndt M, Aspelmeyer M, Bayer M, Giacobino E et al (2016) Quantum technology: from research to application. Appl Phys B 122:1–31. https://doi.org/10.1007/s00340-016-6353-8
62. Repček D, Proschek P, Savinov M, Kachlík M, Pospíšil J, Drahokoupil J, Kamba S et al (2024) Multiferroic quantum criticality in a (Eu, Ba, Sr) TiO 3 solid solution. Phys Rev B 109(21):214109. https://doi.org/10.1103/PhysRevB.109.214109
63. Achary S, Jayakumar O, Tyagi A (2012) Multiferroic materials. Funct Mater: Prep, Process Appl 155-191. https://doi.org/10.1016/B978-0-12-385142-0.00004-0
64. Fiebig M, Lottermoser T, Meier D, Trassin M (2016) The evolution of multiferroics. Nat Rev Mater 1(8):1–14. https://doi.org/10.1038/natrevmats.2016.46
65. Chandra P, Lonzarich GG, Rowley S, Scott J (2017) Prospects and applications near ferroelectric quantum phase transitions: a key issues review. Rep Prog Phys 80(11):112502. https://doi.org/10.1088/1361-6633/aa82d2

66. Scott JF (2015) Searching for new ferroelectrics and multiferroics: a user's point of view. NPJ Comput Mater 1(1):1–9. https://doi.org/10.1038/npjcompumats.2015.6
67. Khorsand Zak A, Hashim AM (2025) Recent advances in $BiFeO_3$-based nanostructures: properties and applications. Coord Chem Rev 523:216297. https://doi.org/10.1016/j.ccr.2024.216297
68. Boddu V, Endres F, Steinmann P (2017) Molecular dynamics study of ferroelectric domain nucleation and domain switching dynamics. Sci Rep 7(1):806. https://doi.org/10.1038/s41598-017-01002-0
69. Scott JF (2015) Searching for new ferroelectrics and multiferroics: a user's point of view. NPJ Comput Mater 1(1):1–9. https://doi.org/10.1038/npjcompumats.2015.6
70. Genenko YA, Glaum J, Hoffmann MJ, Albe K (2015) Mechanisms of aging and fatigue in ferroelectrics. Mater Sci Eng, B 192:52–82. https://doi.org/10.1016/j.mseb.2014.10.003
71. Aso S, Matsuo H, Noguchi Y (2023) Reversible electric-field-induced phase transition in Ca-modified NaNbO3 perovskites for energy storage applications. Sci Rep 13(1):6771. https://doi.org/10.1038/s41598-023-33975-6
72. Bokov A, Ye Z-G (2006) Recent progress in relaxor ferroelectrics with perovskite structure. J Mater Sci 41:31–52. https://doi.org/10.1007/s10853-005-5915-7
73. Jiang X, Huang W, Zhang S (2013) Flexoelectric nano-generator: materials, structures and devices. Nano Energy 2(6):1079–1092. https://doi.org/10.1016/j.nanoen.2013.09.001
74. Zhang H, Zhu Z, Zhang X, Han J (2007) Polarization and dielectric properties of ferroelectric thin film grown compositionally graded substrate: the first-principles study. Solid State Commun 144(7):339–342. https://doi.org/10.1016/j.ssc.2007.08.044
75. Moya-Canul KM, Flores-Ruiz FJ, Leal-Zayas JC, Silva-González R, Tomás SA, Yañez-Limón JM (2025) Influence of nickel incorporation and thermal treatment under atmospheric conditions and controlled oxygen flow on the ferroelectric and electrical properties of BiFeO3 thin films obtained by sol-gel. J Alloy Compd 1010:178015. https://doi.org/10.1016/j.jallcom.2024.178015
76. Supriya S (2024) Research progress, doping strategies and dielectric-ferroelectric anomalies of rare earth-based Bi0.5Na0.5TiO3 perovskites. J Rare Earths 42(11):2003–2017. https://doi.org/10.1016/j.jre.2023.10.009
77. Maier B, Welsch A-M, Mihailova B, Angel R, Zhao J, Paulmann C, Bismayer U et al (2011) Effect of La doping on the ferroic order in Pb-based perovskite-type relaxor ferroelectrics. Phys Rev B 83:134106. https://doi.org/10.1103/PhysRevB.83.134106
78. Morris M, Pendlebury S, Hong J, Dunn S, Durrant J (2016) Effect of internal electric fields on charge carrier dynamics in a ferroelectric material for solar energy conversion. Adv Mater 28. https://doi.org/10.1002/adma.201601238
79. Michel VF, Esswein T, Spaldin NA (2021) Interplay between ferroelectricity and metallicity in BaTiO3. J Mater Chem C 9(27):8640–8649. https://doi.org/10.1039/D1TC01868J
80. Khan MA, Nadeem MA, Idriss H (2016) Ferroelectric polarization effect on surface chemistry and photo-catalytic activity: a review. Surf Sci Rep 71(1):1–31. https://doi.org/10.1016/j.surfrep.2016.01.001
81. Szeremeta A, Macutkevic J, Zubko M, Miga S, Svirskas S, Gruszka I, Molak A et al (2023) Doping influence on structural ferroelectric phase transitions and electrical features of barium calcium titanate. J Eur Ceram Soc 43. https://doi.org/10.1016/j.jeurceramsoc.2023.03.018
82. Ihlefeld J, Jaszewski S, Fields S (2022) A perspective on ferroelectricity in hafnium oxide: mechanisms and considerations regarding its stability and performance. Appl Phys Lett 121:240502. https://doi.org/10.1063/5.0129546
83. Wu Z, Sun Y, Durajski AP, Zheng F, Antropov V, Ho K-M, Wu S (2023) Effect of doping on the phase stability and superconductivity in $\mathrm{La}{\mathrm{H}}_{10}$. Phys Rev Mater 7(10):L101801. https://doi.org/10.1103/PhysRevMaterials.7.L101801
84. Xu R, Zou X, Liu B, Cheng H-M (2018) Computational design and property predictions for two-dimensional nanostructures. Mater Today 21(4):391–418. https://doi.org/10.1016/j.mattod.2018.03.003

85. Schultheiß J, Picht G, Wang J, Genenko YA, Chen LQ, Daniels JE, Koruza J (2023) Ferroelectric polycrystals: structural and microstructural levers for property-engineering via domain-wall dynamics. Prog Mater Sci 136:101101. https://doi.org/10.1016/j.pmatsci.2023.101101
86. Pardo L (2015) 5—Piezoelectric ceramic materials for power ultrasonic transducers. In: Gallego-Juárez JA, Graff KF (eds) Power ultrasonics. Woodhead Publishing, Oxford, pp 101–125. https://doi.org/10.1016/B978-1-78242-028-6.00005-3
87. Scholz T, McLaughlin K, Giuliani F, Clegg W, Espinoza-Beltrán FJ, Swain M, Schneider G (2007) Nanoindentation initiated dislocations in barium titanate (BaTiO3). Appl Phys Lett 91:062903–062903. https://doi.org/10.1063/1.2766846
88. Yudin P, Duchon J, Pacherova O, Klementova M, Kocourek T, Dejneka A, Tyunina M (2021) Ferroelectric phase transitions induced by a strain gradient. Phys Rev Res 3(3):033213. https://doi.org/10.1103/PhysRevResearch.3.033213
89. Haeni J, Irvin P, Chang W, Uecker R, Reiche P, Li Y, Schlom D et al (2004) Room-temperature ferroelectricity in strained SrTiO3. Nature 430:758–761. https://doi.org/10.1038/nature02773
90. Gil-González E, Perejón A, Sánchez-Jiménez PE, Criado JM, Pérez-Maqueda LA (2018) Chapter 16—thermoanalytical characterization techniques for multiferroic materials. In: Vyazovkin S, Koga N, Schick C (eds) Handbook of thermal analysis and calorimetry, vol 6. Elsevier Science B.V, pp 643–683. https://doi.org/10.1016/B978-0-444-64062-8.00010-3
91. Zhang G, Chen Z, Wen J, Hou J, Chen S, Fang Y, Ren Y (2024) Bulk photovoltaic effect in high-temperature lead-halide molecular ferroelectric [C4N2H14][PbI4]. ACS Appl Mater Interfaces 16(40):53873–53880. https://doi.org/10.1021/acsami.4c11534
92. Koutzarova T, Kolev S, Krezhov K, Georgieva B (2022) Phase transitions in magneto-electric hexaferrites. ACS Omega 7(49):44485–44494. https://doi.org/10.1021/acsomega.2c05689
93. Gao Y, Song Z, Hu H, Mei J, Kang R, Zhu X, Lou X et al (2024) Optimizing high-temperature energy storage in tungsten bronze-structured ceramics via high-entropy strategy and bandgap engineering. Nat Commun 15(1):5869. https://doi.org/10.1038/s41467-024-50252-w
94. Zou J, Song M, Zhou X, Chi W, Wei T, Zhou K, Zhang S et al (2024) Enhancing piezoelectric coefficient and thermal stability in lead-free piezoceramics: insights at the atomic-scale. Nat Commun 15(1):8591. https://doi.org/10.1038/s41467-024-53020-y
95. Sabeel K, Al-Greer M, Bashir I (2025) Advancements in vibration testing: effects on thermal performance and degradation of modern batteries. Batteries 11(2):82. https://doi.org/10.3390/batteries11020082
96. Higashitarumizu N, Kawamoto H, Lee C-J, Lin B-H, Chu F-H, Yonemori I, Nagashio K et al (2020) Purely in-plane ferroelectricity in monolayer SnS at room temperature. Nat Commun 11(1):2428. https://doi.org/10.1038/s41467-020-16291-9
97. Shi Q, Parsonnet E, Cheng X, Fedorova N, Peng R-C, Fernandez A, Ramesh R et al (2022) The role of lattice dynamics in ferroelectric switching. Nat Commun 13(1):1110. https://doi.org/10.1038/s41467-022-28622-z
98. Hafner J, Benaglia S, Richheimer F, Teuschel M, Maier FJ, Werner A, Schmid U et al (2021) Multi-scale characterisation of a ferroelectric polymer reveals the emergence of a morphological phase transition driven by temperature. Nat Commun 12(1):152. https://doi.org/10.1038/s41467-020-20407-6
99. Qi H, Hu T, Deng S, Liu H, Fu Z, Chen J (2023) Giant dynamic electromechanical response via field driven pseudo-ergodicity in nonergodic relaxors. Nat Commun 14(1):2414. https://doi.org/10.1038/s41467-023-38006-6
100. Wang J, Shu W, Shimada T, Kitamura T, Zhang TY (2013) Role of grain orientation distribution in the ferroelectric and ferroelastic domain switching of ferroelectric polycrystals. Acta Mater 61(16):6037–6049. https://doi.org/10.1016/j.actamat.2013.06.044
101. Chen B, Tao C, Fan W, Shen B, Ju M, Dou Z, Wang K et al (2024) Composition regulation of potassium sodium niobate thin films through post-annealing under alkali element atmospheres. Nanomaterials 14(3):288. https://doi.org/10.3390/nano14030288
102. Mopoung K, Ning W, Zhang M, Ji F, Mukhuti K, Engelkamp H, Puttisong Y et al (2024) Understanding antiferromagnetic coupling in lead-free halide double perovskite semiconductors. The J Phys Chem C 128(12):5313–5320. https://doi.org/10.1021/acs.jpcc.3c08129

103. Nandy A, Roychowdhury A, Kar T, Das D, Pradhan S (2016) Effect of sodium doping on microstructure, lattice distortion and magnetic properties of GdMnO3 tiny single crystals. RSC Adv 6. https://doi.org/10.1039/C5RA19631K
104. Joseph S, John RE, Saban KV (2019) Unraveling the effects Mn-site doping on structural, magnetic and magnetotransport properties of Nd0.67Sr0.33Mn0.9T0.1O3 (T= Mn, Fe, Cr, Ni). Ceramics Int 45(17, Part A):21249–21262. https://doi.org/10.1016/j.ceramint.2019.07.107
105. Slade TJ, Anand S, Wood M, Male JP, Imasato K, Cheikh D, Snyder GJ et al (2021) Charge-carrier-mediated lattice softening contributes to high zT in thermoelectric semiconductors. Joule 5(5):1168–1182. https://doi.org/10.1016/j.joule.2021.03.009
106. Rowley S, Spalek L, Smith R, Dean M, Itoh M, Scott J, Saxena S et al (2014) Ferroelectric quantum criticality. Nat Phy 10. https://doi.org/10.1038/NPHYS2924
107. Zhang G, Wu H, Jin W, Yang L, Xiao B, Yu J, Chang H et al (2025) Progress and challenges for two-dimensional spin-polarized quantum materials. Cell Rep Phys Sci 6(1):102356. https://doi.org/10.1016/j.xcrp.2024.102356
108. Gong G, Duan Y, Zuo Y, Wang L, Zhou J, Liu D, Su Y et al (2023) Substitution effect of Ni2+ on the magnetism and electrical properties of Ca3CoMnO6 compound. J Alloy Compd 933:167748. https://doi.org/10.1016/j.jallcom.2022.167748
109. Sahasrabudhe M, Patil S, Date S, Adhi K, Kulkarni S, Joy P, Bathe R (2006) Influence of magnetic (Fe +3) and non-magnetic (Ga +3) ion doping at Mn-site on the transport and magnetic properties of La 0.7Ca 0.3MnO 3. Solid state communications—SOLID STATE COMMUN 137:595–600. https://doi.org/10.1016/j.ssc.2006.01.011
110. Dong S, Xiang H, Dagotto E (2019) Magnetoelectricity in multiferroics: a theoretical perspective. Natl Sci Rev 6(4):629–641. https://doi.org/10.1093/nsr/nwz023
111. Zhang L, Xi G, Fang Y-W, Zheng D, Xu S, Li H, Tian J (2024) Shear strain control of multiferroic magnetic ordering for giant ferromagnetism. https://doi.org/10.21203/rs.3.rs-4956109/v1
112. Behr D, Taran LS, Porter DG, Bombardi A, Prabhakaran D, Streltsov SV, Johnson RD (2024) Strain-induced antiferromagnetic domain switching via the spin Jahn-Teller effect. Phys Rev B 110(6):L060408. https://doi.org/10.1103/PhysRevB.110.L060408
113. Wu L, Zhou L, Zhou X, Wang C, Ji W (2022) In-plane epitaxy-strain-tuning intralayer and interlayer magnetic coupling in monolayers and bilayers. Phys Rev B 106(8):L081401. https://doi.org/10.1103/PhysRevB.106.L081401
114. Bastien G, Roslova M, Haghighi MH, Mehlawat K, Hunger J, Isaeva A, Wolter AUB et al (2019) Spin-glass state and reversed magnetic anisotropy induced by Cr doping in the Kitaev magnet. Phys Rev B 99(21):214410. https://doi.org/10.1103/PhysRevB.99.214410
115. Kilanski L, Szymczak R, Dobrowolski W, Szalowski K, Slynko V, Slynko E (2010) Magnetic interactions in spin-glasslike Ge1-x-ySnxMnyTe dilute magnetic semiconductors. Phys Rev B 82:094427. https://doi.org/10.1103/PhysRevB.82.094427
116. Assirey EAR (2019) Perovskite synthesis, properties and their related biochemical and industrial application. Saudi Pharm J 27(6):817–829. https://doi.org/10.1016/j.jsps.2019.05.003
117. Sudarsan V (2012) 8—Optical materials: fundamentals and applications. In: Banerjee S, Tyagi AK (eds) Functional materials). Elsevier, London, pp 285–322. https://doi.org/10.1016/B978-0-12-385142-0.00008-8
118. Mostovoy M (2024) Multiferroics: different routes to magnetoelectric coupling. NPJ Spintronics 2(1):18. https://doi.org/10.1038/s44306-024-00021-8
119. Yadav K, Mukherjee K (2021) Effect of partial substitution of iso-valent Mo at Cr-site on electronic structure and physical properties of Fe2CrAl. Intermetallics 133:107153. https://doi.org/10.1016/j.intermet.2021.107153
120. Li S, Ye Z, Luo X, Ye G, Kim HH, Yang B, Zhao L et al (2020) Magnetic-field-induced quantum phase transitions in a van der waals magnet. Phys Rev X 10(1):011075. https://doi.org/10.1103/PhysRevX.10.011075

121. Fujishita H, Kitazawa S, Saito M, Ishisaka R, Okamoto H, Yamaguchi T (2016) Quantum paraelectric states in SrTiO 3 and KTaO 3: barrett model, vendik model, and quantum criticality. J Phys Soc Jpn 85:074703. https://doi.org/10.7566/JPSJ.85.074703
122. Horiuchi S, Kobayashi K, Kumai R, Minami N, Kagawa F, Tokura Y (2015) Quantum ferroelectricity in charge-transfer complex crystals. Nat Commun 6(1):7469. https://doi.org/10.1038/ncomms8469
123. Starykh OA (2015) Unusual ordered phases of highly frustrated magnets: a review. Rep Prog Phys 78(5):052502. https://doi.org/10.1088/0034-4885/78/5/052502
124. Perren G (2018) Experimental study of pressure-and field-induced quantum phase transitions in clean and disordered quantum magnets. ETH Zurich. https://doi.org/10.3929/ethz-b-000334111
125. Wu Z, Hu C, Wang T, Chen Y, Li Y, Zhao L, Peng X et al (2024) Experimental quantum simulation of multicriticality in closed and open rabi model. Phys Rev Lett 133(17):173602. https://doi.org/10.1103/PhysRevLett.133.173602
126. Shao H, Guo W, Sandvik AW (2016) Quantum criticality with two length scales. Science 352(6282):213–216. https://doi.org/10.1126/science.aad5007
127. Kovács IA (2022) Quantum multicritical point in the two-and three-dimensional random transverse-field Using model. Phys Rev Res 4(1):013072. https://doi.org/10.1103/PhysRevResearch.4.013072
128. Pixley J, Goswami P, Das Sarma S (2016) Disorder-driven itinerant quantum criticality of three-dimensional massless Dirac fermions. Phys Rev B 93(8):085103. https://doi.org/10.1103/PhysRevB.93.085103
129. Parafilo A (2024) Manifestation of Luttinger liquid effects in a hybrid metal-semiconductor double-quantum dot device. Low Temp Phys 50(12):1180–1188. https://doi.org/10.1063/10.0034413
130. Liu X, Gu J, Shen Y, Li J (2016) Crystal metamorphosis at stress extremes: how soft phonons turn into lattice defects. NPG Asia Materials 8(10):e320–e320. https://doi.org/10.1038/am.2016.154
131. Engelkemier J, Fredrickson DC (2016) Chemical pressure schemes for the prediction of soft phonon modes: a chemist's guide to the vibrations of solid state materials. Chem Mater 28(9):3171–3183. https://doi.org/10.1021/acs.chemmater.6b00917
132. Kamba S (2021) Soft-mode spectroscopy of ferroelectrics and multiferroics: a review. APL Materials 9(2). https://doi.org/10.1063/5.0036066
133. Bianco R, Errea I, Paulatto L, Calandra M, Mauri F (2017) Second-order structural phase transitions, free energy curvature, and temperature-dependent anharmonic phonons in the self-consistent harmonic approximation: theory and stochastic implementation. Phys Rev B 96(1):014111. https://doi.org/10.1103/PhysRevB.96.014111
134. Stamp P, Silevitch D, Libersky M, McKenzie R, Geim A, Rosenbaum T (2024) Gallery of soft modes: theory and experiment at a ferromagnetic quantum phase transition. Phys Rev B 110(13):134420. https://doi.org/10.1103/PhysRevB.110.134420
135. Wagué B, Niu G, Dong G, Dai L, Roy P, Saint-Girons G, Vilquin B et al (2020) Structural studies of epitaxial BaTiO3 thin film on silicon. Thin Solid Films 693:137636. https://doi.org/10.1016/j.tsf.2019.137636
136. Hohenberg PC, Krekhov AP (2015) An introduction to the Ginzburg-Landau theory of phase transitions and nonequilibrium patterns. Phys Rep 572:1–42. https://doi.org/10.1016/j.physrep.2015.01.001
137. Thermodynamical Principles and the Landau Theory (2005) In: Fujimoto M (ed) The physics of structural phase transitions). Springer New York, New York, NY, pp 5–29. https://doi.org/10.1007/b138153
138. Pérez-Mato JM, Salje EKH (2000) Quantum fluctuations of order parameters in structural phase transitions and the pressure dependence of transition temperatures. J Phys: Condens Matter 12(3):L29. https://doi.org/10.1088/0953-8984/12/3/101
139. Rohn J, Hörmann M, Genes C, Schmidt KP (2020) Using model in a light-induced quantized transverse field. Physical Review Research 2(2):023131. https://doi.org/10.1103/PhysRevResearch.2.023131

140. Wittek P (2014) 14—Boosting and adiabatic quantum computing. In: Wittek P (ed.) Quantum machine learning. Academic Press, Boston pp 139–151. https://doi.org/10.1038/nature23474
141. Mukherjee S, Son J, Bhattacharya BB (2022) Estimation in tensor Using models. Inf Inference: J IMA 11(4):1457–1500. https://doi.org/10.48550/arXiv.2008.12882
142. Misra PK (2012) Chapter 13—magnetic ordering. In: Misra PK (ed) Physics of condensed matter. Academic Press, Boston, pp 409–449
143. Talanov MV, Shirokov VB, Talanov VM (2023) Spin order on the pyrochlore lattice: magnetic crystallography, landau thermodynamics and emergent phenomena. Acta Mater 259:119187. https://doi.org/10.1016/j.actamat.2023.119187
144. Longa L, Monselesan D, Trebin HR (1987) An extension of the Landau-Ginzburg-de Gennes theory for liquid crystals. Liq Cryst 2(6):769–796. https://doi.org/10.1080/02678298708086335
145. Wang B (2013) Derivation of the Landau-Ginzburg expansion coefficients. In: Wang B (ed.) Mechanics of advanced functional materials. Springer Berlin Heidelberg, Berlin, Heidelberg, pp 321–375. https://doi.org/10.1063/1.5000139
146. Landau D, Binder K (2021) A guide to monte carlo simulations in statistical physics. Cambridge University Press. https://doi.org/10.2277/0521842387
147. Chen W, Legner M, Rüegg A, Sigrist M (2016) Correlation length, universality classes, and scaling laws associated with topological. Phase Transit. https://doi.org/10.1103/PhysRevB.95.075116

Chapter 6
Perovskite Multiferroics as Energy Harvesters: Exploring the Multidimensional Properties for Energy Harvesting Applications

Prateek Mittal, Dinesh Chawla, Sushant, Jimmy Mehta, and Pallav Gupta

1 Introduction

In this chapter, structural characteristics and properties of perovskite multiferroics, importance of the framework, the crystal symmetry, and strain are discussed with reference to energy harvesting. Also, the coupling mechanisms between the ferroic orders, and their dependence on sample composition and external excitation are explored [1, 2]. The generic perovskite structure with the formula ABX_3 provides a maneuverable platform for compositional and structural modifications to engineer properties for various applications [3]. Perovskite multiferroics are a class of materials that have multiple ferroic orders, including ferroelectricity, ferromagnetism, and ferroelasticity, typically at complementary ratios and are highly sought after for device applications where multiple functionalities are needed and can be tuned through electric, magnetic, or mechanical input. According to [3], structural and compositional changes can make the data to contain diverse technical qualities. Perovskite multiferroics are materials that exhibit a combination of ferroic orders such as ferroelectricity, ferromagnetism or ferroelasticity are highly desirable for multifunctional device applications because of their ability to control their physical properties via electric, magnetic or mechanical means.

Multiferroics are being recognized for their ability to provide novel capabilities in devices such as magnetically switched ferroelectricity and electric field-controlled

P. Mittal (✉) · D. Chawla · Sushant
Department of Mechanical Engineering, Manav Rachna International Institute of Research and Studies, Faridabad, India
e-mail: prateekme@gmail.com

J. Mehta
Department of Mechanical Engineering, Manav Rachna University, Faridabad, India

P. Gupta
Mechanical Engineering Department, Madan Mohan Malaviya University of Technology, Gorakhpur, Uttar Pradesh, India

M. Kumar (ed.), *Sustainable Perovskite Multiferroic Materials*, Materials Horizons: From Nature to Nanomaterials, https://doi.org/10.1007/978-981-96-7419-0_6

magnetization. These properties open applications in energy-efficient multifunctional devices, spintronics, and non-volatile memory [4, 5]. Creating their properties for practical applications and understanding the fundamental rules that control their behavior pose several challenges. The perovskite structure consists of a large cation (A-site) surrounded by a 12-fold oxygen coordination, a small cation (B-site) in an octahedral oxygen environment and oxygen anions generating BX_6 octahedra. Nevertheless, the ideal cubic structure is often distorted into lower symmetry phases such as tetragonal, rhombohedral, or orthorhombic due to strain effects, off-centering of cations, or tilting of the BX_6 octahedra [6]. Such distortions are pivotal in giving rise to ferroelectricity and magnetism in multiferroics.

As an example, at ambient conditions, $BiFeO_3$, the most extensively investigated perovskite multiferroic, assumes a rhombohedral R3C structure. Its ferroelectricity, resulting from the stereochemically active lone pair of Bi^{3+} and its antiferromagnetism due to the Fe^{3+} sublattice [7]. Likewise, in strained thin-film form, $PbTiO_3$ has a tetragonal structure with improved ferroelectric properties via c-axis elongation [8].

Apart from the structural distortions, the hybridization between the A-site and B-site cations is pivotal to stabilizing the multiferroic properties. The chemical substitution of A-site or B-site ions with chemically similar species can alter the polar and also magnetic properties of the material, for instance. La or Nd addition was found to improve multiferroic properties through the regulation of structural parameters and reduction of leakage currents [9].

This displacing of the B-site cation with respect to the oxygen octahedron results in a spontaneous polarization and is therefore the source of perovskite ferroelectricity. The presence of d^0 cations on the B-site including Ti^{4+} and Zr^{4+} promotes this behavior [10]. The ferroelectric polarization is significantly increased due to strain engineering in $BiFeO_3$ thin films under compressive epitaxial strain [11]. In perovskite multiferroics, domain wall dynamics is another significant factor that influences ferroelectric switching behavior. Domain walls have special electrical and magnetic properties that improve the overall properties of the material and separate polarization uniform areas. However, recent research in this area suggests that they can also function as nanoscale conductive pathways, suggesting a new avenue for their potential applications [10].

In perovskite multiferroics, the B-site cations with half-filled d-orbitals (e.g., Fe^{3+} and Mn^{3+}) are usually responsible for mediating magnetism. Magnetic order is mediated through super-exchange contact of B-site cations, assisted by oxygen anions. Take $BiFeO_3$, this material has G-type antiferromagnetism but spin–orbit coupling may introduce a weak ferromagnetism by canted magnetic moments [7, 11]. External factors such as mechanical stress, electric fields, and temperature also affect the spin dynamics and magnetic anisotropy of perovskite multiferroics. Due to this dependence, the use of multiferroics lies in tunable magnetic devices, such as sensors and actuators.

Due to the application potential of multiferroics in magneto-electric memory systems, the alignment of ferroelectric and magnetic ordering in multiferroics is a vivid area of research. This coupling is promoted by spin–lattice interactions or strain (elastic coupling). As an example of magneto-electric coupling under applied electric fields, it has been demonstrated in $BiFeO_3$ that the magnetic domain structure is modified by ferroelectric domain switching [12]. Recent developments have focused on understanding the microscopic origin of magneto-electric coupling and devising ways to enhance its strength. For example, nanoscale composites with ferroelectric and magnetic phases show a significant enhancement of couplings, attributed to the interface interactions between two phases [13].

This provides constructive mode of strain engineering to tune the properties of perovskite multiferroics. Thin-film synthesis methods such as molecular beam epitaxy (MBE) and pulsed laser deposition (PLD) enable precise strain tuning through a lattice mismatch between the film and substrate [11]. For instance, a tetragonal-like phase with increased polarization and magnetization is stabilized by compressive strain in $BiFeO_3$ [14]. Reports of bifurcation induced phase changes in perovskite multiferroics have also been made, where the coexistence of multitude of phases is possible with varying ferroic ordering by merely applying a biaxial strain. This phenomenon has an impact on the design of tunable multiferroic devices [15].

Perovskite multiferroics are utilized in energy harvesting devices, sensors, actuators, and non-volatile memory. In particular, magneto-electric random access memory (MeRAM) utilizes electric field-controlled magnetization, enabling high-density and low-power storage [16]. Moreover, multiferroics have been explored in the setting of photovoltaic effects, resulting in above bandgap photovoltaics due to the coupling of the ferroelectric polarization and optical absorption. This unusual phenomenon has potential applications in photodetectors and solar energy conversion [17]. They can also be used as spin filters or injectors in spintronics, another exciting application, because multiferroics have coupled magnetic and electric properties. The combination of multiferroics and conventional semiconductors might give rise to multifunctional spintronic devices [18].

Although tremendous achievements in this aspect had been presented, several issues prevent the widespread use of perovskite multiferroics for practical applications. This requires to overcome issues such as low working temperatures, leakage currents, and the difficulty to achieve significant magneto-electric couplings. In addition, stability and scalability of multiferroic thin films have been of fine concern in device architectures [18]. New promising multiferroic materials with better properties should be discovered, synthesize procedures should be improved, and fundamental mechanisms for their behavior should be understood by future works. Advancements in machine learning and computational modeling may help us discover new multiferroics with properties tailored to our requirements, in a more rapid fashion [19]. Motivated by both structural flexibility and coupled ferroic orders, the perovskite multiferroic is a synthetic landscape that bears decades of creativity and opportunity for means to explore. As the developments in strain engineering, domain control, and material design will continuously evolve their potential for multifunctional device applications. The emphasis will be placed on investigations aimed to characterize and optimize the coupling mechanisms to regulation of these classes of compounds, paving the way for future application of these materials.

1.1 Structure and Properties

Perovskite multiferroics are generally governed by the formula ABX_3, where the anion 'X' is typically oxygen, and 'A' and 'B' are cations of varying sizes. This crystal structure is flexible enough to afford chemical substitution, enabling tuning of their ferroic properties. One of the best-studied perovskite multiferroics that exhibits both ferroelectric and antiferromagnetic order at room temperature is bismuth ferrite ($BiFeO_3$) [1]. The basis of ideal perovskite structure has a cubic symmetry (space group Pm$\hat{h}$3m), yet, fluctuation of the ionic sizes leads to distortions, giving rise to lower symmetries, such as tetragonal, orthorhombic, or rhombohedral phases [3]. Some of the prominent features of the perovskite structure are:

- **Tolerance Factor (τ):** The tolerance factor (τ) is a key parameter in understanding the stability of the perovskite structure, particularly in multiferroic materials. It quantifies the suitability of the ionic radii of the A-site and B-site cations in forming a stable perovskite structure. The tolerance factor is calculated using the formula mentioned in Eq. 1:

$$\tau = \frac{r_A + r_O}{\sqrt{2}(r_B + r_O)} \tag{1}$$

where r_A, r_B, and r_O are the radii of the A-site cation, B-site cation, and oxygen ion, respectively. A tolerance factor approaching 1 indicates a stable cubic structure, while values significantly deviating from 1 lead to distortions, affecting the material's ferroic properties.

- **Flexibility:** One essential feature that adds to the perovskite structure's adaptability in multiferroic materials is its flexibility. Significant distortions are possible in the ABX_3 perovskite framework because of external strain and differences in the ionic radii of the A-site and B-site cations. Lower symmetry phases, such as tetragonal, orthorhombic, or rhombohedral, can be produced by these distortions and affect ferroic characteristics like magnetism and ferroelectricity. The material's electrical, magnetic, and elastic properties can be fine-tuned thanks to the structural flexibility, which also makes doping and ion replacement easier. Perovskite multiferroics can be used in a variety of multifunctional applications because of their versatility.

1.1.1 Ferroelectricity in Perovskite Multiferroics

Ferroelectricity, which allows spontaneous electric polarization that may be reversed with an applied electric field, is one of the main properties of perovskite multiferroics. This effect is primarily caused by the displacement of the B-site cation with respect to the oxygen octahedron in the perovskite structure. In multiferroics, ferroelectric polarization often occurs in cooperation with other ferroic orders, such as magnetism, to give rise to unique coupled behaviors that are widely desired for advanced applications [1]. The perovskite structure is an excellent platform for tuning the ferroelectric properties by replacing the A-site or B-site. By stabilizing the crystal structure and decreasing leakage currents, e.g., rare-earth metal doping may improve the polarization of the material. Another way to improve ferroelectricity is through strain engineering. Due to the change of crystal symmetry and polarization strength under compressive or tensile strain, the ferroelectric response can be enhanced for thin films. Noteworthy are materials exhibiting ferroelectricity at room temperature and large spontaneous polarization, such as $BiFeO_3$, which is the most studied perovskite multiferroic. In particular, candidate materials that are multiferroics based on the perovskite structure have improved memory and sensor functionality after it was discovered that the dynamics of ferroelectric domain walls, the boundaries between regions of uniform polarization, allow for the modulation of electronic and magnetic properties that are novel. Thanks to these special features, perovskite multiferroics are ideal for multifunctional devices of the future. $BaTiO_3$, for instance, undergoes a phase change from high temperature cubic to form ferroelectric tetragonal phase and $BiTiO_3$ is known for high spontaneous polarization and high curie temp.

1.1.2 Magnetism in Perovskite Multiferroics

Magnetism, a hallmark of perovskite multiferroics that significantly increases their versatility for use in memory devices, actuators, and sensors, is the next characteristic. B-site cation (mainly partially filled d-orbitals such as Fe^{3+}, Mn^{3+}, and Co^{3+}) mainly controls the magnetic property in these compounds. The magnetic ordering of perovskites can be one of ferromagnetism, antiferromagnetism, or ferrimagnetism depending on the interactions between the cations. One of the primary mechanisms affecting magnetism of perovskites multiferroics is super-exchange, a quantum mechanical process where magnetic interactions occur through (i.e., mediated by) the oxygen anions that span the neighboring cations. This interaction is particularly salient in materials such as $BiFeO_3$, where the Fe^{3+} ions couple antiferromagnetically via oxygen anions. For these materials, it is often the super-exchange interaction (whereby the oxygen facilitates the exchange of electron spins between the metal cations) that is stabilizing the magnetic structure. However, another important flavor of magnetism comes from double exchange, particularly in perovskites with transition metals sharing an oxidation state. Here, electron migration occurs across cations of varying oxidation state, e.g., Mn^{3+} and Mn^{4+}. This mechanism can lead to ferromagnetic ordering, as seen in minerals like $La_{0}._{7}Ca_{0}._{3}MnO_3$. The double exchange process enhances the magnetic properties of the material and makes it useful for magneto-electric and spintronic devices. The wide range of magnetic degrees of freedom exhibited in perovskite multiferroics due to the super-exchange and double exchange interactions shows promise for the development of unique multifunctional devices. Figure 6.1 shows the Polarization-Magnetic Field curve for calculation results on the relation between polarization and magnetic field for Pr2FeAlO6.

Fig. 6.1 Plot indicating dependence of polarization on magnetic field for Pr_2FeAlO_6 [20] (This image is under creative commons license and no permissions are required for reuse)

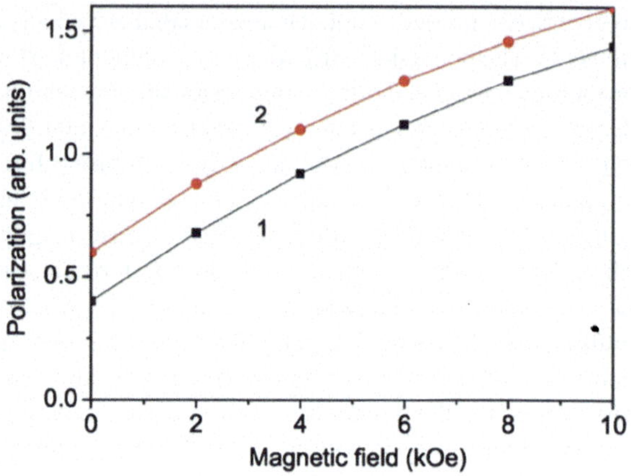

1.1.3 Coupled Ferroic Orders

- Multiferroicity refers to the simultaneous existence of two or more ferroic orders, namely ferroelectricity, ferromagnetism, and ferroelasticity, in a single material, while these orders may interact to influence and couple each other. Perovskite-based multiferroics can, therefore, be considered highly advantageous due to their ability to exhibit such coupled behaviors in advanced applications, including spintronic devices and magneto-electric memory. Perovskite-based materials with ferroelectric and magnetic properties can essentially be divided into the following two categories of multiferroics: Type I and Type II.
  - Type I: Since the ferroelectric and magnetic orders are tightly coupled with each other in multiferroics, an externally applied electric field can directly manipulate their magnetic properties and vice versa. This coupling is normally due to a spin–lattice interaction or shared symmetry where electric polarization is closely related to magnetization. Due to the close association between their ferroelectric and antiferromagnetic orders, compounds such as $BiFeO_3$, the most studied perovskite multiferroic, are known as Type I multiferroics 189. Of interest in these materials is magnetization control by electric fields through polarization, usually directed to modify the magnetic domain topology.
  - Type II: Contrarily, in multiferroics, the connection between magnetic and ferroelectric order is weak where both exist independently. They're still subject to manipulation by factors outside of them, however. In Type II multiferroics, ferroelectricity and magnetism may be present in the same phase, but do not directly interact with each other; hence they are treated as orders in separate channels. These systems are commonly used for applications where it is important to control the electric and magnetic properties separately.

Understanding and controlling these two types of multiferroic features is key to advancing the development of multifunctional materials for next-generation devices.

1.1.4 Dielectric Properties

Dielectric properties of perovskite multiferroics are important to its performance in memory, sensor, and capacitor applications. Because of the strong response of dielectric materials to external electric fields, they have critical significance in energy storage and in the manipulation of the electric field in multipurpose devices. Ferroelectricity and other ferroic properties, such as magnetism, are directly linked to dielectric properties in perovskite multiferroics. Perovskite multiferroics' dielectric behavior is mainly caused by the B-site cation's displacement up in the oxygen octahedra in a perovskite structure. This displacement causes spontaneous polarization which gives rise to ferroelectric behavior. Materials like $BaTiO_3$, $PbTiO_3$, and $BiFeO_3$, possessing a substantial polarization response to the applied electric field, exhibit very high dielectric constants. These materials are well known for their high dielectric permittivity ($\sim$ 50) which is beneficial for energy storage applications. Strain engineering is an effective method to modulate the dielectric properties of perovskite multiferroics. When thin films of perovskite materials are exposed to external strain, such as compressive or tensile strain, their dielectric constants are improved. This is because the strain can modify the domain configuration, polarizability, and crystal symmetry of the material. Compressive strain, for instance, can stabilize certain crystallographic phases which enhance polarization in $BiFeO_3$ simultaneously enhancing the dielectric response. Additionally, fascinating magneto-electric effects where the dielectric and magnetic orders interact are seen in multiferroics. Applying an electric field to these materials can modify their magnetic properties, paving the way for tunable dielectric and magneto-electric devices. The dielectric properties and multifunctional characteristics of perovskite multiferroics render it potential material class for various advanced technologies, such as energy-efficient electronics, actuators, and non-volatile memory.

1.1.5 Magneto-Electric Coupling

Magneto-electric coupling in perovskite multiferroics refers to the phenomenon where the internal orderings of a ferroic material—electrical or magnetic—affect one another while exerting an external field. This coupling phenomena has significant applications in many highly advanced technologies, including magneto-electric memory devices, sensors, and actuators, as it allows for field-controlled magnetism or electric polarization, leading to efficient energy gating in devices. Bismuth ferrite, $BiFeO_3$, and other perovskite multiferroics are a prime example of materials with a magneto-electric interaction. Due to the coupling between the orientations of ferroelectric polarization and the directions of magnetic ordering in these materials, changing the magnetic configuration can affect the polarization, and changing the polarization can affect the magnetic domain structure. This link follows from several mechanisms, including spin–lattice interaction and strain-induced interactions, where electric field-induced mechanical with respect to spin ordering affects the polarization response, while mechanical deformations (due to electric fields) can

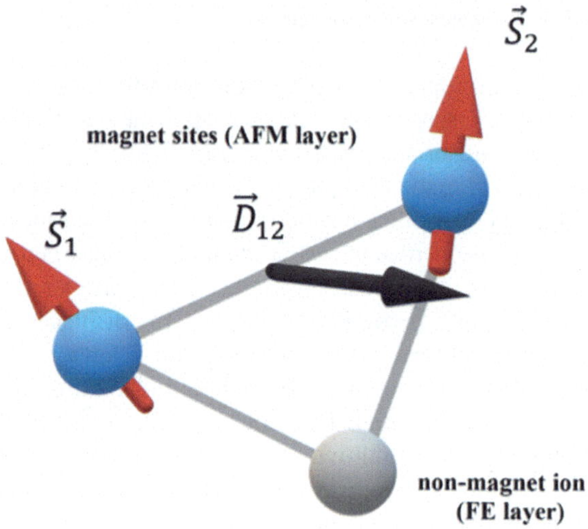

Fig. 6.2 Magneto-ferroelectric bi-layer [21] (This image is under creative commons license and no permissions are required for reuse)

modify the magnetic properties. The strength of the magneto-electric coupling can be modified through different approaches in perovskite multiferroics, such as engineering of the crystal structure, tailoring of the composition, or applying external stimuli like pressure and temperature. Several modifiers, such as changing the A-site and B-site cations or rare-earth elements quenching elements at symmetry locations, can enhance the coupling and amplify the magneto-electric effects. Magneto-electric coupling is one of the exciting mechanisms utilized in the emerging class of non-volatile devices called magneto-electric random access memory (MeRAM), in which information is stored by manipulating the magnetic state with an electric field. Other applications include energy-efficient sensors and actuators, where the coupling provides a low-power means of using electric fields to control of magnetic states [5].

Figure 6.2 represents a magneto-ferroelectric bi-layer, wherein gray spheres represent the non-magnetic ion and blue spheres indicate the spin position.

2 Energy Harvesting Mechanisms

The remarkable functionality of perovskite multiferroics arises from their complex interchange of ferroic orders [3]. The following mechanisms reflect upon the mechanisms involved in energy harvesting through perovskite multiferroics:

2.1 Ferroelectricity

Ferroelectricity is a key property in material science, particularly for perovskite multiferroics with spontaneous polarization tunable by external voltage. This behavior can be characterized by the presence of an internal dipole moment that can be reversed with the application of an electric field in the absence of an external field. Controlling spontaneous polarization enables a variety of applications in energy harvesting, memory devices, sensors, and energy storage. Because of their unique functional parameters and inherent structural abilities, perovskite multiferroics have attracted significant contributions for studying and optimizing ferroelectric behaviors. The generic formula for this class of material is ABX_3, where A and B represent a large cation (A-site) and a smaller cation (B-site), respectively, and where X represents an oxygen anion. The perovskite lattice with the ionic order provides the conditions needed for ferroelectricity. Nevertheless, we know that the presence of spontaneous polarization in ferroelectric materials arises from the generation of an asymmetric charge distribution due to the displacement of the B-site cation with respect to the oxygen octahedra within the lattice. By applying a field, the dipoles in the material can be reoriented in a different direction. Ferroelectricity also has reversible switching properties, which are fundamental to device applications, e.g., efficient devices as well as non-volatile storage.

Due to the structural and chemical versatility of the perovskite framework, ferroelectricity in perovskite multiferroics is often accompanied by other ferroic orders, e.g., magnetism. An illustrative example is $BiFeO_3$, a perovskite which exhibits ferroelectric and antiferromagnetic ordering at room temperature. It is known that the Bi^{3+} cation has a stereochemically active lone pair, which is responsible for displacement of the cations with respect to the lattice, and thus gives rise to spontaneous and ferroelectric polarization in $BiFeO_3$. This material is well-suited for the development of multipurpose devices because it can have the electric and magnetic ordering simultaneously. The ferroelectric properties of perovskite multiferroics can be modified by parameters such as temperature, strain, and cation substitution. Strain engineering is one powerful strategy to enhance the ferroelectric polarizability in thin-film perovskite multiferroics. The polarization response and crystal symmetry can be suitably modified by applying compressive or tensile strain through epitaxial growth on suitable substrates. For example, in $BiFeO_3$, the tetragonal-like phase can be stabilized under compressive strain and, as a consequence, the polarization would be enhanced, which is beneficial for high-performance ferroelectric devices.

Cation substitution is also another method used to modify the ferroelectric properties of perovskite multiferroics. By a different elemental substitution for either the A- or B-site cations it is possible to augment or maintain the polarization. For example, replacing A-site Bi^{3+} ion with La^{3+} in rare-earth elements can suppress leakage currents and improve the overall ferroelectric response of the materials. Subsequently, the ferroelectric polarization was enhanced through B-site doping ions like Mn^{4+} or Zr^{4+}, which simultaneously change the local bonding environment and promote the formation of more stable polarization states.

One of the major advantages of perovskite multiferroics is their ability to couple their ferroelectric and magnetic ordering. The coexistence of these properties within the same material enables the study of magneto-electric effects—where the application of an electric field can influence the magnetic state of the material or vice versa. This coupling is highly relevant for the development of next-generation devices, such as magneto-electric random access memory (MeRAM), in which the magnetic state can be manipulated for non-volatile data storage and the polarization state can be controlled by an electric field. Simultaneous control of electric and magnetic properties enables low-power, high-density memories, and sensors.

Perovskite multiferroics enables spontaneous polarization to be modulated without the need for free carriers, allowing for efficient electro-migration (charge separation) for applications in solar and energy harvesting devices in addition to memory applications. In materials like $BiFeO_3$ that exhibit spontaneous polarization under light, the separation of the photogenerated charge carriers is enhanced, leading to elevated open-circuit voltages and photocurrents. Due to this property, perovskite multiferroics emerge as promising candidates for optoelectronic devices and future solar cells, where efficient charge separation is essentially required to improve the efficiency of devices overall. The ability to perform external electric field regulation on spontaneous polarization in perovskite multiferroics is an important step that greatly improves their performance and expands their possible uses. Using the tunability of these materials, researchers will be able to design multifunctional, energy-efficient devices that exploit the interlinked ferroelectric and magnetic functions. As long as material design, synthesis methods and fabrication procedures develop, perovskite multiferroics should become an important part of future technologies where effective charge separation, non-volatile memory and low-power energy devices are required.

2.2 Magnetism

The main origin of magnetism in the perovskite multiferroics are the B-site cations such as Fe^{3+}, Mn^{3+}, Co^{3+}, and Ni^{2+} with partially filled d-orbitals. These cations couple magnetically, and dictate the overall magnetic ordering of the material. The most common magnetic properties of perovskite multiferroics are antiferromagnetism, ferromagnetism, and weak ferromagnetism. These magnetic interactions are influenced by the crystal structure and cation distribution, as well as external stressors such as strain and electric fields. The exchange interactions between neighboring magnetic ions dictate the nature of magnetic ordering in perovskite multiferroics. These materials are governed by two main processes—super-exchange and double exchange.

Super-exchange refers to the case of an indirect exchange contact between two magnetic ions mediated via an anion like an oxygen ion. This process is frequently observed in wide-band gap perovskite materials that are doped with transition metal ions (Fe^{3+}, Mn^{3+}, Cr^{3+}, etc.). The cause of this super-exchange interaction arises from the overlapping of the d-orbitals of the metal ions and the p-orbitals of the oxygen anions by which the metal ions can transfer electron densities between themselves. The super-exchange mechanism can be ferromagnetic or antiferromagnetic depending on the orbital overlap and the angle formed by the metal–oxygen-metal bond. For example, in $BiFeO_3$ materials, Fe^{3+} ions interact via oxygen anions, resulting in G-type antiferromagnetism (AFM) with weak ferromagnetism due to the canting of the spin, where the super-exchange interaction between Fe^{3+} ions controls the antiferromagnetic ordering, a weak ferromagnetic moment is established by the spin–orbit coupling.

Super-exchange interactions, which allow for the concomitant development of ferroelectric and magnetic orders, play a fundamental role in perovskite multiferroics. In this article, we review the constructive competition between ferroelectric and antiferromagnetic interactions in perovskite structure materials, focusing on $BiFeO_3$, to obtain a multifunctional material that exhibits spontaneous polarization as well as magnetic ordering, paving an avenue to develop a new class of multipurpose devices, e.g., magneto-electric sensors and memory devices. The coupling between ferroelectric polarization and antiferromagnetic order in these materials is responsible for energy conversion in magneto-electric applications.

The double exchange process is another notable exchange interaction occurring in perovskite multiferroics, and becomes particularly important in the case of mixed-valence systems, which can have ions of transition metal cations (e.g., Mn^{3+} or Mn^{4+}). This mechanism is responsible for the hopping of an electron from one metal ion to one that is connected through an oxygen anion. The change of the oxidation states of metal ions from the double exchange interaction results in electrons jumping between the metal ions which ultimately changes the magnetic moments of the metal ions. Mangnese (Mn)-based perovskites systems such as $La_{1-}xCaxMnO_3$ exhibits mixed-valence of the charge carriers (states), e.g., so it shows double exchange hence the ferromagnetic ordering. Inducing a ferromagnetic coupling, the electron hopping between Mn^{3+} and Mn^{4+} ions is due to the overlap of their electron wavefunctions. The double exchange mechanism is particularly important in the case of perovskite multiferroics since it allows for coexistence of electrical conductivity and ferromagnetic ordering. This phenomenon can make energy conversion possible in magneto-electric systems, in which an electric field may control switching of the magnetic state and vice versa.

Among those, perovskite multiferroics have been considered as one of the most promising applications, since their magneto-electric devices based on the coupling between electric and magnetic orders allow energy conversion. This refers to the capability of a material to change its magnetic properties or induce a change in electric polarization under an electric field. This is most evident in perovskite multiferroics, where structural modifications such as doping and strain engineering can strengthen coupling between the electric and magnetic order parameters.

The super-exchange and double exchange pair interactions are the keys to achieve the energy conversion in the magneto-electric devices like magneto-electric random access memory (MeRAM) and magneto-electric sensors. As an example, a device that utilizes an electric field to switch a multiferroic material, through the exchange interactions of the sandwiched material, can both switch the polarization of the material and change the corresponding magnetic state. This enables manipulation of both electric and magnetic properties, which allows for development of efficient sensors and low-power, high-density information storage. This approach can further be generalized to the case of magneto-electric generators that convert magnetic energy into electrical energy by shifting the electric polarization in response to an externally applied magnetic field. This efficient energy conversion is ascribed to coupling of the magnetic and electric orders stemming from the exchange interactions operating in the material.

The magneto-electric effect can be enhanced further by controlling the magnetic ordering, crystal symmetry, and cation substitution in perovskite multiferroics. For example, an improved magneto-electric response can be achieved upon doping the B-site with ions that give rise to mixed-valence states, promoting double exchange interactions. Additionally, the crystal structure can be modified using strain engineering through epitaxial growth techniques, which will enhance the magneto-electric interaction. In fact, positive ferroelectric as well as magneto-electric properties of $BiFeO_3$ have been shown to enhance by applying external strain or doping with rare-earth elements. Coexistence of both ferroelectric and antiferromagnetic orders and tunability of exchange interactions in $BiFeO_3$ makes this material very promising for efficient energy conversion in magneto-electric devices. Super-exchange and double exchange mechanisms in perovskite multiferroics provide magnetism, which is a fundamental requirement for governing energy conversion in magneto-electric devices. The mixed-valence ions are self-magnetized and exhibit ferromagnetic order with antiferromagnetic ordering, which is ne with a weak ferromagnet due to the super-exchange interaction between these mixed-valence ions, while the double exchange mechanism between these mixed-valence ions results in the ordering ferromagnetic again due to the hopping of electrons. Collectively, these procedures allow for building blocks of multifunctional materials suitable for diverse applications in magneto-electric technologies, from memory to energy harvesting devices. This latter area is undergoing active research and remains a goal, as further work remains to be done to probe the applied utility and mission-crossover technologies of these materials. The magneto-electric behavior on crystallographic distortions is illustrated through the arrangement of electric dipole moments of $RCrO_3$ (R = Rare Earth Metal) as represented in Fig. 6.3.

2.3 Piezoelectricity

Due to their capability of coupling different physical phenomena, such as mechanical deformation, electric polarization, or magnetic ordering, these materials are of particular interest for energy harvesting applications. One of the most fascinating applications of perovskite multiferroics concerns the coupling of mechanical strain with electric polarization making them promising candidates for energy harvesting [18, 23]. This thermodynamic coupling enables the conversion of mechanical energy—such as

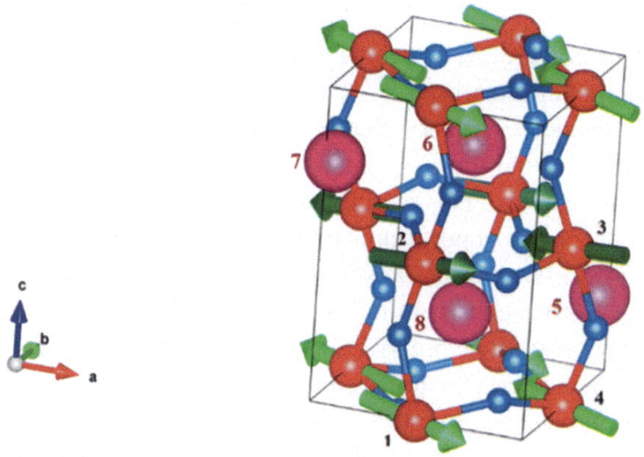

Fig. 6.3 Arrangement of electric dipole moments in the unit cell of $RCrO_3$ [22] (This image is under creative commons license and does not require any permissions for reuse)

pressure or vibrations—into their electrical counterpart, which can be harnessed in a wide range of applications, including sensors, wearable technology and autonomous systems. In this article, the mechanisms involved in energy harvesting by perovskite multiferroics are analyzed and how coupling between strain and electric polarization contributes to the enhancement of the phenomenon. Energy harvesting includes the conversion of ambient energy (e.g., mechanical vibrations or thermal gradients) into electrical energy. For perovskite multiferroics, mechanical strain is among the primary manipulations capable of modulating the material's electric polarization, resulting in potential conversion of mechanical energy to electrical energy. The ferroelasticity and ferroelectricity of perovskite multiferroics make it possible for them to induce mechanical strain in addition to electric polarization under the action of stress or strain. Ferroelectricity and piezoelectricity properties are privileged to perovskite materials like $BiFeO_3$ (BFO), and therefore suit for energy harvest. These materials react to mechanical strain by a process called electric polarization, wherein the application of external stress leads to dipole alignment of usable electric field for energy conversion. Moreover, their very strong piezoelectric effect renders these materials most suitable for mechanical energy harvesting.

Mechanical stress/electric polarization coupling in perovskite multiferroics is mainly determined by the piezoelectric effect and the material ferroelectricity. In detailed, piezoelectric effect refers to the development of an electrical charge in response to mechanical deformation, while ferroelectricity describes the polarization of electric dipoles in the material, resulting in spontaneous polarization. In perovskite multiferroics, applying external mechanical stress—i.e., compressive or tensile strain—triggers a crystal shape change, and subsequent reorientation of electric polarization. The applied strain reoriented the ferroelectric domains in the material and created a net polarization. The polarization, which can be detected as an electrical signal, can be harvested to power devices. In $BiFeO_3$, the spontaneous polarization is the result of the positions of the oxygen and iron ions in the absence of external forces, but if strain is applied, dimensions of the system are changed and thus can be changed the position of the ions. This coupling of the electric polarization to a mechanical strain is thus enhanced to produce an electric field that is accessible. This is achieved by using centrifugal force to convert mechanical energy to electrical energy efficiently.

In addition to ferroelectricity and piezoelectricity, ferroelasticity is also significant in improving the coupling of electric polarization to mechanical strain in perovskite multiferroics. This is often demonstrated in perovskites, ferroelasticity. The ferroelastic nature of perovskite multiferroics suggests that with the application of mechanical stress, a phase transition occurs that alters the lattice structure and thus modifies the ferroelectric nature. Strain modifies the lattice structure, thereby enhancing not only the electric polarization but also enhancing the piezoelectric character, making the energy conversion even more efficient. The ferroelectric property enables dipoles in a typical perovskite ferroelectric, $PbZr1_{-}xTixO_3$ (PZT), to be aligned in the material and results in a large electric field when mechanically deformed through ferroelastic effect. The described ferroelastic contribution can significantly enhance the energy harvesting systems via optimizing the mechanical-to-electrical energy conversion process. The ability of perovskite multiferroics to couple electric polarization to mechanical strain offers diverse opportunities for energy harvesting applications. One of the most significant applications is in vibration energy harvesting, where ambient mechanical vibrations created by equipment, vehicles, or human activity are converted into electrical energy. Translating mechanical vibrations into electrical energy cannot to be more suitable for powering energy devices with low consumption, such as sensors, wearable technology, and medical implants that consume small energy, however, normally a no power source is assigned.

Practical applications of perovskite-based energy harvesting systems can be embedded in smart systems in which the energy harvested is used to energize wireless sensors to detect temperature, pressure, or other environment elements. These devices might be used in industrial applications, where they assist with system maintenance by providing real-time information on equipment condition and performance. As an example, perovskite-based piezoelectric nanogenerators can harness energy from vibrations and converted to useful electric current to power up small devices or charge batteries. In addition to vibration energy harvesting, perovskite multiferroics can find use as strain sensors and actuators in robotics and flexible electronics applications. The ability to respond to strain and produce an electrical signal when mechanically stressed is particularly useful in applications where both sensing and actuation is desirable, such as devices to sense body movement or pressure in wearable sensors.

Thus, perovskite multiferroics are very useful energy harvesting materials, but it still entails several challenges. One of the main obstacles is to develop the materials to be both stable and reliable over long periods, as repeated mechanical strain on the material can lead to its fatigue, and thus, a declining performance as time progresses. Moreover, to be applied, the efficiency of energy conversion process has to be improved, which requires further exploration of the mechanism of strain electric polarization coupling and development of new perovskite compositions with better piezoelectric/ferroelectric properties. Beyond that, whether perovskite-based energy harvesting device can be realized on a large scale is still an important problem, in which the fabrication processes as well as the synthesis of materials should be enhanced to qualify for the large-scale and high-performance devices. Striving for optimal crystallinity, crystallization, and decreased defects will all be vital to the eventual significance of perovskite multiferroics in commercial applications.

Energy harvesting by perovskite multiferroics through the coupled mechanical strain and electric polarization can offer a diversity of potential applications such as self-powered sensors, wearable devices, and cryogenic refrigeration systems. The distinctive characteristics of perovskite multiferroics such as ferroelectricity, piezoelectricity, and ferroelasticity typify the efficient transformation of mechanical energy into electrical energy, thus presenting these materials for self-efficient energy harvesting methods in the near future. Despite challenges surrounding material stability and scalability, further research will likely provide avenues for the implementation of perovskite multiferroics in next-generation energy harvesting devices. Designers would have to look beyond overcoming these practical challenges when ultimately integrating a technology into a device since that marriage will conceivably help all future low-power self-sustaining systems. And as this technology advances, perovskite-based energy harvesters could potentially become a driving factor behind the future of low-power systems contributing to technological advancement-smart energy harvesting technology.

2.4 *Photovoltaic Energy Conversion*

Photovoltaic energy conversion involves using semiconducting materials to convert light energy into electrical energy. In conventional photovoltaic cells, when light is absorbed, it generates electron–hole pairs in the semiconductor, which are separated by an electric field and collected at the electrodes. How effective this process depends heavily on the ability of the material to absorb light, generate charge carriers, and effectively separate and transport those carriers to the electrodes. Based on their design versatility, high charge carrier mobility, and absorption of light, perovskites multiferroics have demonstrated great promise candidates for increased photovoltaic efficiency. Integration of ferroelectric and piezoelectric function onto these materials introduces an additional functionality, that could improve the energy conversion as described in the following points.

- Enhanced Charge Separation: Ferroelectric materials possess an internal polarization, leading to the formation of an internal electric field across the perovskite structure. Under light absorption, the internal field separates the electron–hole pairs and the carriers are driven to the electrodes. This internal field also improves the efficiency of charge collection and reduces recombination loss, a major challenge in traditional solar cells.
- Lower Carrier Recombination: In traditional semiconductors, large fractions of the charge carriers (holes and electrons) recombine before reaching the electrodes which limits the efficiency of the achievable device. However, in the case of perovskite multiferroics, an additional driving force, imparting charge

carrier separation, significantly reduces recombination, thanks to spontaneous polarization.

- Photovoltaic Performance Enhanced by Piezoelectric Effects: Some perovskite multiferroics are also piezoelectric, meaning that mechanical deformation generates an electrical charge. The polarization of the excited states is controllably tunable upon application of mechanical stress to these materials, making the materials more effective in absorbing light and enhancing their photovoltaic performance. This coupling between mechanical and electrical properties opens the door to strain-engineered photovoltaics in which mechanical stimuli from the outside control the properties of the material.
- Bulk Photovoltaic Effect (BPVE)—Perovskite multi-ferromagnetic materials are among the strongest bulk photovoltaic effect (BPVE). Unlike traditional photovoltaic devices where pair creation is promoted by the application of an external electric field, BPVE emerges from and is driven solely by the zeroth order, intrinsic properties of the material itself. For the case of the perovskite multiferroics, the ferroelectricity, in conjunction with the non-centrosymmetric structure of the crystal, is enough to generate photocurrents without an external bias. This feature is especially important because it enables efficient separation of charge that does not rely solely upon traditional device architectures, therefore offering the potential to yield more efficient devices.
- **Enhanced Light Absorption:** · Perovskite multiferroics have higher-light absorbing properties due to their unique crystalline structure and for their chemical composition. Perovskite materials have tunable band gaps, allowing them to absorb a considerable portion of the solar spectrum and be highly efficient at converting light to electrical energy. Furthermore, their high absorption coefficients enable thin-film devices with favorable photovoltaic property. In this regard, using piezoelectric, ferroelectric, or piezoelectric-ferroelectric materials in the architecture design of such materials can further enhance their light absorption capacity and ultimately their photovoltaic cell performance via effective photon harvesting. Here are some explanations to why Perovskite multiferroics are more feasible to be used in photovoltaic energy conversion:

    High Efficiency: In just a few years, perovskite solar cells have shown superior power conversion efficiencies (PCEs) over conventional silicon cells. Material devices (with a multiferroic nature) may lead to higher efficiencies than their single-state counterparts, with enhanced charge separation, decreased recombination and the ability to tune material properties.

    Lightweight + Flexibility: Perovskite materials are a lot more efficient, lightweight, and flexible allowing for applications in flexible solar panels. This versatility enables the incorporation of solar cells into diverse surfaces like windows, textiles, and portable devices, broadening the possible uses for photovoltaic technology.

    One of the major advantages of perovskite based photovoltaic devices is their simple fabrication. Perovskite solar cells may be fabricated using low-temperature, low-cost solution-based techniques (e.g., spin-coating, inkjet printing), in contrast to conventional silicon solar cells that rely on high temperature fabrication processes. This scaling means that perovskite multiferroic solar cells can be designed to be highly repeatable and mass-producible at lower cost.

Tunability of Material Properties: The properties of perovskite multiferroics, like bandgap, polarization and piezoelectric coefficients, can be effectively tuned by changing the chemical composition of the material. Such tunability makes it possible to optimize the properties of the material for different applications such as photovoltaic energy conversion.

The studies on perovskite multiferroics for solar energy conversion are focused on several key areas. Therefore, scientists try to find ways to boost the ferroelectricity and piezoelectricity of perovskite multiferroics to achieve better charge carrier separation and generation. They can maximize polarization and piezoelectricity by doping with certain elements or by ingenious engineering of the crystal structure to create more efficient solar devices. Another promising class of materials are the hybrid perovskites, which are a mixture of organic and inorganic components, and are under research for higher photovoltaic efficiencies. As such, hybrid materials offer more stability and efficiency than pure perovskites and are therefore a strong candidate for commercialization. An emerging field is multilayer or heterostructure perovskite devices. By stacking multiple types of perovskite material, each with different properties, researchers can create devices with enhanced light absorption, charge separation and overall efficiency. Since some perovskite multiferroics have magnetic properties, researchers are also considering the possibility of magneto-electric effects in photovoltaic devices. Incorporating such functionality might provide new pathways to modulate charge transport, as well as to further improve device efficiency.

Despite their promising potential for photovoltaic energy conversion, there are several challenges in designing perovskite multiferroics. Stability is among the most critical challenges since perovskite materials break down over time when exposed to moisture, oxygen, and light. This study is continuing, applying compositional variations and protective layers to improve the long-term stability of the materials. Moreover, scalable fabrication techniques are required for perovskite multiferroic devices for use in both research and commercial applications. Solution-based techniques are low cost, but they need to be optimized for mass production in order to make price comparable to traditional solar technologies.

Multiferroics perovskites lead a paved way for photovoltaic energy conversion device. By leveraging their ferroelectric, piezoelectric and magnetic properties, the perovskite multiferroics allow efficient charge separation and energy conversion, offering efficient, low cost, and flexible solar cells. Compared to conventional materials, two notable characteristics of perovskite multiferroics are the bulk photovoltaic effect and enhanced light absorption. Given that the challenges of material stability and scalability remain (for these devices), perovskite multiferroics may yet play a key role in the future development of next-generation solar energy technology for the world's transition to renewable energy sources [8].

2.5 Thermoelectric Energy Harvesting

Thermoelectric energy harvesting is an emerging technology able to extract heat and transform it into electrical energy potentially giving a sustainable way of recovering energy from waste heat. Because of their remarkable combination of properties, such as ferroelectricity, piezoelectricity, and also thermoelectricity, perovskite multiferroics have emerged as good candidate thermoelectric energy harvesters. The materials demonstrated both high performance in conventional energy conversion processes, but also facilitate hybrid systems, combining different energy conversion pathways to improve overall efficiency. Since the scope of this chapter covers the potential of perovskite multiferroics for the field of thermoelectric energy harvesting techniques, with emphasis on the temperature gradient-induced polarization and the exhibition of hybrid energy harvesting systems [9], this section will be dedicated exclusively to such type of activity.

One of the most important properties that render perovskite multiferroics feasible for thermoelectric energy harvesting is that they can produce a polarization response under thermal gradients. This is because of the coupling between temperature-stimulated lattice distortions and the electric polarization in ferroelectric materials. When a perovskite multiferroic material is under a temperature gradient, the difference in temperature produces disparate lattice expansion or contraction at disparate locations of the material. The differential expansion causes an internal electrostatic field, which, in turn, produces a polarization response. This polarization effect due to temperature gradient is a direct consequence of the thermopiezoelectric effect, in which a thermal gradient creates a mechanical strain that affects the electric polarization of the material. This polarization can be utilized for energy harvesting since the electrostatic field can be used to cause charge separation and collection, thus generating a voltage difference between the hot and cold ends of the material. The created electric field allows the charge carriers to travel toward the electrodes, transferring the thermal energy to electrical energy.

The effectiveness of this process is based on the ferroelectric nature of the material, such as the size of the polarization response, the thermal expansion coefficient, and the thermal conductivity. By optimally tailoring the perovskite multiferroics' composition, one can achieve the best temperature gradient-stimulated polarization effect, and they are useful for thermoelectric energy harvesting.

2.6 Hybrid Energy Harvesting in Perovskite Multiferroics

It appears promising that thermoelectric energy harvesting can be performed from the temperature gradient-induced polarization of perovskite multiferroics, however, the general issues of material stability, efficiency, and high requirement of temperature gradients remain the drawbacks across the board. A novel solution for this state-of-the-art consists of a fabrication of hybrid energy harvesting devices based on combination of thermoelectric conversion with other patterns of energy conversion, such as piezoelectric, photovoltaic, or magneto-electric section. In a hybrid energy harvesting module, the thermoelectric module operates with the remaining energy harvesting mechanisms in a complementary manner to deliver optimal energy conversion efficiency. For instance, you can combine piezoelectric effects with the thermoelectric energy harvesting module and provide more support for efficient energy conversion, especially in scenarios with the availability of mechanical energy and thermal energy at the same time.

- **Thermoelectric and Piezoelectric Hybrid Systems**

Hybridization of thermoelectric and piezoelectric effects, which can be achieved in perovskite multiferroics, is the most promising approach. In these types of materials, mechanical strain or vibrations can be harnessed to generate excess electrical energy, on top of what is produced under a temperature gradient. The hybrid mechanism functions properly in order to make use of applications in which mechanical movement, e.g., vibration or deformation, co-existing with temperature gradients piezoelectric and thermoelectric perovskite multiferroics are capable of efficiently harvesting energy from alternating temperature gradients and applied mechanical stress, making them an attractive candidate for applications in wearable electronics, self-powered sensors, and environmental monitoring. The piezoelectric effect generates a primary voltage that propels charge carriers, which is further enhanced by the thermoelectric process, resulting in increased total harvested energy. Temperature dependence of magnetization for Pr_2FeAlO_6 is displayed in Fig. 6.4

- **Thermoelectric and Photovoltaic Hybrid Systems**

The combination of thermoelectric and photovoltaic energy harvesting in perovskite multiferroics is another possible option. It offers a dual benefit with

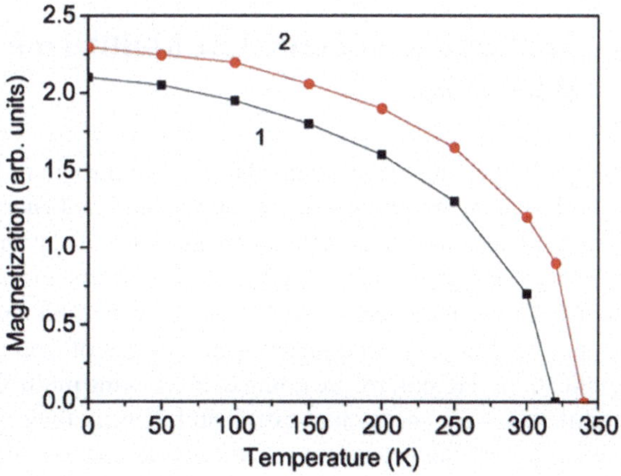

Fig. 6.4 Plot indicating dependence of magnetization on temperature for Pr_2FeAlO_6 [20] (This image is under creative commons license and no permissions are required for reuse)

conversion of energy from two sources sunlight and waste heat via the methods of solar energy conversion and thermoelectric energy harvesting. It can be especially useful for solar cells, thus providing extremely powerful PMS that can be applied to thermal as well as solar energy harvesting. Well-known perovskite materials, in particular, perovskite solar cells, are particularly well-suited to the conversion of solar light into electrical energy. Coupling between the integrated perovskite multiferroic materials and thermoelectric properties, the system would have the capability of harvesting energy efficiently from incoming light as well as the temperature gradient that results from sunlight illumination. This mixed mode enables the device to work properly as specified in varying environmental conditions and enhances harvesting and transformation of energy.

In this regard, perovskite multiferroics hold a continuum of promise for harvesting thermoelectric energy; however, certain challenges still need to be overcome to improve their efficiency and applicability. The high thermal conductivity of perovskite materials is one of the major hurdles, because this decreases the temperature difference across the perovskite material, lowering thermoelectric conversion efficiency. Now, scientists are attempting to fine-tune the thermal properties of these materials through manipulating their ferroic order such that the conversion of heat into electricity can occur more efficiently.

Stability is a major issue, especially in hybrid systems that combine thermoelectric, piezoelectric, and photovoltaic effects. The perovskite materials are often sensitive to UV radiation, oxygen and water, leading to long-term degradation. Advancements in the stability and durability of perovskite multiferroics by way of protective coatings and materials engineering are essential for long lasting operating performance of material in applications to energy harvesting.

Second, the scalability of the perovskite multiferroic devices becomes an issue. Solution-based low-cost synthesis of perovskite materials is feasible, but it may be required to fabricate hybrid structures that amalgamate several energy conversion mechanisms to scale enough for contact on a large scale.

3 Applications of Perovskite Multiferroics in Energy Harvesting

Energy harvesting technologies have come into great prominence over the last few years based on their capability to harness ambient energy from the surroundings and convert it into useful electrical energy. One of the most promising advancements in energy harvesting is with perovskite multiferroics. These materials, which find their defining characteristic in their ferroelectric, piezoelectric, and magneto-electric nature, hold unparalleled promise in the area of energy conversion for a variety of applications. Herein, we are going to discuss the multi-fold applications of perovskite multiferroics for energy harvesting and identify their possibilities in thermoelectric, piezoelectric, magneto-electric, and hybrid energy harvesting.

3.1 Thermoelectric Energy Harvesting

Thermoelectric power harvesting is the process by which temperature gradients across a material are transformed into electric energy. Since perovskite multiferroics are capable of possessing both ferroelectric and thermoelectric functionalities, they hold particular promise for this purpose. When exposed to a temperature gradient, these compounds develop a potential difference as a result of the Seebeck effect, a principle where a voltage develops across a material when a hot end and a cold end are created in the material. In perovskite multiferroics, the temperature gradients are enhanced further by being coupled with the ferroelectric polarization. As the material goes through thermal contraction or expansion resulting from temperature, it causes a polarization response which is capable of driving charge carriers separation. Such a second mechanism of polarization increases the thermoelectric performance enabling such materials to collect waste heat from industrial waste, exhaust gas, or body heat for small electronics such as wearables to power. Through the optimization of the microstructure and composition of perovskite multiferroics, researchers hope to enhance the thermoelectric performance to make them good options for low cost, sustainable energy harvesting from waste heat sources.

3.2 Piezoelectric Energy Harvesting

Piezoelectricity is the creation of electric charge as a result of mechanical stress. Perovskite multiferroics not only are ferroelectric but also piezoelectric in nature, which makes them a prime candidate for energy harvesting due to mechanical vibrations or deformations. They can be employed for harvesting energy from sources such as vibrations in machines, human walking, or even environmental sources such as wind or ocean waves. In a piezoelectric energy harvesting system, the mechanical strain induces a change in the position of the electric dipoles in the material, and an electrical charge is generated. The energy so developed can be stored in batteries or capacitors for future utilization. Perovskite multiferroics, owing to their high piezoelectric coefficients and high electromechanical coupling, are very efficient in the conversion of mechanical energy to electrical energy. These materials have been investigated for a broad range of applications such as self-powered sensors, wearable electronics, and structural health monitoring systems. For instance, perovskite-based piezoelectric devices can be used in smart clothes to power devices such as fitness trackers or medical sensors that record heart rate and body temperature.

3.3 Magneto-Electric Energy Harvesting

Magneto-electric energy harvesting applies the interaction of magnetic and electrical properties to achieve energy generation. Perovskite multiferroics showing ferroelectric and magnetic features are best suited for magneto-electric applications. Such materials are capable of changing mechanical energy to electrical energy based on the interaction between magnetic fields and electric polarization. In a magneto-electric device, the external magnetic field induces a change in the material's electric polarization, which is harvested as electrical energy. Perovskite multiferroics with magneto-electricity enable the direct conversion of magnetic energy to electrical energy and offer a new means to harness energy from magnetic fields. This is especially valuable in applications involving magnetic fields, for example, industrial equipment, sensors, and motors. Magneto-electric energy harvesting efficiency can be further increased by combining perovskite multiferroics with other materials, including piezoelectric and thermoelectric materials, to develop hybrid systems. Such systems would facilitate energy harvesting from various sources, including mechanical, thermal, and magnetic energy, greatly enhancing the overall energy capture efficiency.

3.4 Hybrid Energy Harvesting Systems

Hybrid energy harvesting systems integrate multiple energy conversion mechanisms to achieve optimal energy collection and efficiency. Integrating perovskite multiferroics with thermoelectric, piezoelectric, and magneto-electric characteristics, these materials can collect energy from various ambient sources at the same time. This is highly beneficial in conditions where there are several forms of energy available simultaneously, such as vibrations, temperature gradients, and magnetic fields. For instance, in a thermoelectric-piezoelectric hybrid energy harvesting system, waste heat can be harvested by a perovskite multiferroic material using its thermoelectric behavior and mechanical energy can be harvested using its piezoelectric behavior. Such a combination considerably improves the energy conversion efficiency, particularly in settings with varying temperature gradients and mechanical motion, such as in wearable devices, machinery, and transport systems. Further, systems that integrate magneto-electric effects can also enhance efficiency by tapping into magnetic fields in addition to mechanical and thermal energy. Multifunctional materials like these facilitate the development of self-powered systems that are more practical and functional across a broad range of conditions, from industrial uses to remote and off-grid locations.

3.5 Energy Harvesting for Wearable Electronics

One of the most exciting uses of perovskite multiferroics is in the creation of self-powered wearable electronics. Wearable technology, including fitness monitors, health trackers, and smartwatches, needs to be powered at all times. Conventional battery-driven devices have to be recharged or have batteries replaced on a regular basis, which is inconvenient and eco-unfriendly. By incorporating perovskite multiferroics in wearable devices, it is possible to develop self-sustaining systems that collect energy from environmental sources, like body heat, mechanical movement, or even solar power. For example, perovskite-based thermoelectric generators are able to tap the heat produced by the human body and transform it into electricity, which drives sensors monitoring body temperature, heart rate, and other essential indicators. Concomitantly, piezoelectric energy harvesting can harness power from the motions of the wearer, for instance, walking or running, and use it to drive the device. This process of energy harvesting from the surrounding environment lessens the demand on batteries and makes wearable electronics sustainable and autonomous to use.

3.6 Smart Sensors and IoT Devices

Perovskite multiferroics are also promising to power smart sensors and Internet of Things (IoT) devices. IoT devices and sensors are frequently used in inaccessible or distant areas where it is not possible to install traditional power sources. Using perovskite multiferroics in energy harvesting devices, IoT devices and sensors can be energized using ambient energy sources such as temperature gradients, vibrations, or magnetic fields. For instance, perovskite-based energy harvesters can fuel sensors deployed for environmental monitoring, agricultural applications, and industrial automation. These sensors can gather data on parameters including temperature, humidity, pressure, and movement and send the information wirelessly to a central platform. The presence of perovskite multiferroics guarantees that the devices are autarkic, minimizing maintenance charges and battery replacements.

Perovskite multiferroic applications to energy harvesting are multifaceted and extensive and involve thermoelectric and piezoelectric, magneto-electric and hybrid devices. With its ferroelectric, piezoelectric, and magneto-electric capabilities integrated into it, perovskite multiferroic devices can extract energy from any kind of sources using them for optimal energy capturing and conversion capabilities to suit energy harvesting applications across numerous fields. With ongoing research in this area, perovskite multiferroics are set to transform energy harvesting technologies, opening the door to self-powered devices and green energy solutions in wearable electronics, smart sensors, and IoT devices.

4 Challenges in Developing Perovskite Multiferroic Energy Harvesters

In spite of the immense promise of perovskite multiferroics, there are a number of challenges that engineers and scientists encounter while developing these materials for real-world energy harvesting applications. These challenges encompass material synthesis-related issues, optimization of performance, stability, scalability, and integration into energy harvesting devices.

4.1 Material Synthesis and Composition Control

One of the main challenges in the design of perovskite multiferroic energy harvesters is the challenge of synthesizing materials with the desired multiferroic properties. The perovskite structure is very sensitive to chemical composition and processing conditions. It is challenging to achieve the specific stoichiometry and crystal structure required for maximum ferroelectric, piezoelectric, and magneto-electric properties by controlling synthesis parameters like temperature, pressure, and dopant or additive concentration.

In addition, the use of various additives, like rare-earth metals or transition metals, to improve the properties of the material may create unwanted consequences, like phase instability or secondary phases that affect the multiferroic behavior. This intricacy in material preparation not only complicates the preparation of perovskite multiferroics but also restricts their scalability in large-scale energy harvesting processes.

4.2 Performance Optimization

The performance of materials plays a major role in the efficiency of energy harvesting devices. For perovskite multiferroics, the process of maximizing their ferroelectric, piezoelectric, and magneto-electric properties so that they exhibit high energy conversion efficiency is not an easy process. All three properties must be individually optimized, as in many cases there is a trade-off between them. The piezoelectric response, for instance, might increase at the cost of decreasing magneto-electric coupling, while improvement in the magneto-electric effect could harm the ferroelectric polarization. The difficulty is in tuning these opposing effects to optimize the overall energy conversion efficiency. High electromechanical coupling coefficient, which is necessary for efficient energy harvesting, typically involves microstructure tuning of the material at the nanoscale. Different methods, including doping, nanostructuring, and forming heterostructures, have been attempted by researchers to enhance the performance of perovskite multiferroics. Yet, these methods are still in the experimental stage, and more studies need to be conducted to derive practical strategies for improving performance.

4.3 Stability and Durability

Another significant challenge is the long-term stability and durability of perovskite multiferroics when operating in real-world conditions. Perovskite materials, especially organic–inorganic hybrid perovskites, are susceptible to environmental influences including humidity, thermal cycling, and UV exposure. These environmental factors have been known to cause degradation of material properties with time, which is a critical issue for energy harvesting applications that call for dependable long-term operation. For instance, perovskite solar cells, which have been researched intensively for energy conversion, are plagued by stability problems that prevent them from being commercialized. Similar problems are also expected to occur in perovskite multiferroics employed in energy harvesting devices. The loss of ferroelectric or piezoelectric behavior with time can result in the substantial reduction of energy harvesting efficiency. To get past this, scientists are attempting to create more stable perovskite materials through investigating different compositions as well as enhancing their resistance to environmental stress.

4.4 Scalability and Cost-Effectiveness

Scalability is another problem that hinders the mass application of perovskite multiferroic energy harvesters. Although perovskite-based devices of small scale have exhibited encouraging performances in the lab, scaling these materials to industrial size for practical application is not simple. Fabrication processes of perovskite multiferroics may necessitate sophisticated equipment and conditions that are time and money-consuming. In addition, large-scale production of high-quality perovskite multiferroic materials with the requisite properties is still challenging. Employment of costly raw materials or intricate fabrication processes might restrict the commercial feasibility of perovskite multiferroic energy harvesters. Perovskite multiferroics need to be made commercially viable, cost-effective, and scalable by employing fabrication methods. This might include reducing the synthesis complexity, using available raw materials, or leveraging cheap deposition methods like solution processing or inkjet printing.

4.5 Integration into Energy Harvesting Devices

After perovskite multiferroic compounds are synthesized and optimized, it is the turn of integrating these into functional devices for energy harvesting. Energy harvesting devices usually incorporate intricate device geometries, i.e., electrodes, interconnects, as well as the energy storage. It is highly challenging to incorporate perovskite multiferroics within these devices with no loss in their performance. For instance, the interface between the electrode and the perovskite material has a great influence on the efficiency of the device. Inefficient contact at the interface leads to energy loss and diminishes the power output. Furthermore, perovskite multiferroics' mechanical properties must be carefully analyzed when designing energy harvesters that will be subjected to mechanical stress or vibrations. They have to be mechanically sound so as to resist such stresses without being degraded. In addition, hybrid energy harvesting devices that integrate multiple energy conversion mechanisms (e.g., thermoelectric, piezoelectric, and magneto-electric) can be advantaged by incorporating perovskite multiferroics. Nonetheless, developing such hybrid systems that effectively harvest and convert various forms of energy at the same time is a challenging endeavor. This not only demands the invention of new materials but also the creation of novel device structures that can optimally integrate the best features of each energy conversion mechanism.

4.6 Environmental and Sustainability Concerns

Although perovskite multiferroics can be used for energy harvesting, there are certain environmental issues with their use, especially in view of their widespread application in consumer electronics and bulk energy harvesting devices. The production and utilization of certain elements used in perovskite multiferroics, like rare-earth elements or toxic elements, can have harmful environmental consequences if not controlled properly. In addition, the life cycle of perovskite multiferroic materials, from manufacturing to end-of-life disposal, should be thoroughly assessed to make sure that these materials are environmentally friendly and sustainable. Scientists are more interested in creating environmentally friendly perovskite compositions that do not use toxic elements such as lead or cadmium, which are present in certain perovskite-based materials.

Perovskite multiferroics are outstanding materials for energy harvesting devices due to their singular combination of ferroelectric, piezoelectric, and magneto-electric properties. Nevertheless, there are some challenges that must be overcome before these materials can be effectively developed into useful energy harvesting technologies. These challenges are the challenges in material synthesis and composition control, optimization of performance, long-term stability, scalability, device integration, and environmental sustainability. Increased innovation and research are necessary to bypass these challenges and realize the potential of perovskite multiferroic energy harvesters. With improving material design, fabrication methods, and device integration, perovskite multiferroics could be significant in the evolution of efficient and sustainable future energy harvesting systems.

5 Conclusion

Perovskite multiferroics offer the possibility to design multifunctional materials for energy harvesting purposes because of the special coupling of ferroic orders present in them. Their capability to efficiently transform solar, thermal, and mechanical energy into electricity makes them prime facilitators of self-powered wearable electronics, smart sensors, and renewable energy technologies. Despite their promise, issues like environmental stability, synthesis of materials, and scalability continue to be major hindrances toward extensive applications. Developments in material engineering, strain engineering, and hybrid energy harvesting structures will be central to solving these challenges. Future research and innovation in perovskite multiferroics are anticipated to open up paradigm-shifting possibilities for sustainable energy technologies that support global efforts in energy efficiency and environmental sustainability.

References

1. Catalan G, Scott JF (2009) Physics and applications of bismuth ferrite. Adv Mater 21(24):2463–2485. https://doi.org/10.1002/adma.200802849
2. Eerenstein W, Mathur ND, Scott JF (2006) Multiferroic and magneto-electric materials. Nature 442(7104):759–765. https://doi.org/10.1038/nature05023
3. Hill NA (2000) Why are there so few magnetic ferroelectrics? J Phys Chem B 104(29):6694–6709. https://doi.org/10.1021/jp000114x
4. Ramesh R, Spaldin NA (2007) Multiferroics: progress and prospects in thin films. Nat Mater 6(1):21–29. https://doi.org/10.1038/nmat1805
5. Spaldin NA, Fiebig M (2005) The renaissance of magneto-electric multiferroics. Science 309(5733):391–392. https://doi.org/10.1126/science.1113357
6. Liu H, Yang X (2017) A brief review on perovskite multiferroics. Ferroelectrics 507(1):69–85. https://doi.org/10.1080/00150193.2017.1283171
7. Ding R, Wong MC, Hao J (2020) Recent advances in hybrid perovskite nanogenerators. EcoMat 2(4):e12057. https://doi.org/10.1002/eom2.12057
8. Kumar S, Sharma V, Kumari N, Kaur GA, Saha A, Thakur S, Shandilya M (2024) Recent advances in perovskite materials: exploring multifaceted properties for energy harvesting applications. Ionics 30(9):5159–5188. https://doi.org/10.1007/s11581-024-05658-3
9. Pandey R, Vats G, Yun J, Bowen CR, Ho-Baillie AW, Seidel J (2017) Perovskites for solar and thermal energy harvesting: state of the art technologies, current scenario and future directions. arXiv preprint arXiv:1705.05529. https://doi.org/10.48550/arXiv.1705.05529
10. Jana B, Ghosh K, Rudrapal K, Gaur P, Shihabudeen PK, Roy Chaudhuri A (2022) Recent progress in flexible multiferroics. Front Phys 9:822005. https://doi.org/10.3389/fphy.2021.822005
11. Wei XK, Domingo N, Sun Y, Balke N, Dunin-Borkowski RE, Mayer J (2022) Progress on emerging ferroelectric materials for energy harvesting, storage and conversion. Adv Energy Mater 12(24):2201199. https://doi.org/10.1002/aenm.202201199

12. Ederer C, Spaldin NA (2005) Weak ferromagnetism and magneto-electric coupling in bismuth ferrite. Phys Rev B 71(6):060401. https://doi.org/10.1103/PhysRevB.71.060401
13. Martin LW, Ramesh R (2010) Multiferroic and magneto-electric heterostructures. Acta Mater 58(15):4342–4352. https://doi.org/10.1016/j.actamat.2010.04.001
14. Chu YH, Martin LW, Holcomb MB, Gajek M, Han SJ, He Q, Ramesh R (2008) Electric-field control of local ferromagnetism using a magneto-electric multiferroic. Nat Mater 7(6):478–482. https://doi.org/10.1038/nmat2184
15. Yang SY, Martin LW, Byrnes SJ, Conry TE, Basu SR, Paran D, Ramesh R (2008) Photovoltaic effects in $BiFeO_3$. Appl Phy Let 95(6):062902. https://doi.org/10.1063/1.2966308
16. Spaldin NA, Ramesh R (2019) Advances in magneto-electric multiferroics. Nat Mater 18(3):203–212. https://doi.org/10.1038/s41563-018-0275-2
17. Heron JT, Bosse JL, He Q, Gao Y, Trassin M, Ye L, Ramesh R (2014) Deterministic switching of ferromagnetism at room temperature using an electric field. Nature 516(7531):370–373. https://doi.org/10.1038/nature14004
18. Wang KF, Liu JM, Ren ZF (2009) Multiferroic materials and magneto-electric effects. J Phys: Condens Matter 21(9):1–15. https://doi.org/10.1088/0953-8984/21/9/093202
19. Fiebig M, Lottermoser T, Meier D, Trassin M (2016) The evolution of multiferroics. Nat Rev Mater 1(8):1–14. https://doi.org/10.1038/natrevmats.2016.46
20. Apostolov AT, Apostolova IN, Wesselinowa JM (2024) Multiferroic and phonon properties of the double perovskite Pr_2FeAlO_6. Materials 17(19):4785. https://doi.org/10.3390/ma17194785
21. Gareeva Z, Sharafullin I, Zvezdin A (2023) 2D-perovskite multiferroics: interface-induced magnetoelectric effect in perovskite-based multiferroic superlattices. Crystals 13(9):1404. https://doi.org/10.3390/cryst13091404
22. Gareeva Z, Zvezdin A, Zvezdin K, Chen X (2022) Symmetry analysis of magnetoelectric effects in perovskite-based multiferroics. Materials 15(2):574. https://doi.org/10.3390/ma15020574
23. Catalan G, Béa H, Fusil S, Bibes M, Paruch P, Barthélémy A, Scott JF (2008) Fractal dimension and size scaling of domains in thin films of multiferroic $BiFeO_3$. Phys Rev Lett 100(2):027602. https://doi.org/10.1103/PhysRevLett.100.027602

Chapter 7
Latest Updates on the Enhancement in the Photovoltaic Energy Conversion Efficiency of Sustainable Perovskite Multiferroic Materials-Based Solar Cell Devices

Monika Srivastava and R. C. Singh

1 Introduction

It has been quite evident that to reduce the greenhouse gases, the effective use of renewable sources such as solar energy, wind energy, geothermal energy, biomass energy is the need of the hour. Among these, the solar energy is the most abundant and replenishable source of energy but it is limited by the technology. There has been a number of work emphasizing on harvest and storage of solar energy which seems quite convincing to overcome the energy needs of our planet earth but these techniques are limited by the low efficiency, high costs, and resource issues. Therefore, the major portion of focus of researches based on photovoltaics was focused on conversion of solar energy into electrical energy and storage of this form of energy, with an emphasis on increasing the efficiency and reducing processing and fabrication cost. There have been four different generations of solar cells based on their type of materials used and their fabrication techniques. The first generation was the monocrystalline Si-based solar cells. This generation of solar cells were incorporating crystalline silicon modules of solar cells which gave high efficiency and better stability. The first generation solar cells have occupied about 80% of the commercial market of photovoltaics. But due to high cost of these cells, there was evolution of second generation solar cells which were based thin films of the materials. These solar cells were made up of CdTe, CIGS, and amorphous silicon. They obtained high efficiency but their toxicity and limited availability leads to the evolution of third generation solar cells. This generation is mainly categorized as organic solar cells, perovskite solar

M. Srivastava · R. C. Singh (✉)
Department of Physics and Environmental Sciences, Centre of Excellence in Solar Cell and Renewable Energy, Sharda University, Greater Noida, India
e-mail: rcsingh@sharda.ac.in

M. Kumar (ed.), *Sustainable Perovskite Multiferroic Materials*, Materials Horizons: From Nature to Nanomaterials, https://doi.org/10.1007/978-981-96-7419-0_7

cells, dye sensitized solar cells, and quantum dot solar cells (Fig. 1). Among these, the efficiency of perovskite has ranged from 3.1% to 25.4% in the time span of a decade which has proven that the major research progress has been occurring in the field of perovskite-based photovoltaics. The perovskite materials with the crystal formula of ABX_3 has 'A' as the Hybrid organic inorganic or complete inorganic element, 'B' is the metal cation and 'X' could be oxide or halide anion. The stable perovskite material should have a tolerance factor t with values ranging as $0.813 < t < 1.107$ [1]. For example, the tolerance factor of methyl ammonium lead iodide ($MAPbI_3$) is approximately measured as 0.83. The multiferroic-based perovskite has also gained importance in the field of photovoltaics. The ferroelectric and multiferroelectric materials have regenerated the third generation solar cells by giving good response in terms of photovoltage and photocurrent. There are a number of multiferroics such as Bi_2FeCrO_6 (BFCO), $Pb(Zr,Ti)O_3$ (PZT) and $BiFeO_3$ (BFO) which are used in photovoltaic application. About five decades back this effect was generated but it garnered importance only after the generation of photovoltage of 15 V in BFO thin films [2, 3]. There are a number of multiferroic perovskite which are being used for photovoltaic application. As we know that multiferroics are the materials which have the presence of simultaneous ferromagnetic and ferroelectric order states. Though the multiferroic are typically of the formula ABO_3 for example $BaTiO_3$ and $PbTiO_3$. The non-centrosymmetry of polar perovskite oxides has proved its potential to be used in harvesting in solar energy. The photovoltaic property is generally observed in those multiferroic perovskites which lack center of inversion symmetry, some of such perovskites are $LiNbO_3$, $BaTiO_3$, $KNbO_3$-$Ba(Ni1/2Nb_{1/2})O_3$, $(Pb,La)(Zr,Ti)O_3$ $Pb(Zr,Ti)O_3$, and tungsten-doped $Pb(Mg_{1/3}Nb_{2/3})_{1-x}Ti_xO_3$ (PMN-PT) crystals [4–14]. In the recent studies, it has been found that the photocurrent densities have obtained a range of 10^{-2}–10^{-5} A/m^2 as these multiferroics possess a wide bandgap of the order of 3–4 eV. There has been a number of studies in which enhanced photo conversion efficiencies by optimizing the various parameters such as domain structures, layer thickness, electrode-ferroelectric interface, poling electric field, and band gap. But the base property of the multiferroics which favors its use in photovoltaic is governed by its non-centrosymmetry which cause an easy separation of the charge carriers for the generation of excitons which forms the basis of the photovoltaics. There are some other remarkable properties of multiferroics which makes it suitable for its photovoltaic application such as its room temperature stability, low cost of processing, easy deposition techniques [10]. Among the various multiferroics such as, $BiFeO_3$, $PbTiO_3$, $BaTiO_3$, $PbZrTiO_3$, BFO has the most favored band gap 2.7 eV [15–17].

2 Photovoltaic Phenomenon in Perovskite Multiferroics

The PV effect arises in multiferroics due to their non-centrosymmetry which provides the charge separation due to effective spontaneous polarization. Due to this charge separation, a photo voltage is developed which is governed by the polarization and

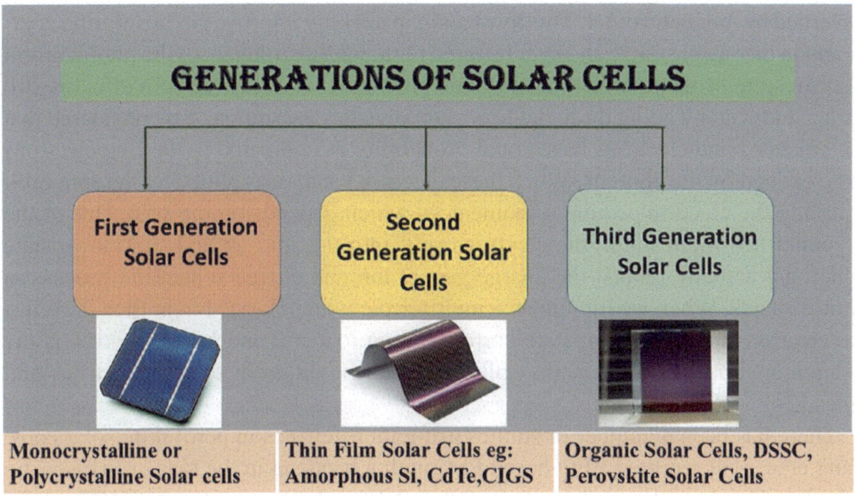

Fig. 1 Generations of solar cells

separation of photo excitons between the electrodes and also on the active layer thickness of the active layer. Generally, these multiferroics show considerably good photovoltage but the photocurrent is comparatively low. However, new techniques such as the concept of domain walls effectively separating the charges, optimization of thin films and tip modified PV effects [18–20] etc., are developed which overcome its limited usage. The photovoltaic phenomenon in multiferroics is different from that of conventional solar cell. In this case, a single material is responsible for generation of excitons instead of forming a p–n junction. The non-centrosymmetry which exist in these multiferroics is responsible for generation of free charges and obtaining considerably high value of open-circuit voltage. But the generation of photocurrent in the multiferroic PSCs is very low due to which the photo conversion efficiency of these solar cells is no way close to that of conventional solar cells. The wide bandgap of the ferroelectrics also reduces their absorption in the visible region of spectrum. Basically in the single crystal samples of some multiferroics such as $LiNbO_3$, $BaTiO_3$, or PZT. The PV effect can be seen as a result of bulk photovoltaic effect which is famously known as photogalvanic effect which generalizes that due to non-centrosymmetry of these materials a steady current is formed [21]. Not quite as much in the case of p–n junction, when light falls on the multiferroic material, the movement of electron is in one particular direction without crossing from one material to another. (Fig. 2). A number of models explains the bulk photovoltaic effect. Some of them are shift current, nonlinear dielectric models, asymmetric carrier, and asymmetric potential well [22]. In addition to the factor of non-centrosymmetry the PV effect in multiferroics is also contributed by the spin property of the electrons and the phenomenon is called as spin photovoltaics. As compared to the ferroelectrics, the multiferroelectrics have a low bandgap due to the electron–electron interaction governed by magnetic order. This low bandgap not only improves its absorption coefficient in the visible range and near infrared region but also enhances the photocurrent [23]. In spite of the fact that photovoltaic phenomenon under the spin effect is still under study but by going through the non-relativistic description, it is considered that the valence band electrons have equal probability as compared to the free electrons to be excited by the photons. When these electron bands are splitted by the spin orbit coupling the electron acquire a momentum which depends on the spin. Due to the asymmetrical momentum, the electrons split into the spin up and spin down state which are actually related the charge generation and charge separation process in photovoltaics. When we illuminate a multiferroic with polarized light then it excites the electrons to momentum state corresponding to $k > 0$ and $k < 0$ where k is the momentum space. However, the splitting could be different for the electrons and holes [24].

There has been a number of studies using multiferroics in perovskite solar cells. Yang et al. [19] reported BFO-based PSC with a bandgap in the range of (Eg≈2.6–2.7 eV). The V_{oc} obtained is about 0.8–0.9 V with an efficiency of 10% under sunlight AM 1.5G with SRO bottom and ITO top electrodes. In another study, BFCO is used with a bandgap of 1.5 eV obtaining a photocurrent of 11.7 m A and PCE of 3.3%. $BaTiO_3$ is a ferroelectric material with bandgap 3.2 eV and it generates a photocurrent of 4 nA (28). Multiferroic $KNbO_3$ Barium nickel niobate (KBNNO) shows a reduction in its bandgap when BNNO is doped with $KNbO_3$ and generates a photocurrent 0.7 mV and photocurrent of 0.1 μA [26]. It has been observed that the BFO-based solar cell gives enhanced efficiency which is attributed to high ordering

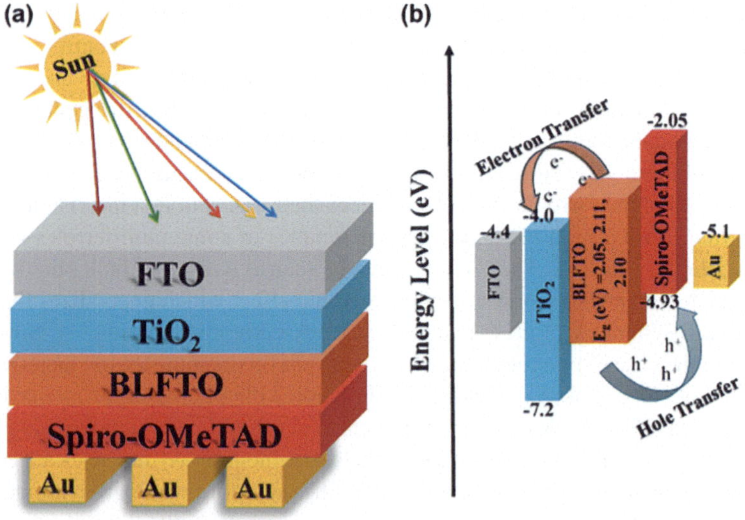

Fig. 2 **a** Schematic diagram of heterostructure PSC device configuration of La and Ti doped BFO. **b** Schematic diagram of energy level diagram of FTO/TiO_2/BLFTO/Spiro-OMeTAD/Au [25]. Copyright 2019. Reused with permission from Elsevier

Table 1 Photoconversion efficiency of multiferroics-based PSCs

Multiferroic-based PSC	E_g (eV)	Photo voltage	Photo current (mA)	Efficiency	References
BFO	2.6–2.7	0.8–0.9	–	10%	[19]
BFCO	1.5	–	11.7	3.3%	[31]
$BaTiO_3$	3.2	–	4×10^{-6}	–	[32]
KBNNO	1.1–3.8	0.7×10^{-3}	0.1×10^{-3}	–	[26]
BFCO (Fe and Cr doped)	1.4–3.2			8.1%	[27]
$KBiFe_2O_5$	1.6	8.8	15×10^{-3}	–	[23]
PFV	–	0.7	20×10^{-6}	–	[28]
Ti-doped BLFO	2.05–2.11	0.91	10.87	5.61%	[29]
$BiFeO_3/BiCrO_3$		1.2	0.013	0.01%	[30]

among the sub lattices of iron and chromium (Fe and Cr). However, this ordering between Fe–Cr can be also be controlled by the effect of magnetic field which results in high efficiency of the upto 8.1% for BFO solar cells. This exceptional work which has obtained marvelous result by effectively tuning of the bandgap by tailoring the order of the cations Fe/Cr [27]. In one of the researches, with the cell structure Sn doped In_2O_3 (ITO)/PFV/ITO, a strong visible-light PV effect is reported, however, the V_{oc} is negative in this case varying from −0.45 to −0.70 V [28]. In another study, lead-free FTO/TiO_2/absorber/Spiro-OMeTAD/Au heterostructure device acquired a (PCE) power conversion efficiency ~ 5.61% with $V_{oc} = 0.91$ V, fill factor (FF) = 55.55%, $J_{sc} = 10.87$ mA/cm^2 where La/Fe doped BTO absorber material with optimized thickness and operating temperature have been used [29]. A power conversion efficiency of ~ 0.01% is obtained by using multistacked films of $BiFeO_3/BiCrO_3$ bilayer used as active layers in a photovoltaic (PV) device [30] (Table 1).

2.1 *Bulk Photovoltaic Effect in Multiferroics Perovskite Solar Cells*

The "bulk photovoltaic effect" (BPVE) is a phenomenon where a material generates a steady photocurrent and voltage directly within its bulk when exposed to light, without the need for a p–n junction, and occurs only in materials lacking inversion symmetry, essentially converting light into electricity through a nonlinear optical process; the first is the asymmetric potential model which takes in to account the electric potential developed due to electric polarization at the light absorbing centers. The model is explained in terms of three energy levels E_0, E_1, and E_2 where electrons are carried from ground state E_0 to E_1 and then to E_2 with respect to the potential well ranging between φ_1 and φ_2. If the excitation energy E_1 is less than φ_1 then the particle is confined in the well and if the excitation energy is higher than φ_2 then the carriers deviate from the absorbing centers, thus reducing the photocurrent. The energy needs to be in between φ_1 and φ_2 for a suitable photocurrent to flow [33]. Another model is the shift current model which is based on the non-centrosymmetry of the FE materials. Due to this, an asymmetry is developed in the momentum of the charge carriers which give rise to steady photocurrent [20, 34].

Another model is the domain wall theory, which is based on the spatial extension of polarization vectors known as domains. Within a FE material domain are identified individually by being separated through the domain walls of 2–3 nm. Since the variation in the structure of potential is more pronounced at the domain walls therefore it become the site for the separation of charge carriers. However, the chance of recombination of the charge carriers is minimum in these domains which result in net diffusion current through each domain. According to this model, the photocurrent does not depend on the direction of light but on the direction of polarization. Thus, the current produced by bulk photovoltaic effect might get added or cancelled depending on the direction of polarization. If the two photocurrents, one by bulk photovoltaic effect and other by domain model are in the same direction, the power conversion efficiency gets enhanced [35]. In Schottky junction effect, a localized electric field is developed at the interface of multiferroic and the electrode, which gives rise to the photocurrent. In this case, the generated photocurrent depends on the Schottky barrier between the top and bottom interfaces of the metal/ferroelectric/metal structure. Thus, it was found that the higher value of the photocurrent largely depends on the difference of top and bottom boundaries of the Schottky barrier [36, 37].

The depolarization model explains the high photovoltage in these PSCs. This model is based on depolarization field which is directly proportional to the strength of polarization. The surface charges are basically governed by the remnant polarization, dielectric effect, charge density of the absorber and the electrode. The external bias voltage easily tunes the polarization which in turn affects the depolarization field [38].

3 Photovoltaic Parameters of Multiferroic PSC

Since the photovoltaic properties of the multiferroics arises due to the ferroelectric polarization therefore its photovoltaic parameters varies from the conventionally operated solar cells. In the case of multiferroics, the current and voltage are governed by the polarization process and built in electric field [39, 40]. The theoretical model of bulk photovoltaic effect gives the expression for photocurrent in terms of a rank 3 tensor β_{ijk} [21, 41]. The formula for PCE is given by PCE $= \frac{\beta_{ijk}}{\sigma} \times J$, where σ is the photoconductivity and J is the current density. As it is known that J–V characteristic is important for evaluating the power conversion efficiency of the PSC. For a traditional solar cell, J–V characteristic is measured under AM 1.5 illumination, i.e., under 1 sun condition with input power as 100 mw/cm^2 at room temperature. The total current of multiferroic perovskite solar cell is contributed by the combined effect of two factors, one is due to photocurrent due to absorption of photons and the other factor is dark current injected from device electrodes.

However, there is only one factor is different from the conventional working of the solar cell that is the separation and the transportation dynamics of the photo generated excitons are governed by the ferroelectric polarization, rest all the process is similar to that of conventional solar cell. Now from the measured J–V curve of the solar cell such as V_{oc}, J_{sc}, and fill factor, the PCE is obtained by the following formula.

$$\mathrm{PCE} = \frac{J_{sc} \times V_{oc} \times \mathrm{FF}}{P_{in}} \quad \text{Where} \quad \mathrm{FF} = \frac{J_{mp} \times V_{mp}}{J_{sc} \times V_{oc}} \tag{1}$$

The significant parameters of the PSC which are required to find its photo conversion efficiency are V_{oc}, J_{sc}, and fill factor (FF). These parameter when applied in the above Eq. 1, they give the value of PCE. It is defined as the ratio of the maximum power output (P_{out}) to the incident light power (P_{in}) under the illumination of the solar cell in one sun condition.

4 Methodologies to Improve the PCE of the Multiferroics-Based Perovskite Solar Cells

Among all the multiferroics BFO is the most frequently used multiferroics for PV applications. This may be due to its low bandgap or due to anti symmetric properties. There has been various models which are proposed for this study such as ferroelectric effect, domain wall model, or P–N junction-based models [42]. In one of the researches, BFO is used in a PV solar cell with the architecture Pt/BFO single crystal/Pt which under the irradiance of $\lambda = 405$ nm attains an open-circuit voltage $V_{oc} = 13$ V and a short-circuit current density $J_{sc} = 0.01$ A/m^2 [43]. The V_{oc} increases to values as good as 50 V with a configuration of ITO/BFO/$SrRuO_3$ under the irradiance of $\lambda = 340$ nm [25]. Another enhanced application of BFO was found in the heterostructure cell with an architecture as ITO/BFO_5Sm ceramic/Au which under the irradiation of 405 nm is successful in achieving a PCE of = 0.37% and an EQE of = 4.1%. In a theoretical P–N junction like model of BFO PSC, there is a band gap variation ranging from 2.24 to 2.15 eV as the material changes from $(Bi_{1-x}Sm_x)FeO_3$ ceramics to $(Bi_{0.9}Sm_{0.1})FeO_3$ (BFO_{10}Sm) [44, 45]. Similarly in one of the studies under the irradiation of $\lambda = 405$ nm a neodymium-doped $BiFeO_3$ with indium-doped tin oxide thin film to generate PCE = 0.8% and EQE = 9% [46]. In the same context, Machado et al. [47] worked on cobalt-doped BFO to be used in the PSC. The study revealed that band gap of the multiferroic reduces when the cobalt concentration was increased in $BiFe_{1-x}Co_xO_3$ ($x \leq 0.3$). The research group of Zhang et al. worked on BFO doped with LA to obtain superior quality multiferroic. The techniques used for the deposition of functional layers in the PSC were magnetron sputtering and pulsed laser deposition with the cell configuration as of FTO/ETLs/La-doped BFO/Au. The PSCs were prepared with three types of ETL, i.e., TiO_2, SnO_2, and ZnO and a comparison of all photovoltaic performances were done. It was found that the PSC with ETL as TiO_2 gave the best efficiency of 5.62% and a $V_{oc} = 1.74$ V. While the PSCs with SnO_2 and ZnO exhibited an efficiency of 5.45% and 4.11%, respectively [48]. Pei et al. [49] used C and Mn-doped BFO and with this doping they not only reduced the bandgap but also improved remnant polarization. Along the same line Renuka et al. used the cell structure of Pt/$CuCr_2$/Ti-doped BFO/WS_2/ITO and obtained a higher V_{OC} of 0.95 V and short-circuit current density (J_{SC}) of 2.60 mA/cm^2. The results also show that the efficiency of Ti-doped BFO-based PSC is 100 times more than that of undoped BFO-based PSCs [50]. In the same continuation, a detailed study based on La and Ti-doped BFO was performed by Raj et al., with the formation of $Bi_{0.80}La_{0.20}Fe_{1-x}Ti_xO_3$ ($x = 0.00, 0.04, 0.08, 0.12$) through the process of solid-state route, to study its structural, optical, and dielectric properties. A PSC with a cell structure FTO/TiO_2/$Bi_{0.80}La_{0.20}Fe_{1-x}Ti_xO_3$/Spiro-OMeTAD/Au was simulated by SCAPS-1D software and its potential as absorber layer in PSCs was further studied.

Furthermore, in one of the simulated studies, the La/Ti modified BFO is incorporated in the cell configuration of the solar cell as the effective absorber layer in the SCAPS-1D simulation software has been used and later on investigated its usefulness in photovoltaic devices. The study even emphasized on investigating the performance of the PSC under variation of temperature (Fig. 3). This study revealed that after consideration of experimental parameters the lead-free doped $Bi_{0.80}La_{0.20}Fe_{1-x}Ti_xO_3$ based PSC exhibited PCE = 5.61%, $V_{oc} = 0.91$ V, FF = 55.55%, and J_{SC} of 10.87 mA/cm^2, however to optimize the thickness of the absorber layer is a crucial requirement for the better performance of the solar cell [51].

The enhancement in power conversion efficiency of the multiferroic perovskite solar cells depend on very crucial factors such as ferroelectric effect, bandgap, interface engineering, etc. The photovoltaic effect in multiferroics is basically governed by two factors. First is APV, i.e., anomalous photovoltaic effect which is attributed by the V_{oc} several orders high than the bandgap of these materials and second is that the direction of photocurrent is dependent on the direction of polarization [52, 53]. As mentioned before also BFO is most used in photovoltaics but its impact in this field is limited by a very factor which is known as the leakage current. There has also been issues such as reduction of Fe^{+3} and volatilization of bismuth [54]. Thus, there are a variety of dopants used to overcome these important issues. In one of the very important studies by Cai et al., Ti-doped BFO is prepared by sol gel method and its photovoltaic properties have been studied. As is shown in this study in the visible region of solar spectrum, i.e., 380–780 nm, the transmittance of Ti-doped BFO thin films is lower than that of BFO thin films means its absorbance is high than that of undoped BFO. This suggests that as we increase the titanium content in BFO, it leads to the decrease in transmittance of the Ti-doped BFO, which shows that addition of Ti^{4+} ions is rather advantageous as it increases the absorption of visible light and ultraviolet light in BFO thin films (Fig. 4a, b). Interface engineering also plays a very crucial role in enhancing the performance of the MPSCs. There are a number of

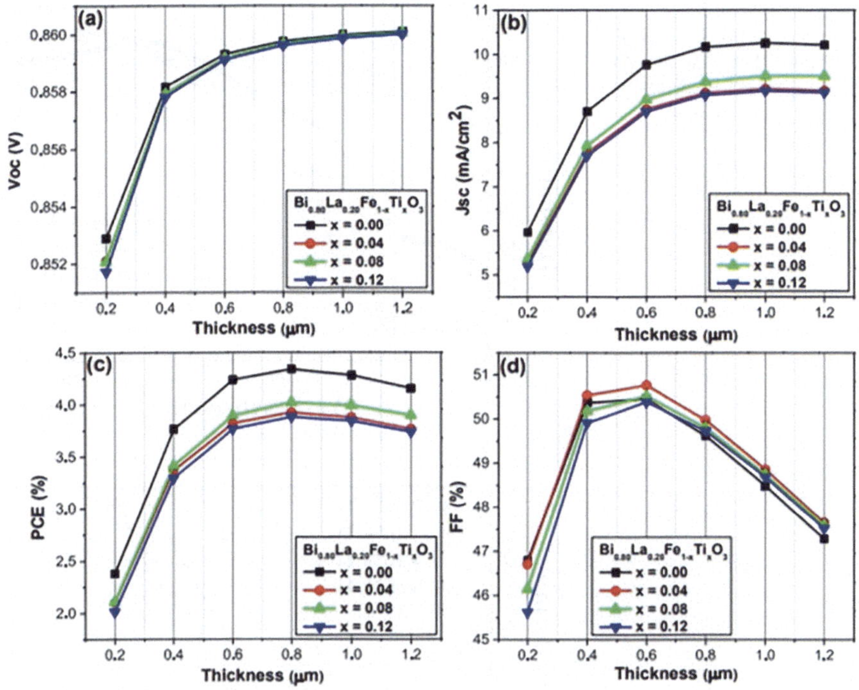

Fig. 3 Variation in device output with thickness of lead-free perovskite absorber $Bi_{0.80}La_{0.20}Fe_{1-x}Ti_xO_3$ ($x = 0.00, 0.04, 0.08, 0.12$) [25]. Copyright 2019. Reused with permission from Elsevier

research work which are based on optical tuning for the usage of bandgap compatible electrodes which provides efficient charge separation and transfer through the respective electrodes. Machado et al. have worked with ZnO electrode which high bandgap matching and easy processability with BCFO-based MPSCs, which reduces the charge recombination, thus enhancing the J_{sc} [37]. Similarly, other materials like WS_2 and reduced graphene oxide are also used for bandgap engineering [45].

The J–V characteristic of the Ti-doped BFO PSC shows a very distinct feature that at first the efficiency decreases and then increases with the increasing content of Ti (Fig. 5). In this case, two opposing factors are dominant at one time. First the high remnant polarization causes the separation of electron and holes on the other hand the stress–strain created due to lattice distortion act as a recombination center for electron hole pair, which reduces the photocurrent. At 2% doping of BFO with Ti, the remnant polarization is dominated by high lattice distortion. This results in the recombination of charge carriers which reduces the photocurrent. However, at 4% and 8% doping, the remnant polarization is higher than the lattice distortion, due to which the photocurrent increases [55]. Chen et al., fabricated a cell with the deposition of 50 nm thick layer of $BaTiO_3$–$BiFeO_3$ (BT–BFO) on the conductive Nb–SrTiO3 substrate to obtain the cell structure of Nb–$SrTiO_3$/Ba TiO_3–$BiFeO_3$/Pt

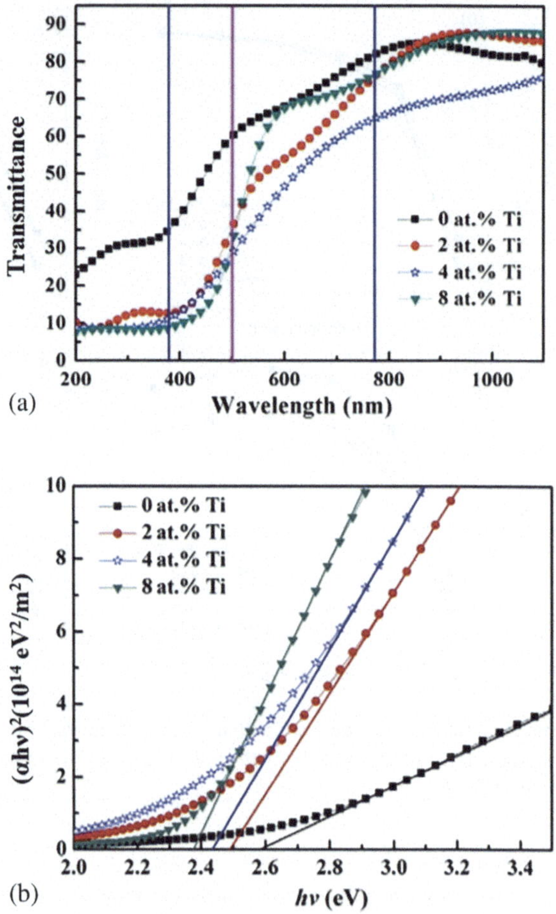

Fig. 4 **a** Transmittance curve of Ti-doped BFO [47]. Copyright 2019. Reused with permission from Elsevier. **b** Absorbance curve of Ti-doped BFO [47]. Copyright 2019. Reused with permission from Elsevier

solar cells. The top electrode is taken as a semitransparent Pt thin film with 20 nm thickness. This study concluded that, with the increase of positive poling voltage, the efficiencies of ferroelectric solar cells keep decreasing [12].

5 Challenges and Opportunities

The PSCs based on multiferroics are the new innovation in the field of multiferroics. They are not only lead free but they are also stable at room temperature, knowing that stability of PSCs is one of the major drawback of the hybrid organic inorganic lead-based perovskite solar cells. Also in comparison with the conventional solar cells, these PSCs are capable of generating high photovoltage though the photocurrent is

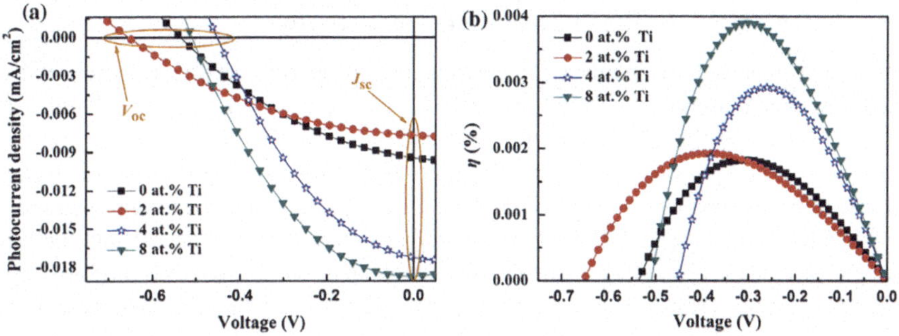

Fig. 5 **a** The variation in photocurrent at different doped concentration of Ti. **b** Efficiency curves of Ti doped BFO [47]. Copyright 2019. Reused with permission from Elsevier

not very effective. In spite of the overwhelming response, the multiferroics-based PSCs face following challenges:

(a) **Solution Processed Fabrication of Multiferroic Layer**

In the hybrid halide-based PSC basically the cost effective and easily processible techniques are applied such as spin coating, Doctor blading followed by annealing at temperature about 500 °C. But the multiferroic material is generally prepared by using epitaxy technology for which the temperature required is 1000 °C. Therefore, fabrication of solution processed multiferroics-based PSCs at temperature less than 500 °C is still a challenge.

(b) **Measurement of J–V Characteristics**

The J–V characteristic of the multiferroic perovskite solar cells (MPSCs) is performed with the same process as done for conventional solar cells. However, the mechanism of working of a multiferroic solar cells is based primarily on the polarization mechanism. Conventionally, in J–V characteristic process an external voltage is applied on the device with the help of two electrodes. But it is not very clear that what is the impact of this voltage on the polarization effect of the multiferroics. If this externally applied voltage is helpful to induce polarization field then the resultant cell will be a multiferroic solar cell of high efficiency.

(c) **Physical Insights**

The correlation of the photovoltaics and polarization effect of multiferroics is still lacking deeper insights. The information about the elaborate processing of generation of charges followed by interfacial injection governed by the multiferroics in solar cells is still lacking the deeper information. The mechanism of defect related charge generation and transport inside the bulk photoferroelectrics perovskite layer is the most important aspect of multiferroics PSCs. This model lacks the better understanding with respect to the photovoltaic behavior of the MPSCs and requires a comprehensive model to effectively describe the whole photovoltaic process of PSSCs.

(d) **Molecule Engineering to Enhance Ferroelectricity**

Later idea was that to enhance the ferroelectric properties of the conventionally used absorber material, i.e., $MAPbI_3$ in PSCs by molecule engineering. This will not only improves the efficiency but it will add on to the already existing champion material for the PSCs. There have been a number of research work which is based on molecule engineering of the halide-based perovskite materials to enhance their ferroelectric property [56, 57]. Later through researches, it was revealed that titanium diisopropoxide bis (acetylacetonate) is very viable material for $MAPbI_3$ as it not only reduces the defect states in the perovskite crystals but rearrange the MA molecules to generate and enhance the in-built electric field [58].

(e) **Possible Efforts to Enhance the Efficiency of MPSCs**

All the multiferroic perovskite solar cells are based on two types of effects, i.e., photovoltaic effect and ferroelectric effect. The appropriate proportion of these two properties in a single material will be the most suitable for solar cell application. The champion material, i.e., $MAPbI_3$ is an efficient absorber material which contains both ferroelectric and piezoelectric properties [59]. If a high portion of the sunlight is absorbed as heat, it affects the photovoltaic performance of the cell. However, it is required that if the greater portion of this sunlight which is absorbed as heat, is converted into electricity then the PCE of these cells could increase due to pyroelectric effect. Moreover, the stress of the flexible substrates is helpful in generating strong polarization effect in ferroelectric materials to obtain high PCE.

6 Conclusion

The chapter concludes that there are certain phenomenon which have identified and applied also to enhance the photo-conversion efficiency of the multiferroics-based perovskite solar cells. Firstly, it emphasized on the possible reasons for the photovoltaic behavior of the multiferroics such as ferroelectricity, non-centrosymmetrical nature which leads to polarization and hence separation of charge carriers, which is an essential process of photovoltaic process. Later, it was emphasized that the multiferroics undergo a phenomenon which is known as bulk photovoltaic effect which is a crucial phenomenon in giving considerable high photovoltage in multiferroics-based PSCs. The detrimental factor relating to low efficiency of MPSCs is mainly due to wide bandgap of multiferroics due to which a large portion of the solar spectrum (about 80%) cannot be utilized by the absorber material of the MPSCs. Therefore, bandgap tuning is one such effective method in improving the absorption coefficient and thus the PCE of these solar cells. The bandgap tuning is mainly achieved by doping of multiferroics with suitable elements so that resultant multiferroic have a bandgap ~ 1.4 eV which is the ideal bandgap as predicted by theoretical studies. The MPSCs generally demonstrates a low value of J_{sc}. This can be improved by interface engineering which means that use of optically suitable and bandgap compatible charge transport layers and electrodes to enhance the separation and transportation of the charge carries which are photogenerated. This improves the device performance to a greater extent. The multiferroics have considerable potential in the field of photovoltaics with suitable bandgap tuning and interface engineering which crucially enhance their performance.

References

1. Assadi MK, Bakhoda S, Saidur R, Hanaei H (2018) Recent progress in perovskite solar cells. Renew Sustain Energy Rev 81:2812–2822. https://doi.org/10.1016/j.rser.2017.06.088
2. Yang SY, Seidel J, Byrnes SJ, Shafer P, Yang CH, Rossell MD, Yu P, Chu YH, Scott JF, Ager III JW, Martin LW, Ramesh R (2010) Above-bandgap voltages from ferroelectric photovoltaic devices. Nat Nanotechnol 5(2):143–147. https://doi.org/10.1038/nnano.2009.451
3. Fridkin VM, Popov B (1978) Anomalous photovoltaic effect in ferroelectrics. Soviet Physics Uspekhi 21(12):981. https://doi.org/10.1070/PU1978v021n12ABEH005722
4. Grekov AA, Malitskaya MA, Spitsina VD, Fridkin VM (1970) Photoelectric effects in $A_5B_6C_7$-type ferroelectrics-semiconductors with low-temperature phase transitions. Kristallografiya 15:500–509
5. Glass AM, Linde DVD, Negran TJ (1995) High-voltage bulk photovoltaic effect and the photorefractive process in $LiNbO_3$. In: Landmark papers on photorefractive nonlinear optics, pp 371–373. https://doi.org/10.1142/9789812832047_0047
6. Brody PS (1973) Large polarization-dependent photovoltages in ceramic $BaTiO_{3+}$ 5 wt% CaTiO3. Solid State Commun 12(7):673–676. https://doi.org/10.1016/0038-1098(73)90310-4
7. Kholkin A, Boiarkine O, Setter N (1998) Transient photocurrents in lead zirconate titanate thin films. Appl Phys Lett 72(1):130–132. https://doi.org/10.1063/1.120663
8. Kholkin AL, Yarmarkin VK, Goltsman BM, Baptista JL (2001) Photoelectric evaluation of polarization and internal field in PZT thin films. Integr Ferroelectr 35(1–4):261–268. https://doi.org/10.1080/10584580108016907
9. Yang YS, Lee SJ, Yi S, Chae BG, Lee SH, Joo HJ, Jang MS (2000) Schottky barrier effects in the photocurrent of sol–gel derived lead zirconate titanate thin film capacitors. Appl Phys Lett 76(6):774–776. https://doi.org/10.1063/1.125891
10. Yarmarkin VK, Gol'tsman BM, Kazanin MM, Lemanov VV (2000) Barrier photovoltaic effects in PZT ferroelectric thin films. Phys Solid State 42:522–527. https://doi.org/10.1134/1.1131242
11. Zheng F, Xu J, Fang L, Shen M, Wu X (2008) Separation of the Schottky barrier and polarization effects on the photocurrent of Pt sandwiched $Pb(Zr_{0.20}Ti_{0.80})O_3$ films. Appl Phys Lett 93(17). https://doi.org/10.1063/1.3009563
12. Chen B, Zuo Z, Liu Y, Zhan QF, Xie Y, Yang H, Dai G, Li Z, Xu G, Li RW (2012) Tunable photovoltaic effects in transparent $Pb(Zr_{0.53},Ti_{0.47})O_3$ capacitors. Appl Phys Lett 100(17). https://doi.org/10.1063/1.4709406
13. Uchino K, Poosanaas P, Tonooka K (2001) Photostrictive actuators-new perspective. Ferroelectrics 258(1):147–158. https://doi.org/10.1080/00150190108235377
14. Qin M, Yao K, Liang YC (2008) High efficient photovoltaics in nanoscaled ferroelectric thin films. Appl Phys Lett 93(12). https://doi.org/10.1063/1.2990754
15. Lu Z, Li P, Wan JG, Huang Z, Tian G, Pan D, Fan Z, Gao X, Liu JM (2017) Controllable photovoltaic effect of microarray derived from epitaxial tetragonal $BiFeO_3$ films. ACS Appl Mater Interfaces 9(32):27284–27289. https://doi.org/10.1021/acsami.7b06535
16. Wang MZ, Feng HJ, Qian CX, He J, Feng J, Cao YH, Yang K, Deng ZY, Yang Z, Yiao X, Shao J, Liu S, Zeng XC (2019) $PbTiO_3$ as electron-selective layer for high-efficiency perovskite solar cells: enhanced electron extraction via tunable ferroelectric polarization. Adv Func Mater 29(1):1806427. https://doi.org/10.1002/adfm.201806427

17. Pérez-Tomas A, Xie H, Wang Z, Kim H-S, Shirley I, Turren-Cruz S-H, Morales-Melgares A, Saliba B, Tanenbaum D, Saliba M, Zakeeruddin SM, Gratzel M, Hagfeldt A, Lira-Cantu M (2019) $PbZrTiO_3$ ferroelectric oxide as an electron extraction material for stable halide perovskite solar cells. Sustain Energy Fuels 3(2):382–389. https://doi.org/10.1039/C8SE00451J
18. Bhatnagar A, Roy Chaudhuri A, Heon Kim Y, Hesse D, Alexe M (2013) Role of domain walls in the abnormal photovoltaic effect in $BiFeO_3$. Nat Commun 4(1):2835. https://doi.org/10.1038/ncomms3835
19. Yang SY, Martin LW, Byrnes SJ, Conry TE, Basu SR, Paran D, Reichertz L, Ihlefeld J, Adamo C, Melville A, Chu Y-H, Yang C-H, Musfeldt JL, Schlom DG, Ager JW, Ramesh R (2009) Photovoltaic effects in $BiFeO_3$. Appl Phys Lett 95(6). https://doi.org/10.1063/1.3204695
20. Alexe M, Hesse D (2011) Tip-enhanced photovoltaic effects in bismuth ferrite. Nat Commun 2(1):256. https://doi.org/10.1038/ncomms1261
21. Butler KT, Frost JM, Walsh A (2015) Ferroelectric materials for solar energy conversion: photoferroics revisited. Energy Environ Sci 8(3):838–848. https://doi.org/10.1039/C4EE03523B
22. Belinicher VI, Sturman BI (1980) The photogalvanic effect in media lacking a center of symmetry. Soviet Physics Uspekhi 23(3):199. https://doi.org/10.1070/PU1980v023n03ABEH004703
23. Zhang G, Wu H, Li G, Huang Q, Yang C, Huang F, Liao F, Lin J (2013) New high T© multiferroics $KBiFe_2O_5$ with narrow band gap and promising photovoltaic effect. Sci Rep 3(1):1265. https://doi.org/10.1038/srep01265
24. Ganichev SD, Ivchenko EL, Bel'Kov VV, Tarasenko SA, Sollinger M, Weiss D, Prettl W (2002) Spin-galvanic effect. Nature 417(6885):153–156. https://doi.org/10.1038/417153a
25. Raj A, Sharma S, Singh DV, Kumar A, Chourasia RK, Siqueiros JM, Kumar M (2024) Photovoltaic energy conversion in multiferroic perovskite absorber-based devices via experiment and theoretical calculations. Physica B 673:415504. https://doi.org/10.1016/j.physb.2023.415504
26. Grinberg I, West DV, Torres M, Gou G, Stein DM, Wu L, Chen G, Gallo EM, Akbashev AR, Davies PK, Spanier JE, Rappe AM (2013) Perovskite oxides for visible-light-absorbing ferroelectric and photovoltaic materials. Nature 503(7477):509–512. https://doi.org/10.1038/nature12622
27. Nechache R, Harnagea C, Li S, Cardenas L, Huang W, Chakrabartty J, Rosei F (2015) Bandgap tuning of multiferroic oxide solar cells. Nat Photonics 9(1):61–67. https://doi.org/10.1038/nphoton.2014.255
28. Wu Z, Zhang Y, Ma K, Cao Y, Lin H, Jia Y, Li2 H (2014) Strong visible-light photovoltaic effect in multiferroic $Pb(Fe_{1/2}V_{1/2})O_3$ bulk ceramics. Phys Status Solidi (RRL)–Rapid Res Lett 8(1):36–39. https://doi.org/10.1002/pssr.201308259
29. Raj A, Sharma S, Singh DV, Kumar A, Chourasia RK, Siqueiros JM, Raymond Herrera O, Anshul A, Kumar M (2024). Photovoltaic energy conversion in multiferroic perovskite absorber-based devices via experiment and theoretical calculations. Phys B Condens Matter 673:415504. https://doi.org/10.1016/j.physb.2023.415504
30. Chakrabartty J, Nechache R, Li S, Nicklaus M, Ruediger A, Rosei F (2014) Photovoltaic properties of multiferroic $BiFeO_3/BiCrO_3$ heterostructures. J Am Ceram Soc 97(6):1837–1840. https://doi.org/10.1111/jace.12837
31. Nechache R, Harnagea C, Li S, Cardenas L, Huang W, Chakrabartty J, Rosei F (2015) Bandgap tuning of multiferroic oxide solar cells. Nat Photonics 9(1):61–67. https://doi.org/10.1038/nphoton.2014.255
32. Koch WTH, Munser R, Ruppel W, Würfel P (1975) Bulk photovoltaic effect in BaTiO3. Solid State Commun 17(7):847–850. https://doi.org/10.1016/0038-1098(75)90735-8
33. Yang SY, Seidel J, Byrnes SJ, Shafer P, Yang CH, Rossell MD, Yu P, Chu YH, Scott JF, Ager III JW, Martin LW, Ramesh R (2010). Above-bandgap voltages from ferroelectric photovoltaic devices. Nat Nanotechnol 5(2):143–147. https://doi.org/10.1038/nnano.2009.451
34. Inoue R, Ishikawa S, Kitanaka Y, Oguchi T, Noguchi Y, Miyayama M (2013) Photocurrent characteristics of Mn-doped barium titanate ferroelectric single crystals. Jpn J Appl Phys 52(9S1):09KF03. https://doi.org/10.7567/JJAP.52.09KF03

35. Tu CS, Chen CS, Chen PY, Wei HH, Schmidt VH, Lin CY, Anthoniappen J, Lee JM (2016) Enhanced photovoltaic effects in A-site samarium doped $BiFeO_3$ ceramics: the roles of domain structure and electronic state. J Eur Ceram Soc 36(5):1149–1157. https://doi.org/10.1016/j.jeurceramsoc.2015.12.019
36. Tu CS, Chen PY, Chen CS, Chien RR, Schmidt VH, Lin CY (2018) Photovoltaic conversion and quantum efficiency in perovskite multiferroic ceramics. Acta Materialia 149:248–255. https://doi.org/10.1016/j.actamat.2018.02.043
37. Machado P, Scigaj M, Gazquez J, Rueda E, Sánchez-Díaz A, Fina I, Coll M (2019) Band gap tuning of solution-processed ferroelectric perovskite $BiFe_{1-x}Co_xO_3$ thin films. Chem Mater 31(3):947–954. https://doi.org/10.1021/acs.chemmater.8b04380
38. Zhang Y, Sun H, Yang C, Su H, Liu X (2019) Modulating photovoltaic conversion efficiency of $BiFeO_3$-based ferroelectric films by the introduction of electron transport layers. ACS Appl Energy Mater 2(8):5540–5546. https://doi.org/10.1021/acsaem.9b00722
39. Chen B, Shi J, Zheng X, Zhou Y, Zhu K, Priya S (2015) Ferroelectric solar cells based on inorganic–organic hybrid perovskites. J Mater Chem A 3(15):7699–7705. https://doi.org/10.1039/C5TA01325A
40. Zheng F, Xu J, Fang L, Shen M, Wu X (2008) Separation of the Schottky barrier and polarization effects on the photocurrent of Pt sandwiched $Pb(Zr_{0.20}Ti_{0.80})O_3$ films. Appl Phys Lett 93:172101. https://doi.org/10.1039/C5TA01325A
41. Fan Z, Sun K, Wang J (2015) Perovskites for photovoltaics: a combined review of organic–inorganic halide perovskites and ferroelectric oxide perovskites. J Mater Chem A 3(37):18809–18828. https://doi.org/10.1039/C5TA04235F
42. Pei W, Chen J, You D, Zhang Q, Li M, Lu Y, Fu Z, He Y (2020) Enhanced photovoltaic effect in Ca and Mn co-doped $BiFeO_3$ epitaxial thin films. Appl Surf Sci 530:147194. https://doi.org/10.1016/j.apsusc.2020.147194
43. Renuka H, Garudapalli A, Raman TA, Venkataraman BH, Ramaswamy K, Raju KJ, Goel S, Kundu S (2021) Enhanced photovoltaic response in ferroelectric Ti-doped BFO heterojunction through interface engineering for building integrated applications. Sol Energy 225:863–874. https://doi.org/10.1016/j.solener.2021.08.002
44. Yang SY, Seidel J, Byrnes SJ, Shafer P, Yang CH, Rossell MD, Ager III JW, Martin LW, Ramesh R (2009) Nat Nanotechnol 5:143. https://doi.org/10.1038/nnano.2009.451
45. Agarwal R, Sharma Y, Mendoza F, Morell G, Katiyar RS (2017) Photovoltaic effect in ITO/rGO/BFO/Au heterostructures. ECS Trans 75(31):1
46. Liu H, Sun Y (2007) Substantially enhanced ferroelectricity in Ti doped $BiFeO_3$ films. J Phys D Appl Phys 40(23):7530. https://doi.org/10.1088/0022-3727/40/23/043
47. Cai W, Fu C, Gao R, Jiang W, Deng X, Chen G (2014) Photovoltaic enhancement based on improvement of ferroelectric property and band gap in Ti-doped bismuth ferrite thin films. J Alloy Compd 617:240–246. https://doi.org/10.1016/j.jallcom.2014.08.011
48. Chen B, Zheng X, Yang M, Zhou Y, Kundu S, Shi J, Priya S (2015) Interface band structure engineering by ferroelectric polarization in perovskite solar cells. Nano Energy 13:582–591. https://doi.org/10.1016/j.nanoen.2015.03.037
49. Lines ME, Glass AM (2001) Principles and applications of ferroelectrics and related materials. Oxford University Press. https://doi.org/10.1093/acprof:oso/9780198507789.001.0001
50. Fridkin VM (2001) Bulk photovoltaic effect in noncentrosymmetric crystals. Crystallogr Rep 46:654–658. https://doi.org/10.1134/1.1387133
51. Fridkin VM (1977) The possible mechanism for the bulk photovoltaic effect and optical damage in ferroelectrics. Appl Phys 13:357–359. https://doi.org/10.1007/BF00882610
52. Chynoweth AG (1956) Surface space-charge layers in barium titanate. Phys Rev 102(3):705. https://doi.org/10.1103/PhysRev.102.705
53. Seidel J, Fu D, Yang SY, Alarcón-Lladó E, Wu J, Ramesh R, Ager JW III (2011) Efficient photovoltaic current generation at ferroelectric domain walls. Phys Rev Lett 107(12):126805. https://doi.org/10.1103/PhysRevLett.107.126805
54. Cao D, Xu J, Fang L, Dong W, Zheng F, Shen M (2010) Interface effect on the photocurrent: a comparative study on Pt sandwiched $(Bi_{3.7}Nd_{0.3})Ti_3O_{12}$ and $Pb(Zr_{0.2}Ti_{0.8})O_3$ films. Appl Phys Lett 96(19). https://doi.org/10.1063/1.3427500

55. Schafranek R, Payan S, Maglione M, Klein A (2008) Barrier height at (Ba,Sr)TiO_3/Pt interfaces studied by photoemission. Phys Rev B Condens Matter Mater Phys 77(19):195310. https://doi.org/10.1103/PhysRevB.77.195310
56. Li L, Liu X, He C, Wang S, Ji C, Zhang X, Luo J (2020) A potential Sn-based hybrid perovskite ferroelectric semiconductor. J Am Chem Soc 142(3):1159–1163. https://doi.org/10.1021/jacs.9b11341
57. Hu Y, Florio F, Chen Z, Adam Phelan W, Siegler MA, Zhou Z, Guo Y, Hawks R, Jiang J, Feng J, Zhang L, Wang B, Wang Y, Gall D, Palermo EF, Lu Z, Sun X, Lu T-M, Zhou H, Ren Y, Wertz E, Sundararaman R, Shi J (2020) A chiral switchable photovoltaic ferroelectric 1D perovskite. Sci Adv 6(9):eaay4213. https://doi.org/10.1126/sciadv.aay4213
58. Weng Y, Shen Z, Guo M, Wu F, Li F, Zhu L, Chen C (2019) Electric dipole moment-assisted charge extraction and effective defect passivation in perovskite solar cells by depositing a PCBM: TIPD blend film on a $CH_3NH_3PbI_3$ layer. J Mater Chem C 7(37):11559–11568. https://doi.org/10.1039/C9TC03881G
59. Liu S, Zheng F, Grinberg I, Rappe AM (2016) Photoferroelectric and photopiezoelectric properties of organometal halide perovskites. J Phys Chem Lett 7(8):1460–1465. https://doi.org/10.1021/acs.jpclett.6b00527

Chapter 8
Sustainable Perovskite Multiferroic Materials for Spintronics Applications

Anuja Dhingra, O. P. Thakur, and Raghvendra Pandey

1 Introduction

1.1 Perovskite Structure

Perovskite materials with chemical formula, ABX_3 where A and B are cations and X is an anion are called after the mineral perovskite, i.e., $CaTiO_3$. Here, A is a big cation generally of alkaline earth or an alkali metal, B is a small cation generally of transition metal, and X is an anion generally oxygen (Fig. 1b). The compositions of perovskite structure is highly tunable due to diversity in incorporation of cations at both the A and B positions. The tolerance factor (t), an important characteristics which is determined by spatial configuration of the ion provides information about the perovskite material's stability. Mathematically, it can be given as,

$$t = \frac{r_A + r_X}{\sqrt{2}(r_B + r_X)} \quad (1)$$

where r_A, r_X, and r_B represents the ionic radii of the A, X, and B sites, respectively. The value of tolerance factor between 0.8 and 1.0 indicates stable perovskite structure. The flexibility allows for cation substitution at both sites, namely A and B, therefore results in altered optical, magnetic and electronic properties. This makes perovskite materials an essential candidate for the development of multiferroic compounds.

A. Dhingra · O. P. Thakur
Materials Analysis and Research Lab, Department of Physics, Netaji Subhas University of Technology, New Delhi, India
e-mail: opthakur@nsut.ac.in

R. Pandey (✉)
Green Energy Physics and Solid State Ionics Lab, Department of Physics, ARSD College, University of Delhi, New Delhi, India
e-mail: raghvendra@arsd.du.ac.in

M. Kumar (ed.), *Sustainable Perovskite Multiferroic Materials*, Materials Horizons: From Nature to Nanomaterials, https://doi.org/10.1007/978-981-96-7419-0_8

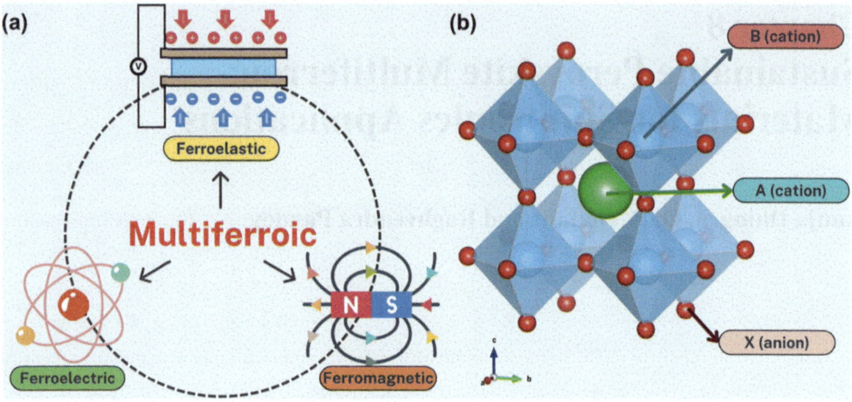

Fig. 1 **a** Schematic illustration of magnetism, electricity, and elasticity coupling in multiferroic and **b** crystal structure of ABX_3-type multiferroic perovskite material

1.2 Multiferroic Materials

Multiferroic materials are the materials which exhibit two or more ferroic properties such as ferroelectricity, ferromagnetism, and ferroelasticity (Fig. 1a). Multiferroic materials are particularly appealing in the devices which employs simultaneous use of electric and magnetic functionalities, like in spintronic devices. The key features of multiferroic materials include are ferroelectricity, ferromagnetism, and ferroelasticity. Ferroelectricity is the property wherein electric dipoles are already oriented and can be displaced by external electric field. Ferromagnetism is the characteristic of magnetic moments of electrons to spontaneously align along the same direction and therefore, leading to net magnetization. This alignment of magnetic moment happens below the Curie temperature and Neel temperature for ferromagnetic antiferromagnetic materials, respectively. Ferroelasticity is the ability of materials to change its shape or form via application of stress, and allows for change in position of atoms. The unique behavior of multiferroic materials is their magnetoelectric coupling, in which magnetic properties are influenced by the application of an electric field and electric properties are influenced by the application of magnetic field. This coupling allows new opportunities of functionalities that are not possible to achieve with traditional ferromagnetic materials alone [6].

1.3 Importance for Spintronics

Spintronics or spin transport electronics is a field that manipulates the electron spin, to process, transport and store data. The main advantage of spintronics is its ability to significantly reduce consumption of energy along with faster processing and storage speed. In perovskite multiferroic materials, the magnetoelectric coupling enables electron spin control via application of field, which can read as well as write data in spintronic devices. Perovskite-based multiferroics provide a means to control electric as well as magnetic properties, therefore results in smaller size, energy-efficient, and capable of complex operations. Perovskite multiferroics can be potentially used in spin-based transistors, magnetic memory including ferroelectric random access memory and magneto-resistive random access memory or ferroelectric random access memory and spin filters in which perovskite materials can block or transmit selective spins in a controlled way.

2 Properties of Perovskite Multiferroics

2.1 Magnetoelectric Coupling

Magnetoelectric coupling is a phenomenon in which magnetization of material can be tuned by application of electric field and polarization of material can be tuned by application of magnetic field. This interaction can take place in two different forms, i.e., direct and indirect coupling [11]. Direct coupling arises when the electric field directly affects the magnetic alignment via interaction of the electric and magnetic dipoles. In few cases, change in the direction of magnetization or orientation of the spin can be induced by the application of electric field. Indirect coupling arises due to intermediate coupling within the ferroelectric and magnetic orders via strain. Electric field causes a change in the crystal lattice, which affects the spin's magnetic alignment. Magnetoelectric coupling mediated via charge generates because of the electrically modified spin-polarized charge present at the interface of ferromagnetic and ferroelectric material. Exchange bias is another form of indirect coupling which is due to uncompensated ferromagnetic spins with ferromagnetic spins at the interface. This causes a shift in the M-H loop and enhancement in coercivity during cooling below Neel's temperature in the presence of magnetic field. Different perovskite such as $BiFeO_3$, $YMnO_3$, and $LuMnO_3$ has depicted magnetoelectric coupling via exchange bias [19, 34, 41, 44]. Table 1 compares room-temperature magnetoelectric coupling coefficient and type of magnetoelectric effect in different perovskite material. A schematic illustration of different types of indirect coupling is shown in Fig. 2.

For perovskite multiferroics, such as bismuth ferrite, the magnetoelectric coupling is quite strong due to the hybridization of the 3d orbitals of the Fe ions with the 2p orbitals of the O ions, which generates a strong interaction within the spin and the charge density.

Table 1 Room-temperature magnetoelectric coupling coefficient at room temperature and type of magnetoelectric effect in perovskite material

Compositions	Magnetoelectric coupling coefficient (mV cm^{-1} Oe^{-1})	Type	References
$GdFe_{0.94}Mg_{0.06}O_3$	1.67	Direct	Shah and Kotnala [31]
$Na_{0.49}Bi_{0.51}Ti_{0.98}Fe_{0.02}O_3$	7.38	Direct	Ruth et al. [29]
$BiFeO_3$ (Thin film)	3000	Direct	Wang et al. [40]
$Pb(Fe_{0.5}Ta_{0.5})_x(Zr_{0.53}Ti_{0.47})_{1-x}O_3$	12	Direct	Rahman et al. [26]
$CoFe_2O_4$–$BiFeO_3$	78	Indirect via strain	Rafique et al. [25]
$Pb(Fe_{0.5}Nb_{0.5})O_3/Ni_{0.65}Zn_{0.35}Fe_2O_4/Pb(Fe_{0.5}Nb_{0.5})O_3$	383	Indirect via charge	Pradhan et al. [24]
$0.7Pb(Mg_{1/3}Nb_{2/3})O_3$–$0.3PbTiO_3$/metglass	751×10^3	Indirect via strain	Rajaram Patil et al. [27]
$PbFe_{0.5}Nb_{0.5}O_3$–$Co_{0.6}Zn_{0.4}Fe_{1.7}Mn_{0.3}O_4$	26.78	Indirect via strain	Bhoi et al. [2]
$[Pb(Mg_{1/3}Nb_{2/3})O_3]_{0.7}$-$[PbTiO_3]_{0.3}$/MGO/FeGa	4.5×10^3	Indirect via strain	Begue and Ciria [1]

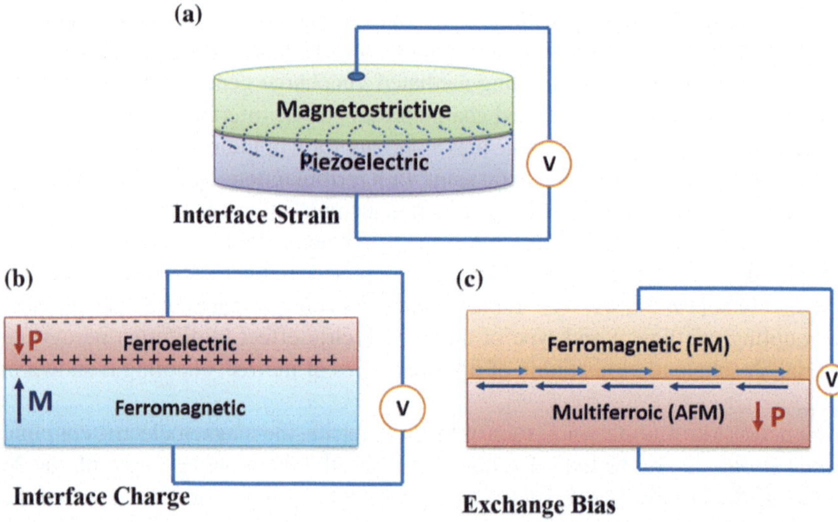

Fig. 2 Schematic illustration of **a** strain-mediated coupling in magneto-strictive and piezoelectric interface, **b** charge-mediated coupling in ferroelectric and ferromagnetic interface and **c** exchange bias coupling in ferromagnetic and anti-ferromagnetic interface. Reprinted with permission from Ref. [11]

2.2 Challenges in Coupling

Achieving high-quality magnetoelectric coupling in perovskite multiferroics is a significant challenge [18]. Different limiting factors for effective coupling in these materials are (i) **Temperature Stability**: Perovskite multiferroics generally shows magnetoelectric coupling at very low temperatures, but maintaining a strong coupling at room temperature is a critical limitation for practical applications [30]. The Neel temperature and Curie temperature limit the operational range of temperature, making it difficult for practical applications. (ii) **Separation of Phase**: In few perovskite multiferroics, the coexistence of ferromagnetic and ferroelectric phase is not be uniform. This results in separation of phase which results in regions with distinct electric and magnetic characteristics. Such non-uniformity may degrade the coupling and limit the performance of the material. Achieving a homogeneous phase so that both characteristics interact efficiently requires a precise control during the synthesis process. (iii) **Material Defects**: The presence of various defects, like oxygen vacancies or grain boundaries, may significantly impact the coupling phenomenon. Local polarization variations can be introduced via defects, which changes the behavior of material and results in effecting the magnetoelectric coupling behavior. To minimize defects, careful synthesis or doping should be done to improve the coupling behaviour [28] and (iv) **Strain and Lattice Mismatch**: Strain inside the crystal lattice, often due to the thermal expansion plays an important role in affecting the magnetoelectric coupling. Strain may modify the alignment of the magnetic moments and ferroelectric dipoles, resulting in altered coupling strength or even complete decoupling of both the properties. Managing strain via epitaxial growth or through designing composite materials is a current limitation.

To overcome these limitations, different strategies have been explored by researchers [47] such as doping with foreign elements in order to stabilize the coupling, engineering nanostructures so as to reduce improve material homogeneity and defects, synthesizing thin-film and multilayered structures to control strain.

3 Green Chemistry Approaches for Synthesis for Perovskite Multiferroics

Green chemistry is a sustainable field that focuses on the use of environmentally friendly as well as sustainable methods for the synthesis. Different green chemistry approaches are now being applied to the synthesis of perovskite multiferroics, including:

3.1 Sol–Gel Method

The sol–gel process is a most commonly used technique for synthesizing powders as well as thin films of perovskite multiferroics. In this methods, precursor chemicals are dissolved in a solvent to produce a sol, which is then heated and subsequently converts to gel. It is then calcined to produce a solid material. When synthesized in an eco-friendly way using solvent as water or non-toxic medium, for instance, the sol–gel method can be more sustainable as compared to other high-temperature synthesis methods. This method allows for control over size of the particle, homogeneity of structure and microstructure which results in maximizing the magnetic and electric properties of perovskites multiferroics [12].

3.2 Hydrothermal or Solvothermal Method

The hydrothermal or solvothermal synthesis route include the material growth within water or other solvents under high temperature and pressure inside a sealed autoclave. It is a cleaner, low energy, and free from toxic solvents, therefore ideal for the synthesis of multiferroic complexes [4]. For perovskite multiferroics, hydrothermal methods make it possible to grow quality crystals and also limits the toxic solvents and reagents. Moreover, method of hydrothermal synthesis can be applied effectively for industrial scale synthesis which makes it a optimal candidate for environment friendly synthesis of perovskite multiferroic materials.

3.3 Room-Temperature Fabrication

Synthesizing at room temperatures is a concept that is attracting significant attention into the field of green chemistry. This encompasses the synthesis of perovskite multiferroics without using of high temperatures, therefore significantly reducing the consumption of energy. Room-temperature chemical vapor deposition, chemical solution deposition, and other similar methods are currently being tested for high-quality film growth without the use of high temperatures or harsh chemicals.

By reducing the need of energy as well as toxic materials, these methods contribute significantly to the sustainability of the perovskite multiferroic synthesis process [7].

4 Characterization Techniques for Perovskite Multiferroics

Characterization is an important part to understand the properties and behavior of perovskite multiferroics. As these materials shows both magnetic and electric characteristics, it is essential to employ different number of techniques to study their structural, microstructural, electric, magnetic, and coupling behaviors. This section will provide information about the prominent techniques available for the analysis of behavior of these materials, providing information about their crystal structure, bonding characteristics, morphology, magnetic characteristics, ferroelectric behavior, and magnetoelectric coupling. The structural characterization employing XRD identifies the perovskite phase and also distinguishes between different phases, like rhombohedral, tetragonal, or orthorhombic phases, which can alter the ferroelectric and magnetic properties of perovskite material. Shifting in diffraction peaks or changes in intensity of peak phase transitions, impurities or lattice strain, which can affect the performance of material [5]. Use of Raman spectroscopy and X-ray photoelectron spectroscopy (XPS) provide information about the bonding characteristics, which influence both magnetism and ferroelectricity. The study of surface morphology using Scanning Electron Microscopy, allows to observe the surface features including grain size, porosity, and crack formation which are crucial to study the mechanical properties and stability. Magnetic characterization using Vibrating Sample Magnetometry or Superconducting Quantum Interference Device (SQUID) Magnetometry provides crucial insights of magnetic properties of perovskite multiferroics, including their ferrimagnetic, ferromagnetic, or antiferromagnetic behaviors. The magnetic hysteresis loop provides information related to retentivity coercivity as well as saturation magnetization which is crucial for studying magnetic memory devices. Mössbauer Spectroscopy is an effective method to study the local magnetic environment of specific elements in perovskite multiferroics. It depends on the resonant absorption of gamma rays by certain isotopes (e.g., Fe-57) and also provides detailed information about the hyperfine interactions between the nuclear spins and the electron spins. The polarization–electric field loop is an important technique to analyze the ferroelectric property of perovskite multiferroics [32]. The P-E loop shape can indicate the ferroelectric properties, like the ability of material to sustain a stable state of polarization without any significant fatigue. Dielectric measurements provide information about ability of material to store electrical energy and its response to an applied electric field. It helps to determine the Curie temperature, i.e., the material's ferroelectric phase transition temperature and its dielectric stability under different conditions, such as changing temperatures or applied fields [8]. Piezoelectric tests investigate the perovskite multiferroics ability to generate an electrical charge with the application of mechanical stress. As many perovskite materials exhibit both ferroelectricity as well as piezoelectricity, piezoelectric measurements provide information about the strain-electric coupling [46]. Techniques such as d33 piezoelectric measurements (where the piezoelectric charge coefficient is determined) and laser interferometry are used to analyze the material's response to external mechanical forces. These measurements are critical for applications in sensors, actuators, and energy harvesting. Magnetoelectric coupling is a crucial technique for perovskite multiferroics, to understand how the magnetic and electrical properties influence each other. Different methods are used to study the coupling behavior like magnetoelectric susceptibility. It refers to the response of material to changes in the applied magnetic field and electric field. By measuring how the magnetization or polarization changes with the application of a electric field or magnetic field, it is possible to determine the nature and strength of the magnetoelectric coupling in a given perovskite material. Another important method is to study magnetoelectric coupling is the Ferroelectric Polarization Reversal in the Presence of Magnetic Fields. It determines how the polarization reverses with the application of external magnetic field. By monitoring the polarization switching with varying magnetic fields, the coupling strength can be determined between the magnetic and ferroelectric parameters. This often includes measuring of P-E loops with the application of both electric and magnetic fields, which helps to determine the potential for spin-polarized current generation as well as multiferroic memory devices.

5 Spintronics and the Role of Multiferroics

Spintronics, refers to the field that manipulates spin of electron for information processing, offering development of novel next-generation devices. Conventional electronics materials, relies on the electronic charge for storing and processing of information, whereas spintronics make use of the intrinsic angular momentum of electrons, giving rise to several advantages, including improved energy efficiency, enhanced processing of data and reduction in heat generation. Perovskite multiferroics, because to their coupled electric and magnetic characteristics, makes them a superior candidate for spintronic applications. This section includes overview of spintronics fundamentals, potential of multiferroic perovskite in spintronic applications, and enhancement of device functionality using magnetoelectric coupling.

5.1 Spintronics Overview

Spintronics, known as spin transport electronics, exploits quantum mechanical characteristic of electronic spin and manipulate it for storage, processing, and transmission of data. Traditional electronics works on the principle of electron movement, i.e., charge for conveying information, whereas spintronics makes use of electron spin, which can be oriented in either up or down state. The state of spin can be manipulated and therefore, employed as a binary bit, similar to charge but with several advantages like lower consumption of power, high speed, non-volatility, integration, and miniaturization. Conventional electronic materials require a continuous movement of electron, i.e., generation of current to function, which causes loss of energy due to heating and resistance. While, spintronic devices manipulates spin of electron causing significant reduction in dissipation of energy, causing lower consumption of power. Spin-based phenomena occurs rapidly on timescales of femto- to pico-seconds, therefore, spintronic devices have enormous potential for ultra-data processing and storage of information. Spintronic devices retain data even in the absence of power, therefore making them an ideal candidate for memory storage devices. This also results in instant-on and less boot-up period of time. Spintronic devices can be integrated with traditional semiconductor technologies, therefore resulting in further miniaturization and improved functionality, specifically in artificial intelligence and quantum computing applications.

The spintronics devices include materials having high spin polarization, (ferromagnetic metals), spin injectors, i.e., devices that introduce spin into a system, spin filters, i.e., devices that control polarization of spins, and spin detectors, i.e., devices that read states of spin.

5.2 Potential of Multiferroic Perovskite in Spintronic Applications

Multiferroic materials with perovskite structures, possess enormous potential for application in spintronic devices due to their ability of coupling magnetic and electrical field. In multiferroic materials, the magnetic and electric magnetic properties are linked through magnetoelectric effect, that means magnetic field can influence polarization, and vice versa. This coupling is possess enormous opportunities for spintronic devices that needs the manipulation of spin as well as charge.

5.2.1 Spin Filters

Spin filters devices allow selective electrons with a certain orientation of spin (for example, up or down) to pass through and blocks electrons with opposite orientation of spin. In perovskite multiferroics, the electrical properties can be manipulated by the applying magnetic field, which results in polarization of spin, therefore acting like a spin-dependent transport medium. For example, ferromagnetic perovskites are employed in spin valve devices, in which the relative orientation of two ferromagnetic layers determines the device resistance. By employing multiferroic materials, tunable spin filtering can be achieved depending on the applied electric field, therefore enhancing the efficiency and versatility of the device.

5.2.2 Memory Storage

Multiferroic perovskites are suitable the application of non-volatile memory storage devices. The ability to multiferroic perovskite to write and read data using both polarization as well as magnetization results for enhanced storage densities and relatively faster time compared to conventional traditional magnetic memory materials. Magnetoelectric random access memory (MeRAM) where the magnetization can be switched by the application an electric field, while maintaining the stable spin state, offers both non-volatility as well as low consumption of power consumption. Kosub et al. [17] discovered an antiferromagnetic magnetoelectric memory in a Cr_2O_3/Pt heterostructure where Pt layer affects the surface magnetization of Cr_2O_3, which can be measured using Hall measurement (Fig. 3). Surface magnetization was determined by the antiferromagnetic domain state, which can be altered by applying an electric and magnetic field simultaneously.

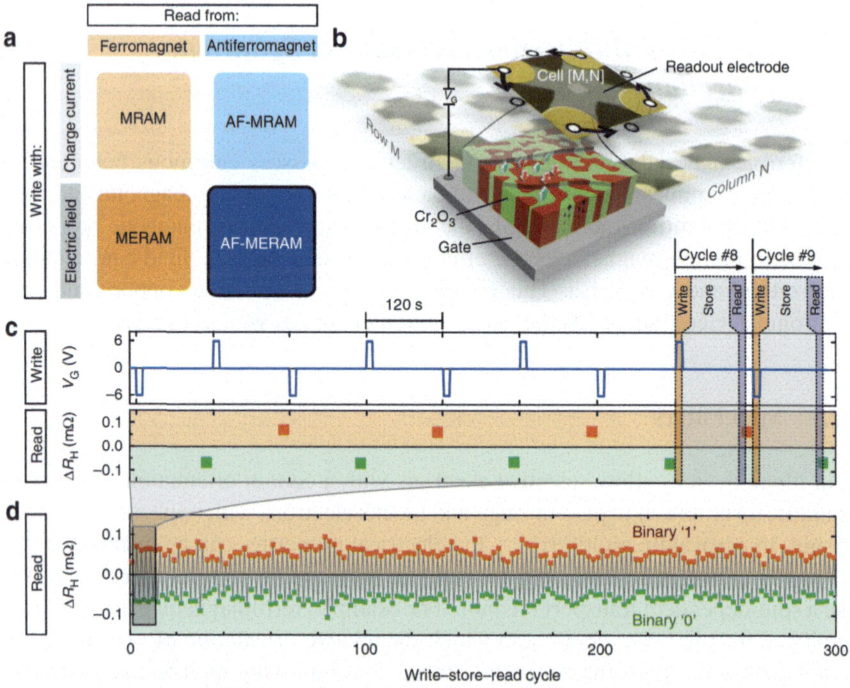

Fig. 3 **a** Different types of non-volatile magnetic random access memory elements, **b** illustration of memory cell within device matrix. Here arrow represents contact permutation for obtaining offset free Hall readings, **c** random access memory in which information is written using voltage pulse and stored in antiferromagnetic order. The magnetic state is read after writing stimulus is removed and **d** Behavior of device after 300 write-store-read cycles. Reprinted with permission from Ref. [17]

5.2.3 Logic Gates

Logic gates, the fundamental building blocks of digital electronics, can also benefit from the unique properties of multiferroics. The magnetoelectric coupling allows for the production of novel logic gates that perform computations based on the interaction between charge, spin, and electric polarization. For instance, a spin-polarized field-effect transistor (FET) based on a multiferroic material could potentially operate using both the spin and charge of electrons, resulting in new opportunities for energy-efficient computation. Li et al. [21] proposed spin valve device based on two-dimensional $CrI_3/BiFeO_3$ heterostructure. (Fig. 4). The bias electrodes (Gate 1 and Gate 2) read the electrical resistance, and gating electrode (Gate 3) regulates the electron polarization.

5.3 Spin–Orbit Coupling and Magnetoelectric Effects

The performance of spintronic devices can be significantly enhanced by spin–orbit coupling, a phenomenon that arises from the interaction between an electron's spin and its orbital motion in an electric field. Spin–orbit coupling plays an important role in spintronic devices, including the spin current generation and the manipulation of spin polarization. In perovskite multiferroic with strong spin–orbit coupling, spin currents can be manipulated and controlled by the electric field application. In addition to spin–orbit coupling, the magnetoelectric effects in multiferroics provide an additional functionality. Therefore, making them suitable for magnetoelectric memory devices, spin-polarized transport, and reversible magnetization switching.

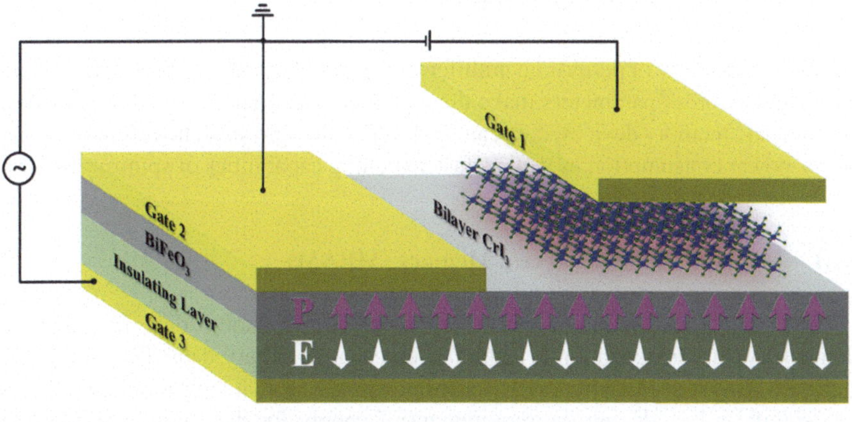

Fig. 4 Spin valve device based on two-dimensional $CrI_3/BiFeO_3$ heterostructure. The bias electrodes (Gate 1 and Gate 2) read the electrical resistance, and gating electrode (Gate 3) is used to the regulate the electron polarization. Reprinted with permission from Ref. [21]

The combination of magnetoelectric effects and spin–orbit coupling in multiferroics perovskite leads (i) **Spin Hall Effect (SHE)**: In SHE, electric current generates a transverse spin current due to the interaction of spin–orbit. The interplay between the magnetic field and electric field can improve the SHE, therefore resulting in efficient generation of spin current and manipulation. (ii) **Rashba Effect**: Rashba effect causes a spin state splitting in materials with strong spin–orbit coupling, which can be manipulated by application of electric field. The Rashba effect can be utilized for enhanced spin manipulation in spintronic devices like or spin FETs or spin valves [20]. (iii) **Magnetoelectric Switching**: The magnetoelectric effect results in switching of the electrical state using a magnetic field. When magnetoelectric effect is combined along spin–orbit coupling, spin polarization can be manipulated in the material, resulting their applications in spin logic gates, spin-based memory devices, and spin FETs.

6 Applications of Sustainable Perovskite Multiferroics in Spintronics

Sustainable perovskite multiferroic materials possess enormous potential for the next-generation energy-efficient spintronic devices. The magnetoelectric coupling, in perovskite materials offer enormous potential for numerous applications in spintronics, including magnetic memory devices, nano-oscillators, logic devices, and sensors (Fig. 5).

6.1 Spintronic Memory Devices

The unique property of perovskite multiferroics particularly ability to couple electric and magnetic order parameters make them an ideal candidate for memory devices. In spintronic memory devices, data can be stored in the spin state, therefore resulting in low-power consumption as well as fast switching capabilities of spintronics [45].

6.1.1 Magnetic Random Access Memory (MRAM)

MRAM is a non-volatile memory technology which stores data in magnetic elements, with magnetization orientation representing binary data. The perovskite multiferroics-based MRAM devices can offers improved switching efficiency, therefore significantly reducing the requirement of energy for data writing in MRAM cells. This makes them a suitable candidate for large-scale memory devices, where less power consumption of individual memory cells plays a critical role. Another

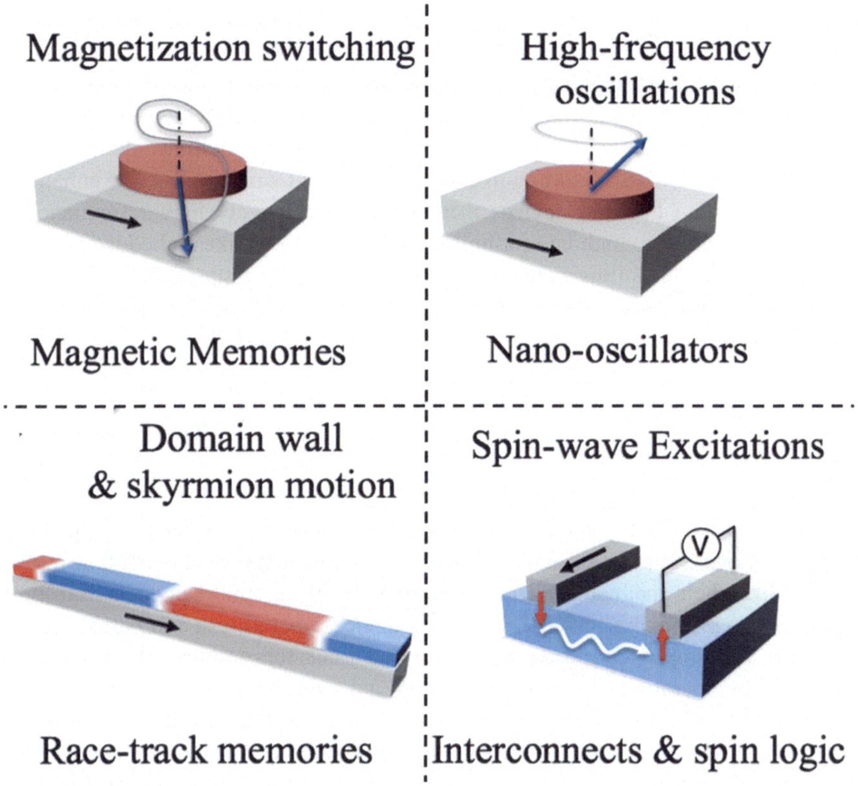

Fig. 5 Application of sustainable perovskite multiferroics in spintronics. Adapted from Ref. [23]

advantage is enhanced writing speed due to fast magnetization reversal on the application of external electric field, therefore resulting in high-speed data writing. The rapid switching behavior is particularly essential in high-performance computing applications where fast memory access is needed. Perovskite-based MRAM devices retain information even after power is switched off like traditional MRAM devices. The non-volatile behavior makes these MRAM memory more stable and reliable. Synthesis techniques, including solvent-free methods and the use non-toxic and abundant precursors, adds advantage to the development of sustainable MRAM technology. Therefore, perovskite multiferroics-based MRAM devices significantly enhance the speed, efficiency, and sustainability, positioning them as a leading material for future memory devices.

6.1.2 Ferroelectric Random Access Memory (FeRAM)

Ferroelectric random-access memory (FeRAM) utilizes the polarization state of ferroelectric materials for data storage. The perovskite multiferroics-based FeRAM devices offers reduced power consumption due to their ability of switching magnetic moment and ferroelectric polarization employing a single external field. This results in energy-efficient reading and writing operations, therefore reduced overall power consumption in contrast to traditional FeRAMs that particularly depends on ferroelectric polarization switching. Another advantage is faster switching of both magnetization and polarization, resulting in enhanced read and write cycles. The enhanced speed can lead to FeRAM devices that operate at relatively higher frequencies, making them suitable candidate for high-speed applications like embedded systems and mobile. By controlling both the magnetic and electric states, these materials based FeRAM devices ensures that data remains intact under different environmental conditions, therefore improving the overall reliability of FeRAM devices.

6.2 Logic Devices

Spintronics possess enormous potential to revolutionize logic circuits by enabling energy-efficient computation and switching. The unique features of perovskite multiferroics like their magnetoelectric coupling as well as spin-dependent transport makes them a promising candidates for logic devices operating at comparably lower power and with improved speed [36]. Spin-based transistors employs electron spin to perform logic operations, in contrast to traditional semiconductor transistors, which depends on the electron charge [10]. The perovskite multiferroics-based spin-based transistors offers advantage of energy-efficient switching as the ability to manipulate spin of electron with the application of electric fields in multiferroics leads to low-energy switching in spin-based transistors. This leads to the fabrication of sustainable logic circuits due to low-power consumption in large-scale computing devices. The coupling between the magnetic and electric and states allows for fast electron-spin manipulation, therefore improving the speed of logic gate operation. Spin-based transistors that use multiferroic materials can have the potential to achieve speeds that are on par with or even surpass traditional semiconductor transistors. The non-volatile behavior of perovskite multiferroics ensures that spin-based transistors retain their state without requiring continuous power, therefore leading to the development of ultra-low-power logic devices. Further, perovskite multiferroics-based spin-based transistors leads to further miniaturization of logic circuits, therefore enabling the creation of more compact and efficient devices for next-generation electronics industry.

6.3 Spin–Orbit Torque (SOT) Devices

Spin–orbit torque (SOT) devices make use of the interaction between spin of electron and its orbital motion to manipulate the magnetization of a material. Perovskite multiferroics are suitable candidates for use in SOT devices because of their strong spin–orbit coupling and magnetoelectric effects [15]. Perovskite multiferroics can efficiently switch the magnetization of a ferromagnetic layer in SOT devices, therefore requiring relatively low voltage and current. This leads to the development of energy-efficient SOT devices and memory cells. Further, integration of perovskite multiferroics with traditional semiconductor technologies leads to the development of hybrid systems which combines the advantages of both materials leading to novel device architectures which can combine the scalability of semiconductors with the high speed of spintronics. Therefore, incorporation of perovskite multiferroics into SOT devices is a promising approach for the fabrication of high-performance and energy-efficient spintronic logic devices.

6.4 Sensors and Actuators

Perovskite multiferroics have also emerged as a leading candidate in applications including sensors and actuators, where the magnetoelectric property can be utilized to fabricate highly sensitive and low-power devices [37]. Magnetoelectric sensors implies magnetoelectric effect, where the external electric field influences the magnetic properties of material, resulting in detection of external magnetic fields. Perovskite multiferroics are ideal choice for magnetoelectric sensors because of their strong magnetoelectric coupling. The large magnetoelectric response in results in detection of even small magnetic fields, which is necessary for a variety of applications such as biosensing, magnetic field monitoring, and geophysical measurements. Perovskite magnetoelectric sensors can operate with minimum input energy, making them a suitable choice for portable or battery-powered devices where low-power consumption is necessary. The unique characteristics of perovskite multiferroics enable the miniaturization of magnetoelectric sensors, resulting in efficient and compact that can be employed in a wide range of sensing applications, such as consumer electronics or industrial monitoring systems.

6.5 Microactuators

Microactuators are devices which convert electrical energy into mechanical motion at the microscale [38]. The presence of magnetoelectric coupling in perovskite multiferroics makes them a promising candidate for microactuator applications. The ability to control polarization as well as magnetization with electric fields results in precise and tunable mechanical actuation at the nano-scale, which is necessary for applications in microelectronics, MEMS (Microelectromechanical Systems), and nano-robotics. Microactuators based on multiferroic materials operate with low input power, making these materials an ideal choice for portable and energy-efficient applications, like wearable technology or implantable medical devices. The perovskite multiferroics integration with microactuators may enable their use in combination with different electronic systems, like sensors, to generate fully integrated micro-electromechanical systems for advanced applications.

7 Limitations

Despite the enormous potential of perovskite multiferroics in the field of spintronic, different significant limitation exists in practical integration and material development. Addressing these limitations is necessary for widespread adoption of perovskite multiferroics in real-world device application. In this section, we present some of the key limitations, such as issues of scaling and integration, material stability, energy efficiency versus performance trade-offs, and tuning of material properties along with the solution.

7.1 Scaling and Integration

While perovskite multiferroics show immense potential for spintronic applications, one of the key limitation is scaling up the synthesis method for large-scale production of devices without compromising the quality of material. Scaling the production of these materials without compromising with their properties poses different challenges such as (i) **synthesis of good-quality thin films**: The synthesis of good-quality thin films and their heterostructures, which are necessary for device integration, is particularly difficult. Different techniques such as pulsed laser deposition or sputtering are generally used for the synthesis of thin films of perovskites [13], but maintaining purity, uniformity, and crystallinity at mass production scales can be difficult. Inconsistent thickness of film, defects, and crystallographic alignment may lead to performance variations. Another limitation is (ii) **large-area device fabrication**: For perovskite multiferroics-based spintronic devices to be commercially available, large-area synthesis techniques must be developed. Current synthesis techniques for thin-film deposition are generally not cost-effective for large-scale production, and the cost associated with scaling these processes for industrial application can be prohibitive. Another limitation is (iii) **Integration with existing technologies**: Integrating perovskite multiferroics materials into existing spintronic and semiconductor device architectures is still a challenge. The compatibility of these materials with current synthesis techniques, such as those used for complementary metal–oxide–semiconductor (CMOS) or other semiconductor devices, must be ensured. The requirement for precise alignment and compatibility with a variety of materials at the nanoscale presents a significant barrier for successful integration.

7.1.1 Solutions and Future Directions

Developing more scalable deposition techniques, like chemical vapor deposition or solution-processing techniques, may reduce costs and therefore, improves the scalability of perovskite synthesis. Research on hybrid integration strategies, in which perovskite multiferroics are combined with conventional semiconductor materials, may lead to more efficient device with high performance. Exploring the use of roll-to-roll techniques for large-area device manufacturing may reduce costs and increase the scalability of perovskite-based spintronic devices.

7.2 Material Stability

The stability of sustainable perovskite multiferroic materials is a major concern, especially when these materials are employed in devices which operate over extended periods like varying operational and environmental conditions [16]. Perovskites, particularly with complex compositions, may experience degradation or transition of phase that degrade their performance. Perovskite multiferroics are highly sensitive to different environmental factors like temperature, humidity, and oxygen exposure. For example, a number of perovskite-based materials are highly prone to moisture-induced degradation, which results in the loss of ferromagnetic and ferroelectric properties. Similarly, exposure to high temperature range may cause phase transitions which affects the magnetoelectric coupling strength. The multiferroic materials synthesis generally leads to phase separation or defect formation which degrade the stability of material. Inhomogeneous distribution of metal ions as well as lattice distortions may cause localized regions where the polarization and magnetism are weak or even absent. These defects may adversely affect the magnetoelectric coupling, making it harder to achieve reliable and consistent performance in devices. For spintronic devices that depend on perovskite multiferroics, should ensure long-term reliability. However, perovskite multiferroics materials may experience fatigue with repeated cycling in devices like MRAM or FeRAM, where the switching of polarization or magnetization is a frequent phenomenon. Prolonged cycling may cause degradation of the ferromagnetic or ferroelectric properties, leading to a decrease in performance.

7.2.1 Solutions and Future Directions

Improving the stability of perovskite multiferroics via use of protective coatings or encapsulation techniques may help shield them from environmental stress. Researchers are focusing on the development of stable perovskite compositions that can withstand high humidity and temperatures, ensuring better long-term performance and durability.

7.3 Energy Efficiency Versus Performance

Balancing the sustainability goals of energy efficiency and low environmental impact along-with the high-performance requirements of spintronic devices is a critical limitation in the development of perovskite multiferroics-based devices. Perovskite materials are attractive because of their potential for low-power consumption in devices and low-energy synthesis, but there are trade-offs in performance that need to be carefully considered such as (i) **Power Consumption Versus Switching Speed**: In spintronics, fast switching speeds are necessary for high-performance devices. While perovskite multiferroics offer lowest switching energy requirements among present materials, achieving ultra-high-speed operation may require higher input power or more complex device architectures. Balancing between energy efficiency and the demands for high-speed operation remains a critical challenge. Another limitation is the (ii) **Toxicity of Precursors and Materials**: Many perovskite materials, specifically lead-based perovskite material, consist of toxic elements which may pose risk to environmental and health during manufacturing, use, or disposal. Although different sustainable synthesis techniques have already been developed for some of the perovskite materials, eliminating toxic elements like lead while maintaining high performance remains a key limitation [9]. (iii) **Device Performance Versus Environmental Impact**: While perovskite materials show high potential for low-carbon footprint production, the trade-off with-in achieving high performance in spintronic devices as well as reducing the environmental impact of material synthesis is still an active subject of research. The complexity of achieving both goals simultaneously makes the development of truly sustainable spintronic devices difficult.

7.3.1 Solutions and Future Directions

Research is being conducted to find lead-free perovskite materials and alternative environmentally friendly materials without compromising with the performance. For example, bismuth and calcium-based perovskites are being explored as alternatives to lead-based perovskites. Different techniques in low-power spintronic devices, like the development of more efficient switching mechanisms such as spin–orbit torque or the incorporation of energy-efficient ferroelectric materials, may help in maintaining a better balance between performance and power consumption. Developing novel device architectures that may leverage the unique properties of perovskite multiferroics to lessen power consumption without compromising performance could pave the way for high-performance, greener spintronic devices.

7.4 Tuning Material Properties

Another significant challenge in the development of sustainable perovskite multiferroics is the difficulty in tuning their properties, like Curie temperatures, band gap, and strength of magnetoelectric coupling. The performance of perovskite-based spintronic devices is highly dependent on the precise control of these properties, which may vary greatly depending on fabrication conditions, material composition, and external factors. The strength of magnetoelectric coupling in perovskite multiferroic material is a critical factor in determining performance of device. Achieving a stable and strong coupling requires control over the material's composition and crystal structure. Small variations in the synthesis conditions or introduction of defects may lead to significant variations in the strength of coupling, making it challenging to achieve consistent performance. In most of the perovskite multiferroic materials, the band gap plays an important role in determining their optical and electrical properties. Tuning the band gap is necessary for optimizing device performance, especially for applications in sensors and logic devices. However, controlling the band gap in perovskite materials is not easy, as it is often sensitive to a number of factors such as crystal size, doping, and strain [14]. Another parameter that must be carefully controlled is the Curie temperature, which defines the transition from ferromagnetic to paramagnetic behavior. Achieving high Curie temperatures in perovskite multiferroics is necessary to ensure the reliability and stability of spintronic devices, particularly under high-temperature operating conditions [42]. However, this can be challenging to achieve, and most of the perovskite multiferroics exhibit relatively low Curie temperatures compared to traditional ferromagnetic materials.

7.4.1 Solutions and Future Directions

The development of novel doping strategies or heterostructures formation may enhance the magnetoelectric coupling strength or improving thermal stability may help to optimize these materials for spintronic applications. Advances in computational materials design may provide better insights into how to fine tune the properties of perovskite multiferroics for optimal results, enabling more precise control over their crystal structure and composition. Exploring high-throughput screening techniques to identify new multiferroic compositions with optimized properties may accelerate the discovery of materials with tailored performance for specific spintronic applications.

8 Future Directions and Emerging Trends

As perovskite multiferroics continue to gain prominence in spintronic applications, future research will focus on discovering new materials, engineering hybrid structures, utilizing nanotechnology, and exploring integration with quantum technologies. These emerging trends offer exciting possibilities for overcoming the current limitations and advancing the practical use of perovskite multiferroics in next-generation electronic devices. This section explores these promising directions in detail.

8.1 New Materials Discovery

The search for novel perovskite compositions is a crucial avenue of research aimed at improving the fundamental properties of multiferroic materials, particularly in terms of high Curie temperatures, strong magnetoelectric coupling, and long-term stability. While existing perovskite materials show significant promise, further advances in material discovery are needed to optimize performance and tailor materials to the specific needs of spintronic applications [22]. Most of the perovskite multiferroics possess relatively low Curie temperatures, limiting their range of operation. The development of novel materials with higher Curie temperatures would allow these materials to function in a wider operational range like high-temperature environments. Researchers are exploring novel compositions, doping strategies, and different synthetic techniques to increase the Curie temperatures of perovskite multiferroics. For example, perovskite structures based on titanium, bismuth, or other alternative metals may offer higher magnetic ordering temperatures without compromising with the ferroelectricity. Strong magnetoelectric coupling is necessary for spintronic devices that require efficient conversion between electric and magnetic states. Although some perovskites show reasonable coupling, the performance of multiferroic materials could be further improved. Novel perovskite compositions that exhibit enhanced magnetoelectric coupling such as those incorporating rare-earth elements or transition metals could help unlock new functionalities. Combinations of metal ions with different oxidation states could be explored to engineer the desired coupling effects. Some of the multiferroic perovskites shows complex phase transitions between different ferroic states. Controlling these transitions is necessary to develop multiferroics with more robust and tunable properties. By carefully adjusting the ionic composition, synthesis conditions, and strain, researchers may discover new materials with more desirable phase transitions temperatures that occur at or under specific conditions.

8.2 Hybrid Materials

The development of hybrid materials, like combining organic and inorganic elements or integrate other ferroelectric and ferromagnetic materials into perovskite structures, is an emerging trend with great significance for enhancing multiferroic properties. Hybrid perovskite materials may offer synergistic effects that enhance performance in a number of ways that single-phase materials may not be able to achieve. The integration perovskite multiferroics with organic materials may result in materials with enhanced flexibility, tunable properties, as well as lower processing costs. Organic materials may provide additional degrees of design freedom, resulting in fine-tuning of properties like magnetic behavior, electrical conductivity, and overall stability [35]. Additionally, the use of organic molecules may improve the processability of perovskite films, making these materials more suitable for large-area applications such as flexible electronics. Another way is hybridization between ferromagnetic and ferroelectric material. Combining perovskite materials with different ferromagnetic or ferroelectric materials may enhance the magnetoelectric coupling and stability of hybrid systems [39]. For example, integrating perovskite multiferroics with different ferromagnetic oxide materials, like yttrium iron garnet (YIG), may result in composite materials with higher saturation magnetization, improved magnetization control, and enhanced polarization switching properties. These hybrid materials could also possess tailored properties that are ideal for spintronic applications. The combination of perovskite materials with two-dimensional (2D) materials such as transition metal dichalcogenides (TMDs) or graphene could result in hybrid structures that may benefit from the unique electronic, magnetic, and optical properties of 2D materials [3]. These materials may facilitate faster electron and spin transport, improved magnetoelectric coupling, and lessen energy losses, opening up novel possibilities for high-performance, scalable spintronic devices.

8.3 Nanoengineering

Nanoengineering of perovskite multiferroics offers a pathway to overcome some of the shortcomings of bulk materials. By controlling the morphology, size, and interface properties of the materials at the nanoscale, researchers can optimize magnetoelectric coupling strength, improve efficiency of device, and enhance the scalability of these materials for practical applications. Nanoscale engineering of perovskite materials, includes the fabrication of nanoparticles, quantum dots, nanowires, and thin films, which can significantly improve their magnetoelectric coupling strength. As nanoscale material exhibit enhanced surface-to-volume ratios, therefore leads to more efficient coupling between the electric and magnetic domains. Furthermore, nanostructures may also reduce energy losses by reducing defect formation and phase separation, which may occur in bulk materials. At the nanoscale, quantum confinement effects come into picture, altering the magnetic, electronic, and ferroelectric properties of materials [43]. By designing perovskite multiferroics in the nanometer scale, it is possible to fine-tune material properties for a varied number of applications. For example, quantum dots or nanorods of perovskite materials may exhibit improved switching speeds or more robust magnetoelectric coupling as compared to bulk counterparts. The interfaces between different perovskite nanostructures, as well as between other materials and perovskites may influence the overall performance of the material. Understanding and controlling these interfaces is necessary for optimizing the magnetoelectric coupling strength and achieving high-performance real-life working devices. Hybrid perovskite-nanomaterial heterostructures may also be engineered to exploit unique interfacial effects, leading to more efficient spintronic devices with lower power consumption.

8.4 Integration with Quantum Technologies

Another rapidly emerging field in spintronics is quantum spintronics, which harness quantum mechanical properties, like superposition and entanglement, to improve the capabilities of spin-based devices. Perovskite multiferroics, with their unique ability to couple magnetic and electric fields, play a pivotal role in this field, offering novel opportunities for the development of quantum technologies. Perovskite material can be integrated in a variety of ways such as in (i) **Quantum Memory Devices**: Perovskite multiferroics may be employed in quantum memory devices, where the coupling between electric and magnetic fields is leveraged to store quantum information. The ability to control both the charge and spin states of perovskite materials may provide an ideal platform for quantum bits (qubits), which are the fundamental units of information in quantum computing. (ii) **Quantum Sensors**: Perovskite-based materials may also be employed in the field of quantum sensors, particularly for the detection of weak magnetic fields or electromagnetic radiation with high sensitivity [33]. Quantum sensors, which leverages the advantage of quantum coherence and entanglement, may benefit from the magnetoelectric properties of perovskites to detect signals in a wide range of applications, including fundamental physics to medical diagnostics. (iii) **Quantum Spintronic Circuits**: The integration of perovskite multiferroics into quantum spintronic circuits may lead to the development of hybrid devices that combine quantum and classical information processing. These devices may perform operations that uses both the classical spintronic effects and the quantum mechanical properties, leading to more powerful and energy-efficient quantum computers.

9 Conclusion and Future Perspectives

In this chapter, the significance of sustainable perovskite multiferroic materials in the context of spintronics applications has been explored. These materials, with their unique characteristics to couple magnetic and electric properties, hold enormous potential for revolutionizing next-generation electronic devices. Perovskite multiferroics offer a platform for logic circuits, energy-efficient memory devices, and sensors because of their inherent multiferroic properties including ferromagnetism, ferroelectricity, and magnetoelectric coupling. This coupling allows for the manipulation of electron spins, which is a necessary aspect of spintronics.

The chapter highlighted the different synthesis strategies for these materials, with a focus on environmentally friendly and sustainable approaches. Green synthesis techniques, such including solvent-free and energy-efficient methods, are necessary for making perovskite multiferroics not only more effective but also more sustainable. We have also discussed the challenges associated with magnetoelectric coupling. Despite these limitations, the potential benefits including low-power consumption, high speed, and scalability make sustainable perovskite multiferroics a highly promising candidate for spintronics applications.

Additionally, we also delved into the future directions in hybrid materials, material discovery, nanoengineering, and integration with quantum technologies. Advances in these areas are also discussed to further enhance the performance of perovskite multiferroics, overcome current limitations and unlock new functionalities for both classical and quantum spintronic devices.

Sustainable perovskite multiferroics materials are well-positioned to contribute significantly to the high-performance, energy-efficient devices, such as quantum computing, memory storage, and sensors. As we continue to explore novel compositions, improve present synthesis methods, and integrate these materials with advanced technologies, perovskite multiferroics will become increasingly vital in the development of scalable, smart, and environmentally friendly electronic devices.

In conclusion, sustained research into the field of synthesis, characterization, and application of sustainable perovskite multiferroics is essential to unlock their full potential for spintronics. Upon inscribing the current limitations, these materials have the potential to significantly shape the future of electronics, driving us toward more sustainable, powerful and efficient technologies.

References

1. Begue A, Ciria M (2021) Strain-mediated giant magnetoelectric coupling in a crystalline multiferroic heterostructure. ACS Appl Mater Interfaces 13(5):6778–6784
2. Bhoi K, Mohanty HS, Ravikant, Abdullah MF, Pradhan DK, Babu SN, Singh AK, Vishwakarma PN, Kumar A, Thomas R (2021) Unravelling the nature of magneto-electric coupling in room temperature multiferroic particulate ($PbFe_{0.5}Nb_{0.5}O_3$)–($Co_{0.6}Zn_{0.4}Fe_{1.7}Mn_{0.3}O_4$) composites. Sci Rep 11(1):3149

3. Chen Y, Gao C, Yang T, Li W, Xu H, Sun Z (2022) Research advances of ferroelectric semiconductors of 2D hybrid perovskites toward photoelectronic applications. Chin J Struct Chem 41(4):2204001–2204011
4. Dhingra A, Thakur OP, Pandey R (2024) Enhanced magnetic properties of single domain Al-substituted strontium hexaferrite ($SrFe_{12-x}Al_xO_{19}$) synthesized via hydrothermal method for permanent magnets. J Alloys Compd 175132. https://doi.org/10.1016/j.jallcom.2024.175132
5. Dhingra A, Thakur OP, Pandey R (2024) Magnetic properties, microwave absorption and EMI shielding study of Al-doped strontium hexaferrites. Inorgan Chem Commun 113810. https://doi.org/10.1016/j.inoche.2024.113810
6. Dhingra A, Thakur OP, Pandey R (2025) Polyol-mediated synthesis of strontium hexaferrite nanoparticles and their potential application for resistive switching and non-volatile memory storage devices. Emergent Mater. https://doi.org/10.1007/s42247-025-00996-y
7. Dhingra A, Thakur OP, Pandey R (2025) Structural, optical and magnetic properties of polyol-mediated synthesized Al-substituted Ca-Sr Hexaferrites for data storage and microwave absorption applications. Ceram Int
8. Dhingra A, Thakur OP, Pandey R (2025) Structure–property relationship and electronic structure calculation of cubic YSZ solid electrolyte for electrochemical applications. J Electron Mater 1–12
9. Ding G, Zheng Y, Xiao X, Cheng H, Zhang G, Shi Y, Shao Y (2022) Sustainable development of perovskite solar cells: keeping a balance between toxicity and efficiency. J Mater Chem A 10(15):8159–8171
10. Gradauskaite E, Meisenheimer P, Müller M, Heron J, Trassin M (2021) Multiferroic heterostructures for spintronics. Phys Sci Rev 6(2):20190072
11. Gupta R, Kotnala RK (2022) A review on current status and mechanisms of room-temperature magnetoelectric coupling in multiferroics for device applications. J Mater Sci 57(27):12710–12737
12. Hench LL, West JK (1990) The sol-gel process. Chem Rev 90(1):33–72
13. Hu W, Huang W, Yang S, Wang X, Jiang Z, Zhu X, Zhou H, Liu H, Zhang Q, Zhuang X, Yang J, Kim DH, Pan A (2017) High-performance flexible photodetectors based on high-quality perovskite thin films by a vapor-solution method. Adv Mater 29(43):1703256
14. Hu Z, Lin Z, Su J, Zhang J, Chang J, Hao Y (2019) A review on energy band-gap engineering for perovskite photovoltaics. Solar RRL 3(12):1900304
15. Ji G, Zhang Y, Chai Y, Nan T (2024) Recent progress on controlling spin-orbit torques by materials design. Npj Spintronics 2(1):56
16. Kim H, Seo J, Park N (2016) Material and device stability in perovskite solar cells. Chemsuschem 9(18):2528–2540
17. Kosub T, Kopte M, Hühne R, Appel P, Shields B, Maletinsky P, Hübner R, Liedke MO, Fassbender J, Schmidt OG, Makarov D (2017) Purely antiferromagnetic magnetoelectric random access memory. Nat Commun 8(1):13985
18. Kreisel J, Kenzelmann M (2009) Multiferroics-the challenge of coupling magnetism and ferroelectricity. Europhys News 40(5):17–20
19. Laukhin V, Skumryev V, Martí X, Hrabovsky D, Sánchez F, García-Cuenca MV, Ferrater C, Varela M, Lüders U, Bobo J-F, Fontcuberta J (2006) Electric-field control of exchange bias in multiferroic epitaxial heterostructures. Phys Rev Lett 97(22):227201
20. Leng K, Li R, Lau SP, Loh KP (2021) Ferroelectricity and Rashba effect in 2D organic–inorganic hybrid perovskites. Trends Chem 3(9):716–732
21. Li P, Zhou X-S, Guo Z-X (2022) Intriguing magnetoelectric effect in two-dimensional ferromagnetic/perovskite oxide ferroelectric heterostructure. NPJ Comput Mater 8(1):20
22. Liang J, Wu T, Wang Z, Yu Y, Hu L, Li H, Zhang X, Zhu X, Zhao Y (2022) Accelerating perovskite materials discovery and correlated energy applications through artificial intelligence. Energy Mater 2(3):200016
23. Manchon A, Železný J, Miron IM, Jungwirth T, Sinova J, Thiaville A, Garello K, Gambardella P, Manchon A (2019) Current-induced spin-orbit torques in ferromagnetic and antiferromagnetic systems. Rev Mod Phys 91(3):035004

24. Pradhan DK, Kumari S, Vasudevan RK, Strelcov E, Puli VS, Pradhan DK, Kumar A, Gregg JM, Pradhan AK, Kalinin SV, Katiyar RS (2018) Exploring the magnetoelectric coupling at the composite interfaces of FE/FM/FE heterostructures. Sci Rep 8(1):17381
25. Rafique M, Herklotz A, Dörr K, Manzoor S (2017). Giant room temperature magnetoelectric response in strain controlled nanocomposites. Appl Phys Lett 110(20)
26. Rahman RAU, Jain Ruth DE, Chakravarty S, Schmid-Beurmann P, Duraisamy D, Venkatesan GN, Zhou P, Srinivasan G, Murugan R (2019) Room temperature magnetoelectric coupling and relaxor-like multiferroic nature in a biphase of cubic pyrochlore and spinel. J Appl Phys 126(4)
27. Rajaram Patil D, Chai Y, Kambale RC, Jeon B-G, Yoo K, Ryu J, Yoon W-H, Park D-S, Jeong D-Y, Lee S-G, Lee J, Nam J-H, Cho J-H, Kim B-I, Kim KH (2013) Enhancement of resonant and non-resonant magnetoelectric coupling in multiferroic laminates with anisotropic piezoelectric properties. Appl Phys Lett 102(6)
28. Ricca C, Aschauer U (2022) Mechanisms for point defect-induced functionality in complex perovskite oxides. Appl Phys A 128(12):1083. https://doi.org/10.1007/s00339-022-06210-8
29. Ruth DEJ, Rahman RAU, Dhamodaran M, Lakshmanan V, Balasubramanian S, Schmid-Beurmann P, Zhou P, Srinivasan G, Ramaswamy M (2020) Room temperature magnetoelectric coupling in Fe-doped sodium bismuth titanate ceramics. J Alloy Compd 830:154679
30. Scott JF (2013) Room-temperature multiferroic magnetoelectrics. NPG Asia Mater 5(11):e72–e72. https://doi.org/10.1038/am.2013.58
31. Shah J, Kotnala RK (2012) Room temperature magnetoelectric coupling enhancement in Mg-substituted polycrystalline $GdFeO_3$. Scripta Mater 67(4):316–319
32. Shang M, Zhang C, Zhang T, Yuan L, Ge L, Yuan H, Feng S (2013) The multiferroic perovskite $YFeO_3$. Appl Phys Lett 102(6)
33. Shellaiah M, Sun KW (2020) Review on sensing applications of perovskite nanomaterials. Chemosensors 8(3):55
34. Skumryev V, Laukhin V, Fina I, Martí X, Sánchez F, Gospodinov M, Fontcuberta J (2011) Magnetization reversal by electric-field decoupling of magnetic and ferroelectric domain walls in multiferroic-based heterostructures. Phys Rev Lett 106(5):057206
35. Stylianakis MM, Maksudov T, Panagiotopoulos A, Kakavelakis G, Petridis K (2019) Inorganic and hybrid perovskite based laser devices: a review. Materials 12(6):859
36. Sun B, Zhou G, Sun L, Zhao H, Chen Y, Yang F, Zhao Y, Song Q (2021) ABO_3 multiferroic perovskite materials for memristive memory and neuromorphic computing. Nanoscale Horizons 6(12):939–970
37. Surowiak Z, Bochenek D (2008) Multiferroic materials for sensors, transducers and memory devices. Arch Acoust 33(2)
38. Tuchiya T, Itoh T, Sasaki G, Suga T (1996) Preparation and properties of piezoelectric lead zirconate titanate thin films for microsensors and microactuators by sol-gel processing. J Ceram Soc Jpn 104(1207):159–163
39. Wang J, Chen A, Li P, Zhang S (2021) Magnetoelectric memory based on ferromagnetic/ferroelectric multiferroic heterostructure. Materials 14(16):4623
40. Wang N, Luo X, Han L, Zhang Z, Zhang R, Olin H, Yang Y (2020) Structure, performance, and application of $BiFeO_3$ nanomaterials. Nano-Micro Lett 12:1–23
41. Wei L, Hu Z, Du G, Yuan Y, Wang J, Tu H, You B, Zhou S, Qu J, Liu H, Zheng R, Hu Y, Du J (2018) Full electric control of exchange bias at room temperature by resistive switching. Adv Mater 30(30):1801885
42. Wei Z, He J, Shi Y, Li Q, Niu Q, Hou Y, Yang S (2023) Understanding the temperature sensitivity of the photovoltaic parameters of perovskite solar cells. Sol Energy 264:112040
43. Wright AD, Volonakis G, Borchert J, Davies CL, Giustino F, Johnston MB, Herz LM (2020) Intrinsic quantum confinement in formamidinium lead triiodide perovskite. Nat Mater 19(11):1201–1206
44. Wu SM, Cybart SA, Yu P, Rossell MD, Zhang JX, Ramesh R, Dynes RC (2010) Reversible electric control of exchange bias in a multiferroic field-effect device. Nat Mater 9(9):756–761

45. Yakout SM (2021) Spintronics and innovative memory devices: a review on advances in magnetoelectric $BiFeO_3$. J Supercond Novel Magn 34(2):317–338
46. Yao M, Marzouki A, Hao S, Salmanov S, Otonicar M, Loyau V, Dkhil B (2023) Grain size and piezoelectric effect on magnetoelectric coupling in BFO/PZT perovskite-perovskite composites. J Alloy Compd 948:169731
47. Zhang J, Xie Y, Ji K, Shen X (2024) Perspective on 2D perovskite ferroelectrics and multiferroics. Appl Phys Lett 125(23):230503. https://doi.org/10.1063/5.0235723

Chapter 9
Perovskite Multiferroic Materials for Sensors and Transducers Applications

Sidharth D. Nair, Birundhavi Palanisamy, S. Alden Jenish, and Sunil Kumar Pradhan

1 Introduction

Ferroelectric systems with a magnetic order parameter are known as multiferroic materials. Many of these systems have been successfully synthesized in the recent years with the intention of utilizing their various characteristics for device applications. $BiFeO_3$ (BFO), $BiMnO_3$ (BMO), and Bi_2FeCrO_6 (BFCO) are notable examples [1–4]. They exhibit more than one of the primary properties in the same ferroic phase such as:

- Ferromagnetism—A magnetization that is switchable by an applied magnetic field.
- Ferroelectricity—An electric polarization that is switchable by an applied electric field.
- Ferroelasticity—A deformation that is switchable by an applied stress.

Although photovoltaics (PV) is thought to be one of the most promising renewable energy technologies for the ensuing decades, significant advancements are still needed to increase efficiency and lower the cost of fabricating current materials. Even though ferroelectric-based solar cells' PV performance has increased quickly, significant work is still needed to determine the best device configuration and optimize carrier generation and collection before commercial technologies can be implemented. Solar cells made of crystalline silicon, which have the highest efficiency of roughly 18% in single junction devices, now dominate the PV market. Although this technology is well-established, continued growth is nevertheless constrained by its relatively high cost in comparison with carbon fossil fuel-based energy [5].

The primary causes of this are the high price of crystalline silicon as well as more costly and intricate manufacture and processing procedures. To solve these

S. D. Nair · B. Palanisamy · S. A. Jenish · S. K. Pradhan (✉)
School of Electronics Engineering, Vellore Institute of Technology, Chennai, India
e-mail: sunilkumar.pradhan@vit.ac.in

M. Kumar (ed.), *Sustainable Perovskite Multiferroic Materials*, Materials Horizons: From Nature to Nanomaterials, https://doi.org/10.1007/978-981-96-7419-0_9

problems, a number of initiatives are being made, specifically to (i) lower the cost of fabricating silicon, (ii) boost cell efficiency, and (iii) create new generations of solar cells using different materials. Solar technologies that use simpler components, designs, fabrication, and processing to achieve high conversion efficiencies are still needed.

In order for a material to be classified as multiferroic, it must meet a number of conditions. To pass spontaneous electric polarization, for example, ferroelectric material needs to be non-centrosymmetric; there are 31 point groups that permit this. Among these, both features can exist simultaneously in 13 point groups. It seems doubtful that symmetry will have a significant impact on whether a material is multiferroic or not because these point groups could result in a huge number of potential materials displaying multiferroics [6].

1.1 Ferromagnetism

Certain materials, like iron, have a property called ferromagnetism that causes a noticeable magnetic permeability and, frequently, a large magnetic coercivity, enabling the substance to become a permanent magnet. Because of their high magnetic permeability, ferromagnetic materials exhibit a pronounced attraction to a magnet.

A material's induced magnetization as a result of an external magnetic field is known as magnetic permeability. Different kinds of material magnetism have been identified in physics. The strongest kind, ferromagnetism (and its related effect, ferrimagnetism), is what causes the ubiquitous magnetism of daily life. A refrigerator magnet is an example of a permanent magnet made of ferromagnetic material [7, 8].

1.2 Ferroelectricity

Ferroelectricity, a property of some materials with spontaneous electric polarization that may be reversed by applying an external electric field, is used in physics and materials science. With the added characteristic of having reversible electrical polarization by nature, all ferroelectrics are both piezoelectric and pyroelectric. The phrase is used to compare a material's permanent magnetic moment to ferromagnetism. Capacitors with changeable capacitance can be created using ferroelectric materials' nonlinear nature. Generally, a ferroelectric capacitor is just made up of two electrodes encasing a ferroelectric layer [5, 9]. In addition to being tunable, ferroelectrics typically have very high permittivity, particularly near the phase transition temperature. Ferroelectric capacitors are hence smaller in size than dielectric (non-tunable) capacitors with comparable capacitance.

1.3 Ferroelasticity

Ferroelasticity, the mechanical counterpart of ferroelectricity and ferromagnetism in the context of ferroics, is a phenomenon where a material may show spontaneous strain. Applying a stress or an electric field greater than a critical value can reliably flip between the two or more stable orientational states (also known as remanent states) that a ferroelastic crystal possesses when there is no mechanical stress or electric field. As the system moves back and forth across an energy barrier, hysteresis results from the application of opposing fields. The energy dissipated by this transition is equivalent to the area that the hysteresis loop encloses. The crystal symmetry usually decreases as the crystal's parent structure transitions to one of its stable ferroelastic stresses. Other observable characteristics including birefringence, optical absorbance, and polarizability may also spontaneously change in tandem with the strain and crystal structure.

Ferroelasticity manifests as super elasticity and the shape memory effect. Depending on the nickel-to-titanium ratio, the typical ferroelastic alloy nitinol (nickel titanium) can exhibit either super elasticity or the shape memory effect at room temperature.

2 Historic and Present Developments

Particularly significant early works included the finding of large ferroelectric polarization in epitaxially grown thin films of magnetic $BiFeO_3$, the observation that ferroelectricity is caused by non-collinear magnetic ordering in orthorhombic $TbMnO_3$ and $TbMn_2O_5$, and the coexistence of magnetism in hexagonal manganite $YMnO_3$. The first recorded mention of magnetoelectricity was in the 1959 edition of Landau & Lifshitz's Electrodynamics of Continuous Media. The following statement marks the end of the piezoelectricity section: "Let us note out two further phenomena, which, in principle, could exist." A linear coupling between a deformation and a magnetic field in a material is the basis of piezo magnetism, which is comparable to piezoelectricity. For instance, a magnetization proportional to an electric field would result from a linear connection between magnetic and electric fields in a medium. Both of these phenomena may be present for specific magneto crystalline symmetry classes.

A year later, I. E. Dzyaloshinskii used symmetry considerations to show that the material Cr_2O_3 should behave in a linear magnetoelectric manner. D. Astrov promptly verified his prediction. Over the ensuing decades, research on magnetoelectric materials continued in a number of European groups, most notably in the former Soviet Union and in H. Schmid's group at U. Geneva. Magnetoelectric Interaction Phenomena in Crystals (MEIPIC) is an East–West conference series that took place from 1973 in Seattle until 2009 in Santa Barbara. The phrase "multi-ferroic magnetoelectric" was really first used by H. Schmid in the proceedings of the 1993 MEIPIC conference in Ascona [10–13].

Since multiferroics have the potential to revolutionize and greatly increase the efficiency of contemporary memory storage systems, they are the subject of much research. One excellent option is barium monoferrite (also known as γ-$BaFe_2O_4$ or BaFeO), which has gained popularity because of its beneficial multiferroic properties, especially when it is at room temperature.

A recent study used the pulsed electron deposition (PED) method to produce pure and crystalline γ-$BaFe_2O_4$ thin films. Two primary mechanisms governing the kinetics of BaFeO deposition were identified by analysis of the plasma plume: an incongruent low-energy evaporation process and a congruent ablation process that is far from thermodynamic equilibrium [14]. These results demonstrate PED's capability for creating superior γ-$BaFe_2O_4$ polycrystalline thin films. Another important finding was made by Chinese and Australian researchers who used symmetry analysis to integrate a coupling layer concept with multiferroics and demonstrate the layer Hall effect (LHE). The Bloch electrons in one valley undergo a substantial Berry curvature as a result of valley physics and time-reversal symmetry breakdown.

The resulting LHE is ferroelectrically controlled and reversible, the researchers showed. They confirmed this process and its anticipated effects in the bilayer of the multiferroic material Co_2CF_2 using first-principles calculations. This discovery provides a fresh perspective on LHE research and the contemporary use of multiferroic 2D materials science [15–18].

Multiferroics are not new, yet there are still a lot of problems. Strengthening the magnetoelectric coupling of new room-temperature multiferroics is essential. For ferroelectric/magnetoelectric switching to have a major effect, the voltage needed should be less than 100 mV. One of the main goals is still to create an electric field-controlled multiferroic device, particularly at low voltages, room temperature, and with fast switching. Multiferroics will surely become a necessary material of the future with further improvements.

3 Multiferroic Materials

Materials with several ferroic orders are called multiferroics, and the term "magnetoelectricity" describes the reciprocal connection of electricity and magnetism. Multiferroicity is one of the most active fields in condensed matter physics and materials science, and it has never been as active as it is in the first ten years of the twenty-first century. We now have a much more thorough and profound understanding of multiferroic physics thanks to a number of significant developments and consistent advancements over the last ten years, which is further advancing the horizon of this fascinating field of study. More multiferroic materials and enhanced magnetoelectric performance will soon be available, bringing applications closer to reality. Current research on flexible pressure sensors is illustrated in Fig. 1.

Transition metal oxides, which are compounds composed of (often 3D) transition metals with oxygen and frequently an additional main-group cation, make up the majority of multiferroic materials discovered to date [19].

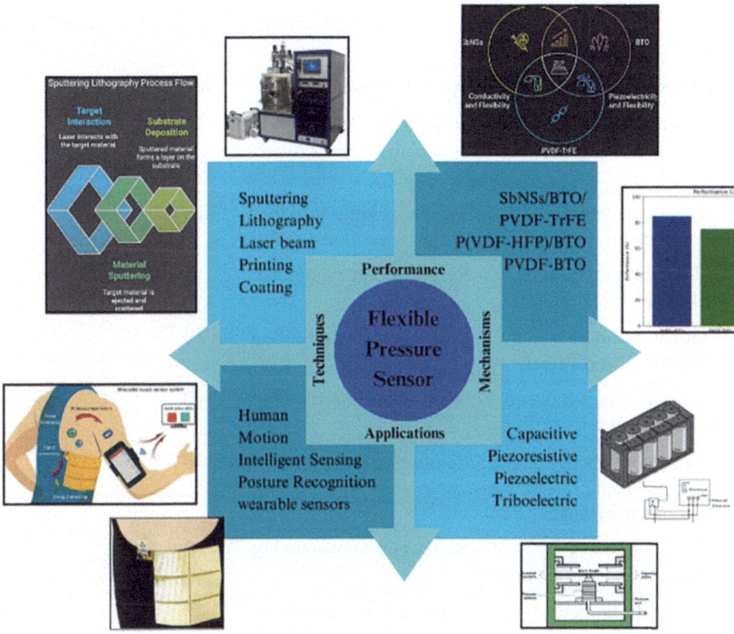

Fig. 1 Current research on flexible pressure sensors

A few factors make transition metal oxides an advantageous class of materials for multiferroic identification:

If the transition metal's localized 3D electrons are partially filled with electrons, they are typically magnetic. Because its bonds with transition metals are neither too covalent like its neighbor nitrogen, N nor too ionic like its neighbor fluorine, F, oxygen occupies a "sweet spot" on the periodic table. Consequently, ferroelectricity benefits from its relatively polarizable bonds with transition metals [20–22].

Oxygen and transition metals are generally stable, non-toxic, abundant on Earth, and safe for the environment.

Some of the most researched multiferroics are listed in Table 1 along with their ferroelectric and magnetic ordering temperatures.

Table 1 Different types of magnetic field sensors based on their operating principles, along with their approximate detection limits

Sensor type	Working principle	Special considerations	Field detection limit (T)
Inductive coil	Electromagnetic induction (Faraday's Law)	Suitable for AC magnetic fields only	~ 10^{-4}
Flux gate	Magnetic induction mechanism	Typically, large in size	~ 10^{-8}
Magneto-optic	Polarization rotation (Kerr/Faraday effect)	Requires precise optical alignment	~ 10^{-12}
Squid	Quantum Interference via Josephson Effect	Needs cryogenic cooling	~ 10^{-15}
Hall sensor	Lorentz Force-Induced Hall Effect	No special requirements	~ 10^{-5}
AMR	Resistance variation due to anisotropic magnetoresistance	Requires DC biasing	~ 10^{-3}
GMR	Spin-dependent scattering at interfaces	No special requirements	~ 10^{-3}
TMR	Quantum tunneling dependent on spin states	No special requirements	~ 10^{-3}
CMR	Double exchange mechanism with phase transition	Highly sensitive to temperature	~ 10^{-2}
Rubidium gas	Spin-exchange relaxation process	Requires cryogenic cooling	~ 10^{-14}
Multiferroic	Magneto-electric coupling phenomenon	Requires DC biasing	~ 10^{-12}

4 Possible Applications of Multiferroics

4.1 Magnetic Field Sensors

Magnetic field sensors detect changes or the presence of magnetic fields without making physical contact. Direction, presence, rotation, angle, speed, electrical currents, temperature, and other parameters are used to determine the information about the magnetic fields. These sensors are used in many applications, including metrology, microscopy imaging, geo-positioning, security and detection, data storage, medical imaging, speed measurement, and many more.

The market demand for these sensors in 2010 was 3.67 billion units (a value of $1.1 billion). The industry's total revenue is $4.4 billion in 2022. By 2023, the magnetic field sensors industry will have a $12.4 billion overall market share.

There is notable non-linearity in the relationship between magnetic strain and the applied DC magnetic field. When external magnetic field applied is zero, strain

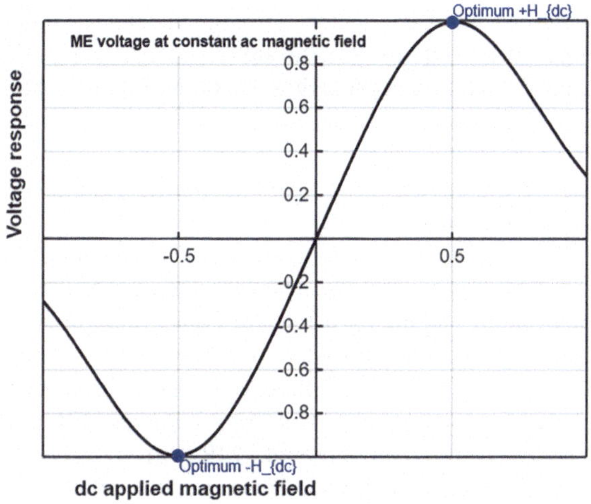

Fig. 2 Typical behavior of a composite multiferroic under a constant AC magnetic field (fixed amplitude and frequency) when subjected to a DC magnetic field

remains unchanged. As the magnetic field increases, strain rises sharply until it reaches a specific value known as the saturation field. At this point, the magnetic strain becomes constant, and any further increase in the applied field has no additional effect on the strain. This saturation field approximately aligns with the saturation magnetization observed in the material's magnetic hysteresis loop [23–27].

The non-linear nature of magnetic strain is reflected in the piezomagnetic coefficient, which is defined as the derivative of strain for the applied field. Consequently, this non-linearity is also transferred to the magnetoelectric coupling coefficient. Figure 1 illustrates a typical voltage response of a multiferroic composite sensor under varying DC magnetic fields while maintaining a constant AC magnetic field amplitude and frequency. Figure 2 depicts typical behavior of a composite multiferroic under a unchanging AC magnetic field when subjected to a DC magnetic field.

Multiferroic materials provide a linear response to an applied magnetic field amplitude, which is crucial for the sensor's proper functioning. This property makes them suitable for magnetic field sensors [28].

Composite multiferroics provide better sensing capabilities than single-phase multiferroics. Single-phase multiferroics have small magneto-electric effects and operate at cryogenic temperatures. Composite multiferroics offer string effects of magneto electrics at ambient temperature. Composite multiferroics can be used to meet the industry's growing demands.

4.2 Multiferroic Microwave Resonators Tuned Electrically

High-frequency excitations applied to ferromagnetic materials, exhibit a phenomenon known as ferromagnetic resonance (FMR) [29]. This occurs when disturbances in the vectors of magnetization or magnetic moments cause them to relax back to equilibrium through a precessional motion. FMR is observed when the precession frequency matches that of the external excitation field, typically occurring in the microwave frequency range for ferrite materials. As a result, soft ferrite-based FMR devices are widely used in microwave signal processing applications, including resonators, band-pass and band-stop filters, and phase shifters operating within the 1–70 GHz range.

For these devices to function correctly and adjust their frequency, a DC magnetic bias field is necessary. Nevertheless, there are a number of disadvantages to magnetic field tuning, including a long response time, high noise levels, substantial power consumption, and restricted miniaturization. Modern communication and radar systems require quick frequency changes with low power consumption. To address these limitations, replacing ferrite materials with multiferroic bi-layer composites enables electric field tuning of FMR [30]

Experimental demonstrations have shown that a YIG/PZT multiferroic composite can achieve electric field-controlled FMR. In a specific study, a YIG layer with a thickness of 15 μm and lateral dimensions of 1×2.2 mm^2 was bonded to a PZT layer measuring 4 mm thick, measuring 4×0.5 mm^2 on the side. To create the composite structure, silver electrodes were positioned on both sides of the PZT. A 50 μm thick, 3 mm long microstrip line that was put onto an alumina substrate was another feature of the device [30].

Under experimental conditions, a signal input of 2–10 GHz with 0.1 mW power was applied to the microstrip line, along with a 1.12 kOe in-plane DC magnetic bias field and no external electric field. FMR absorption was observed at approximately 5 GHz with a linewidth of 3.4 MHz. When a 10 kV/cm electric field was applied, the resonance frequency shifted upward by 16 MHz, whereas applying a -10 kV/cm electric field caused a downward shift of 18 MHz. Importantly, while the FMR peak position changed, the overall shape of the response remained unaltered.

The multiferroic bi-layer's strain-mediated magnetoelectric interaction is responsible for this frequency shift. By causing piezo-strain in the PZT layer, the applied electric field modifies the internal magnetization of the YIG. This effect can be understood as the FMR response being shifted by the creation of an internal magnetic field. According to estimates, an internal magnetic field of roughly 5.77 Oe is induced by an electric field of 10 kV/cm [31, 32].

4.3 Multiferroic Microwave Phase Shifter

In phased array antenna applications, radar systems, and telecommunications, microwave phase shifters are essential components. Usually, semiconductors, ferroelectrics, or ferrite materials serve as the foundation for conventional phase shifters. Ferrite-based phase shifters operate by using electromagnetic waves' Faraday rotation in magnetized ferrites; nevertheless, this approach is frequently impracticable because it necessitates the application of strong magnetic bias fields. An alternative method for phase shifting based on ferrite involves the propagation of spin waves in planar ferrites, which offers significant magnetic tunability across a broad frequency range of 1–26 GHz.

Unlike ferrite-based devices, ferroelectric phase shifters provide advantages such as rapid electric tunability and low power consumption, though they suffer from high signal loss, particularly at frequencies exceeding 5 GHz. To overcome these limitations, researchers have explored multiferroic composite structures, which integrate ferrite and ferroelectric materials to enable dual-phase tunability, extending operability beyond 5 GHz. Theoretical studies have suggested the feasibility of such hybrid devices, and experimental advancements have been demonstrated by Ustinov through his work [33].

Ustinov et al. created a bi-layer multiferroic phase shifter in their work. It consists of a Yttrium iron garnet (YIG) ferrite layer 5.7 μm thick and a $Ba_{0.6}Sr_{0.4}TiO_3$ (BST) ferroelectric layer 500 μm thick. Under a DC magnetic bias field of 1413 Oe, experimental tests showed that applying voltages of 0, 500, and 1000 V to the ferroelectric layer—corresponding to applied electric fields of 0, 10, and 20 kV/cm—produced differential phase shifts. Further investigations were conducted at various DC magnetic bias fields while maintaining the same electric field conditions, successfully tuning the operating frequency range from 3.5 to 8 GHz. Additionally, electrically controlled microwave phase shifters based on stress-mediated magnetoelectric coupling have been reported in YIG/PZT multiferroic composites, utilizing a device similar to the one described in the following section [34].

4.4 Multiferroic Microwave Signal Delay Line

Typically, two microstrip microwave transducers are sandwiched by a thin layer of ferrite for microwave delay lines. The signal source is the first transducer, and the output port is the second transducer. These devices are frequently employed in the processing of high-frequency signals. In this setup, the input signal propagates through the ferrite material to the output transducer, exciting spin waves within the ferrite. Since the wavelength and velocity of these spin waves are about two orders of magnitude smaller than those of the electromagnetic waves at the same frequency, it is possible to induce a delay of up to 500 ns with a few millimeters of ferrite material.

In order to tune delay times using magnetic fields, current devices rely on the fact that spin wave velocity is highly dependent on the magnetic bias field. Despite its effectiveness, this approach has drawbacks, including a sluggish tuning speed, a high magnetic bias field power consumption, and size restrictions that make it difficult to integrate and miniaturize such devices into solid-state chips.

In their new microwave delay line design, Fetisov and Srinivasan substituted a composite multiferroic bilayer structure—more precisely, ferrite/ferroelectric—for the ferrite element. They used Yttrium iron garnet (YIG) ferrite and lead magnesium niobate-lead-titanate (PMN-PT) ferroelectric materials in their experiment. The YIG layer had a thickness of 4.1 μm and lateral dimensions of 2×25 mm^2, while the PMN-PT ferroelectric layer was 500 μm thick and measured 4 × 4 mm. Gold microstrip transducers, each 50 μm wide and 3 mm long, were deposited onto alumina substrates. The distance between the transducers was 8 mm [35].

The ability to adjust the delay time by subjecting the multiferroic element to an electric field is the main benefit of this device. The dielectric permittivity of the ferroelectric material in the multiferroic composite is dependent on the electric field, which enables this. This electrical tuning technique allows for further device downsizing and is quick and power-efficient. When the applied electric field was adjusted from 0 to 8 kV/cm, the authors noticed a difference in the delay time of 10–25% [36].

4.5 Multi-state and Multiferroic Random Access Memory

Random-access memory is an important device in modern computers manufactured today. They can then broadly be classified into volatile and non-volatile RAM. Volatile RAM requires constant power supply to retain data, and any power outage will completely erase the information. Non-volatile RAM retains information even when electrical power is switched off. However, volatile RAM must be used for the installation of operating systems, applications and to furnish the processor with memory capacities since volatile RAM has the capability to provide an extremely fast pace of access on the data compared to the non-volatile one [31].

Dynamic random-access memory (DRAM) is one of the most commonly used types of RAM. An integrated circuit stores each bit of information in a capacitor-transistor element "one capacitor and one transistor (1C-1T)." Charged capacitors are used to represent a "zero" bit in dynamic random-access memory, while discharged capacitors represent a "one" bit. These capacitors are so tiny, and due to their nano-scale dimensions, it becomes inevitable for them to lose charge gradually with time. As a result, data must be periodically refreshed, which is where the term "dynamic" comes from. Unlike DRAM, static random-access memory (SRAM) does not require periodic refreshing and uses either 4–6 transistors per bit in its architecture. Although SRAM is faster and more stable, its structural complexity makes DRAM more suitable for higher data densities.

In 1952, ferroelectric-based non-volatile random-access memory (RAM) technology was introduced (FRAM). Computers may boot up rapidly without reloading software because non-volatile RAM, in contrast to volatile RAM, keeps data when powered down. FRAM, which entered mass production in 1992, has a similar architecture to DRAM (1T-1C) but uses the remanent polarization states of a ferroelectric capacitor to store binary data. However, FRAM has limitations, such as requiring destructive read and reset operations [37].

Magneto-resistive random-access memory, or MRAM, is another type of non-volatile memory technology that uses magnetic states and magneto-resistance characteristics to encode data. A data bit in MRAM is stored or conceptualized as existing in one of two resistance states within a multilayered magnetic structure: a binary "1" state at high resistance and a binary "0" state at low resistance. The data is read by measuring the resistance state, while the write process involves changing the magnetization of a free magnetic layer relative to a pinned layer. This process relies on currents through bit and word lines, which are power-intensive and generate heat, limiting MRAM's efficiency and scalability [38].

To address these limitations, multiferroic random access memory (MFRAM) has been proposed. MFRAM leverages materials that exhibit both magnetic and electric polarization properties. By combining features of FRAM and MRAM, MFRAM can encode data in four states: two based on ferroelectric polarization and two based on magneto-resistance. This design uses a multiferroic layer with electric polarization states for data encoding and a magnetic free layer separated by a non-magnetic or dielectric spacer. Magnetization changes in the free layer are induced via strain-mediated magneto-electric coupling, triggered by applied voltages. This method reduces power consumption and improves memory density while maintaining non-volatility.

MFRAM operates through two distinct write processes: an electric write process to switch ferroelectric polarization and a magnetic write process to rotate the free layer's magnetization. The magnetic write process requires a voltage below the coercive field, ensuring that the ferroelectric state remains unchanged. MFRAM's advantages include all-electric read/write operations, higher memory density, and compatibility with existing memory architectures.

Several researchers have proposed innovative designs incorporating multiferroics into RAM. One design utilizes a multiferroic layer with ferroelectric and antiferromagnetic properties, enabling magnetization reversal through voltage control rather than current-generated magnetic fields. This approach addresses the power consumption and thermal stability issues associated with MRAM. Another study demonstrated the feasibility of using multiferroic materials as quantum tunnel barriers in memory systems. These systems exhibit four resistance states based on the orientation of magnetic and electric polarization, although practical implementation may require overcoming challenges such as operating at cryogenic temperatures [39].

Further advancements, such as those using nano-scale multiferroic thin films, show promise for room-temperature operation and future multiferroic memory devices. These developments represent significant steps toward achieving efficient, non-volatile, and high-density memory solutions.

4.6 Read Heads for Multiferroic Magnetic Recording

Known as "perpendicular magnetic recording media" or PMRM, modern magnetic hard disk drive (HDD) technology stores data in tiny areas of a magnetic thin film that is magnetized perpendicularly. Since IBM's first HDD was introduced in 1956, areal storage densities have increased steadily, roughly doubling every two years, thanks to the switch from longitudinal to perpendicular recording. Following Toshiba's initial commercial launch of PMRM in 2005, Seagate, Hitachi, Western Digital, and Fujitsu serialized the software in 2006. This technology has reached the record areal densities of 500–700 Gb/in^2 in current commercial HDDs. Data are read off these media with the help of magneto-resistive (MR) read heads operating on AMR, GMR, or TMR effects. These MR effects lead to a change in the resistance of the sensor as it interacts with the stray magnetic fields from the recorded bits. The variation in resistance is used to generate a read signal via the voltage change $\Delta V = I \times \Delta R$.

As recording densities increase, it is necessary to scale down the size of the read head sensor to effectively read smaller magnetic bits. However, this reduction presents challenges, especially in constructing the sensor stack, which now requires at least 15 layers with a total thickness of less than 20 nm (about twice the width of a data bit at 1 Tb/in^2 density).

A potential solution lies in a new reader technology based on the magneto-electric effect in multiferroic materials suggested in 2007 and proven experimentally in 2008. A tri-layer Ferromagnet (FM)/Ferroelectric (FE)/Ferromagnet (FM) multiferroic composite structure is used in this concept, which detects AC magnetic fields through the magneto-electric effect. As the head moves along the magnetic recording track, the field from a recorded bit provides the AC excitation for the sensor, and a DC magnetic bias field is required to induce a strain-mediated magneto-electric effect. A self-biasing technique that makes use of the exchange bias effect between the FM and anti-FM layers is the best method for biasing the sensor. Other techniques that rely on demagnetization states, including self-biasing with hysteretic ferromagnets, are less reliable [40].

An innovative method in sensor design is the creation of self-biased multiferroic thin-film devices using the exchange bias effect. This method was first introduced in the same paper and later experimentally confirmed in 2010. Currently, it is employed in the production of self-biased multiferroic composites-based AC magnetic field sensors with excellent sensitivity. The sensor stack is made up of seven layers: Seed/AFM/FM/FE/FM/AFM/Cap, which ensures that both FM electrodes are properly biased.

Theoretical calculations for the proposed multiferroic read head predict the open-circuit voltage output as a function of various material properties, layer thickness ratios, and the applied AC magnetic field frequency and amplitude. For an areal density of 1 Tb/in^2, numerical estimates show an output of 5.7 μV at low frequencies. At the electromechanical resonance frequency (around 10.7 GHz for a sensor with a 120 nm stripe height), 1.57 mV, or a 3.14 mV peak-to-peak signal, is the output voltage. These calculations are based on parameters like the reader width, bit length, shield-to-shield spacing, and sensor stack thickness, all of which contribute to a higher operating frequency compared to conventional MR sensors [41].

This multiferroic read head technology offers several advantages over traditional MR readers, such as reduced power consumption, enhanced thermal performance (due to the capacitive nature of the multiferroic reader), increased frequency of operation, and a smaller sensor consisting of only seven layers, which enables magnetic recording densities of 1 Tb/in^2 and greater.

4.7 Photovoltaic Multiferroic Solar Cells

With the sun emitting energy across almost the entire electromagnetic spectrum, it is also imperative to realize that human sight perceives a range of wavelengths between 400 and 750 nm, known as the visible spectrum. The segment is estimated to cause about 45% of the solar energy emitted from the Sun. Solar radiation provides approximate calculations in a manner whereby 46.3% can be made up by the infrared, and 9.1% occurs in the ultraviolet region.

Solar energy stands out as an abundant and renewable energy source capable of meeting a good portion of the energy demand on our planet. The solar constant is the average solar power density at the Earth's surface, which is roughly 1360 W/m^2. As far as the mass of the planet is concerned, with respect to a radius of about 6350 km, it is estimated that nearly 173,000 terawatts (TW) of solar energy reaches our planet. World energy consumption today is close to only 16 TW and is still expanding because of the population and economic growth. When comparing various renewable sources, solar energy vastly exceeds alternatives like wind power (2–4 TW), hydropower (2–4 TW), geothermal resources (20–30 TW), and biomass—offering between 7 and 10 TW from half of arable land—which solidifies its role as a promising contender for satisfying global energy demands [42].

To accommodate today's global requirement of around 16 TW through solar means necessitates an area coverage of roughly 11,764 km^2 registered with perfect photovoltaic conversion efficiency. However, when considering a more realistic efficiency estimate of merely 10%, this demand escalates to about 17,640 km^2—a surface area comparable to half that of the United Kingdom yet just a small fraction compared to Sahara Desert's sprawling 4619,260 km^2 territory. Such calculations highlight how feasible it could be to meet planetary needs using solar solutions as advancements continue in photovoltaic technology.

The operation principle behind photovoltaic effects in solar cells hinges on two primary processes: firstly, involves generating electron–hole pairs followed by their distinct separation which ultimately leads to electric current production. Conventional solar cell technologies consist of p-n semiconductor junctions whereby photons with energies exceeding the band gap excite electron–hole pairs that are separated by an electric field in the depletion layer. However, efficiency of operation is constrained by restricted band gap energy and narrow depletion region. Open-circuit voltage can be maximized to around 1 V only; a small fraction of the solar spectrum is well exploited in these devices. This trade-off, commonly known as the "solar power dilemma," is due to the fact that while semiconductors with smaller band gaps will produce higher currents, they will also produce lower voltages, and conversely for larger band gaps [43].

An alternative approach to overcome these limitations is the bulk photovoltaic effect observed in ferroelectric materials. Unlike the interfacial charge separation in p–n junctions, the bulk photovoltaic effect relies on the internal electric field associated with the remanent polarization of ferroelectric crystals. This internal field enables charge separation throughout the bulk material rather than being confined to a narrow depletion layer. While it has been reported that ferroelectric materials, such as lead zirconate titanate (PZT), barium titanate ($BaTiO_3$), and lithium niobate ($LiNbO_3$), show promise, with photovoltages reported as high as 1600 V under certain conditions, further studies reveal that these materials, with large bandgaps in the range of 3–6 eV, suffer from poor efficiency and limited light absorption, critically restricting their absorption only to the ultraviolet region of the solar spectrum [15].

Multiferroics, materials which display more than one order parameter (electric and magnetic ordering, for example), present a promising solution to these challenges. Such materials could effectively reduce the band gaps and broaden the absorption range of the solar spectrum. For example, bismuth ferrite ($BiFeO_3$) and bismuth chromate ($Bi_2Cr_3O_{12}$) show reduced band gaps (ranging between 1.4 and 2 eV) with an enhanced performance in photovoltaics. Studies have reported photovoltages of 0.6 V and conversion efficiencies of approximately 6% in thin films of Bi_2CrFeO_6 under red light illumination.

The coupling between the electrical, magnetic, and elastic order parameters in multiferroic materials allows their properties to be tuned using external fields or intrinsic strain. This tunability can improve the photovoltaic performance and also broaden the range of applications for multiferroics in opto-electro-mechanical implementations, beyond energy generation. Though this is still in the beginning stage of research, multiferroic photovoltaic materials represent a great hope for tackling global energy concerns and advancing sustainable energy technologies. 2D illustration of perovskite lattice structure of $BiFeO_3$ is shown in Fig. 3.

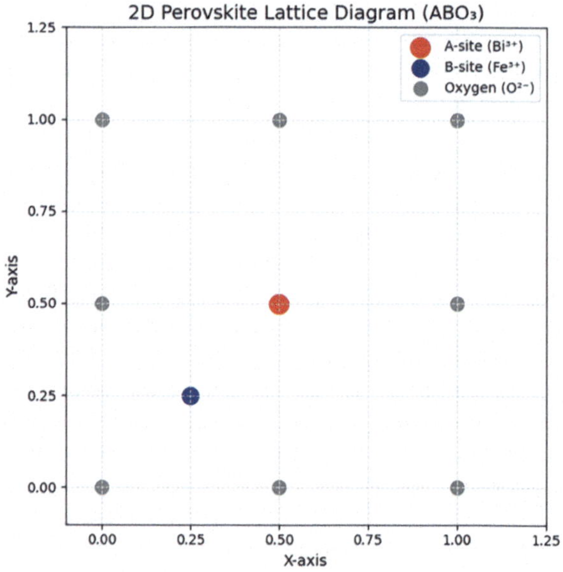

Fig. 3. 2D illustration of perovskite lattice structure of $BiFeO_3$

5 Perovskite Multiferroic for Sensors

The emerging field of perovskite multiferroic materials is set to transform sensor technology because of their distinctive functional and structural features. Perovskites, generally written as ABO_3, have a flexible lattice structure, generally with a large cation at the A-site, while the B-site has smaller transition metal cations, both coordinated with oxygen. Figure 3 shows the 2D perovskite lattice structure diagram (ABO_3) for $BiFeO_3$, showing the arrangement of A-site and B-site cations with oxygen atoms coordinating them. These materials have structural flexibility that allows for the coexistence of multiple ferroic orders, such as ferroelectricity, ferromagnetism, and ferroelasticity, which allows them to be attractive candidates for advanced sensing applications [32]. These materials exhibit many special properties of electron spins, early, and later coupling. Such electromagnetic features of materials make them promising candidates for data storage, information processing, and even neuromorphic computing applications in the field of artificial intelligence.

5.1 *Principles and Mechanisms of Sensing*

As such perovskite multiferroics use different sensing mechanisms owing to their ferroelectricity and compatibility, these mechanisms can be included in the paradigms to show how high-performance sensors can come about across different fields.

Perovskite multiferroics, in particular, rare-earth perovskites ABO_3, show an unusually strong degree of magnetoelectric coupling, making them strong candidates for magnetoelectric sensing applications. Such coupling enables fine manipulation of electron spins, which is pivotal in developing ultrahigh-performance spintronic sensors. For example, in $BiFeO_3$-like materials, external magnetic fields will induce electric polarization due to the interaction between magnetic and electric dipoles. It has been utilized in a precision magnetic field sensor important in biomedical diagnostics such as magnetoencephalography, where detection of subtle changes in the magnetic field produced by the brain is crucial. In addition, magnetoelectric sensors are used in navigation systems and geophysical exploration that enable weak magnetic anomalies to be detected without needing a cryogenic cooling arrangement, a huge advantage in comparison with their traditional superconducting quantum interference device (SQUID) counterparts. In the realm of spintronic devices, ABO_3-type multiferroic rare-earth perovskites have shown great prospects [44].

In the realm of piezoelectric and strain sensing, perovskite multiferroics utilize their high piezoelectric coefficients to effectively convert mechanical stress into electrical signals. Given the piezoelectricity of every FE material the application of an external voltage to a heterostructure induces dimensional variations within the FE layer through the converse piezoelectric mechanism. Chen et al. reported a novel multiferroic heterostructure of a PMN-PT substrate with piezo-strain properties. They fabricated MgO-based MTJs on PMT-PT substrates demonstrating a high TMR ratio of 235%. By applying electric fields to the substrate, the researchers achieved electrical control over the magnetic easy axis orientation within the MTJ, leading to tunable TMR. This technological advancement presents a significant way for developing memory devices that can be written electrically and read magnetically. It is also crucial for applications in structural health monitoring systems, where real-time detection of stress in critical infrastructure—such as bridges and aircraft—is vital. Additionally, wearable sensors that incorporate flexible perovskite thin films can track biomechanical movements, presenting significant potential in fields like rehabilitation and sports science [35, 44, 45].

Pyroelectric sensing takes advantage of temperature-dependent polarization of multiferroics. Materials like $BaTiO_3$ exhibit a high pyroelectric coefficient, thus generating electric current in response to temperature variations. This property is critical in infrared detectors used for thermal imaging and night vision, as well as in industrial thermal monitoring systems. The high sensitivity and low thermal noise of perovskite-based pyroelectric sensors ensure accurate and reliable temperature detection over a broad range of operating conditions. Capacitive sensing in perovskite multiferroics involves measuring changes in capacitance resulting from variations in the dielectric constant under external stimuli. $BiFeO_3$-based capacitive sensors are highly effective in humidity and pressure detection due to their high dielectric tunability. In such sensors, the dielectric constant changes proportionally to the external stimulus, resulting in a measurable capacitance shift. This mechanism finds extensive use in environmental monitoring and industrial process control. Table 2 presents key technological applications of perovskite multiferroics, comparing performance metrics with conventional sensor technologies [35].

Table 2 Presents summary of the applications

Application	Performance metrics	Key mechanism	Advantage
Biomedical imaging	5 fT/$\sqrt{}$Hz sensitivity @ 300 K	Spin-phonon coupling	Cryogen-free operation
Inertial navigation	0.001°/h bias instability	Strain-mediated spin control	Zero-power shielding
Structural health monitoring	0.1 $\mu\varepsilon$ resolution	Domain wall conductivity	Self-powered operation
Gas sensing	0.017 ppm LOD for SO_2 @ RT	Oxygen vacancy adsorption	Multi-gas discrimination
Energy harvesting	12.5 V open-circuit voltage	Piezo-phototronic coupling	Dual light/strain response

5.2 Advantages of Perovskite Multiferroics in Sensor Applications

We know that, uniquely, perovskite multiferroics show coupled ferroic orders, which open up the possibilities for multifunctional sensors. Such materials can provide simultaneous detection of magnetic fields, mechanical stress, and temperature variation. High magnetoelectric and piezoelectric coefficients allow such sensors significantly improved sensitivity, enabling the detection of weak signals that conventional approaches would be unable to detect. The compatibilities of perovskite multiferroics with modern microelectromechanical systems (MEMS) technology ease the integration of such sensors within compact and portable equipment. Many perovskite multiferroic sensors operate with low power consumption, which is critical for applications involving wearable electronics and remote environmental monitoring [34].

5.3 Challenges in Perovskite Sensors

Perovskite multiferroics face significant challenges in while achieving long-term operational and environmental stability, which are critical for real-world sensor applications. ABO_3 perovskites, such as $BaTiO_3$, transition from monoclinic to tetragonal to cubic between 303 and 393 K, leading to dielectric anomalies ($\pm 15\%$ shift in ε_r/K). This destabilizes temperature-sensitive sensors. Comparably, an increase in pressure can induce structural transitions-from rhombohedral to orthorhombic-and results in hysteresis effects. An electric field induces a Low Spin (LS) $\rightarrow$ High Spin (HS) transition, affecting ferroelectric polarization (decreasing by 30%) and changing magnetic ordering, vital for spintronic sensor applications. Hybrid halide perovskites decompose, under conditions of humidity, into volatile organics. Selectivity of gas sensors toward NO_2 drops from approximately 95 to 72% after undergoing a 500-cycle thermal test due to the clustering of oxygen vacancies. Likewise, the dielectric losses ($\tan \delta > 0.1$) that result from vacancy formation decrease the sensitivity of electrochemical H_2O_2 detection. Radiation from UV light affects optoelectronic properties in halide perovskites. Photooxidation degrades quantum dots (QDs) for 90% of the photoluminescence quantum yield (PLQY) in 100 h, thereby limiting their applications to O_2/VOC sensing [46].

5.4 Mitigation Strategies for Challenges

A meticulous approach of development is needed to address issues with perovskite multiferroic sensors. Co-doping with transition metal and rare-earth ions lowers oxygen vacancy formation energies by 1.2 eV, enhancing thermal stability up to 450 °C. Moisture resistance of hybrid halides can be improved by alloying them with inorganic cations which reduce ionic conductivity by 50% under 85% RH. Layered 2D/3D heterostructures mitigate water ingress by 70% thereby extending the sensor life in humid conditions. Graphene interlayers reduce interfacial trap densities by two orders of magnitude improving magneto-electric coupling; while atomic layer electrodes lower the Schottky barrier height from 1.2 to 0.7 eV lowering the leakage currents. Encapsulation via atomic layer-deposited Al_2O_3 preserves 95% photoluminescence quantum yield under 85% RH for 30 days, and flexible substrate engineering improves piezoelectric coefficient retention (>80%) after 10^4 bending cycles. Pulsed electric field poling realigns ferroelectric domains, maintaining 95% sensitivity after 10^6 cycles, and real-time impedance spectroscopy enables predictive maintenance by detecting early-stage degradation through resonance frequency shifts (>0.1%) [47].

6 Perovskite Multiferroic Materials for Transducers

A transducer transforms one form of energy to another. Perovskite multiferroics are ideal candidates of such devices since it exhibits both piezoelectric and magneto-electric effects. These materials have gained significant attention in the transducer design and development due to their multifunctional properties and strong coupling between multiple orders.

6.1 Operating Principles

Perovskite multiferroic transducers use the coupling of ferroic orders to convert energy. Transducers of magnetoelectric type use the magnetoelectric phenomenon to transform magnetic fields into electrical impulses, and vice versa. In $BiFeO_3$, an applied alternating magnetic field interacts with electric dipoles, resulting in an alternating electric field. This approach has important applications in magnetic field detection and wireless energy transmission. In piezoelectric transducers, mechanical energy is converted into electrical energy, thanks to the non-centrosymmetric crystal structure of perovskite materials. The mechanical stress imposed will displace the charged ions from their equilibrium positions, resulting in the creation of electric dipole moments. The efficiency of the process factors such as intrinsic lattice anisotropy and crystallographic orientation. In perovskite materials, fluctuations in temperature induce a variation in net dipole moment due to expansion or contraction of the lattice and shift in ionic positions. The pyroelectric transducers operate based on this temperature-dependent variation of the dipole moment [38].

Recent advances in perovskite multiferroic transducers highlight the role of strain-mediated coupling in composite systems, where interfacial interactions between ferromagnetic and ferroelectric phases amplify magnetoelectric coefficients. Key efficiency determiners are interfacial lattice coherence, domain wall mobility, and defect engineering. Advanced synthesis techniques like pulsed laser deposition enable atomic-scale control over crystallographic alignment, critical for ultra-sensitive magnetometers. These systems now power contactless energy transfer in biomedical implants, achieving 68% wireless efficiency at 5 cm distances via resonance coupling [48–50].

6.2 Applications

The unique properties of multiferroic perovskites enable a broad spectrum of transducer applications. They are capable of high-resolution imaging in medical ultrasonics, while allowing high sensitivity along with a broad bandwidth for accurate imaging and diagnosis. The hybrid transducers in energy harvesting applications convert mechanical vibrations and thermal energy to electricity through their unique piezoelectric and pyroelectric properties. Highly sensitive magnetic field sensors based on magnetoelectric effects are employed in diverse fields, including geophysics and biomedical diagnostics. Perovskite multiferroics like PMN-PZT and $BiFeO_3$ composites enable high-performance piezoelectric micromachined ultrasound transducers (PMUTs) for medical diagnostics. Recent designs include annular-shaped PMUT arrays with frequencies up to 20–100 MHz, achieving acoustic pressures of 40 kPa at 2.66 MHz for intravascular imaging [51].

Multiferroic composites exploit a synergism of piezoelectric-magnetostrictive coupling for hybrid energy harvesting, for example, Pr_2FeAlO_6 double perovskites feature magnetodielectric response ($\frac{\Delta P_s}{P_s} = 5.71\%$), in which dual mechanical/magnetic energy can be converted. Flexible Au/$BiFeO_3$/mica films on PDMS retain 97% polarization after bending cycles, ideal for biomechanical energy scavenging in implantable devices. Recent advances in core–shell Pb(Zr,Ti)O_3-$NiFe_2O_4$ composites show 4× enhanced flexoelectric coefficients, optimizing low-frequency vibration harvesting. Pr_2FeAlO_6 and $BiFeO_3$-$CoFe_2O_4$ heterostructures demonstrate room-temperature magnetoelectric coupling, outperforming traditional $Bi_{0.5}Na_{0.5}TiO_3$-$La_{0.67}Sr_{0.33}MnO_3$ composites by 124%. These sensors achieve picotesla-level resolution for geophysical exploration and biomagnetic diagnostics [52]. Flexible Fe/$BaTiO_3$/PDMS transducers exhibit ferromagnetic hysteresis at bending radii less than 5 mm, allowing conformal mapping of the magnetic field in wearable systems [23].

Core–shell lanthanide-upconversion nanoparticles were incorporated with lead halide perovskites and show 75% NIR-to-visible light conversion efficiency. $CsPbBr_3$/BNNO 2D/3D heterolayers have a pyroelectric coefficient of 350 $\mu C \cdot m^{-2} \cdot K^{-1}$ and are capable of uncooled infrared thermal imaging for monitoring in industrial processes. $CoFe_2O_4$-$BiFeO_3$ multiferroic tunnel junctions show non-volatile resistive switching with 10 ns response times, suitable for ultra-low-power MRAM. 2D organic–inorganic perovskites (e.g., PEA-Fe) exhibit strain-tunable magnetization shifts via spin–orbit coupling, enabling AFM-based logic devices. Room-temperature magnetoelectric coupling in $PbZr_{0.53}Ti_{0.47}O_3$-$PbFe_{1/2}Ta_{1/2}O_3$

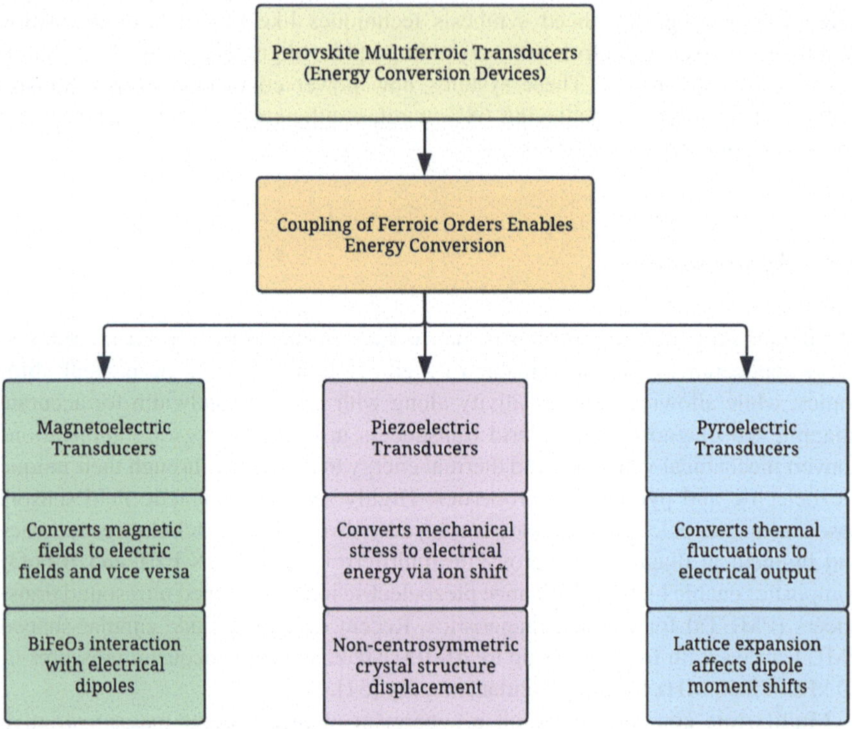

Fig. 4 Summary of operating principles of transducers

thin films allows electric-field control of magnetic anisotropy, advancing spintronic sensors. Summary of operating principles of transducers is depicted in Fig. 4.

6.3 *Material Optimization and Challenges*

Perovskite multiferroics face intrinsic and synthesis-related limitations that hinder their transducer applications, including hysteresis losses of 12–18% due to anisotropic strain broadening in cyclic magnetoelectric switching, temperature instability where piezoelectric coefficients degrade by ~ 30% at 150 °C due to domain wall pinning, and interface coupling inefficiencies limiting composite multiferroics to 40–60% of theoretical magnetoelectric coupling due to lattice mismatch. Optimization strategies include A/B-site co-doping, where La-Mn codoping in perovskites reduces leakage currents to < 10^{-8} A/cm^2 by mitigating oxygen vacancies, nano-structuring through core–shell architectures that enhance flexoelectric coefficients by 4× via strain engineering, and epitaxial growth of 2D/3D hybrid perovskites, improving thermal stability up to 250 °C [17].

The recent advances in operando characterization show that gradient doping, that is through the layered A-site substitutions (La^{3+}/Sr^{2+}), reduces thermal depolarization by stabilizing oxygen octahedral tilts up to 300 °C [53, 54]. Mixed with machine learning-guided combinatorial screening for low-hysteresis composition through selective suppression of 71° domain walls, $(Bi_{0.9}Sm_{0.1})(Fe_{0.7}Mn_{0.3})O_3$ presented with < 8% loss. The interface coupling is ameliorated through ALD HfO_2 buffer layers with the reduction of lattice mismatch in $BiFeO_3/CoFe_2O_4$ composites yielding 89% of magnetoelectric coupling efficiency, brought about by strain transfer. Difficult scaling problems exist: sol–gel synthesis of core–shell nanoparticles (<30% monodispersity), and pulsed laser deposition of 2D/3D hybrids is prohibitively expensive (>450 $/cm^2). Lead-free alternatives such as $KNbO_3$-$BaTiO_3$ display promising d33 coefficients (~120 pC/N) but phase segregate at elevated temperatures (>180 °C). Therefore, ionic liquid defect passivation (for example, [EMIM][BF_4]) during spin-coating produced a 63% decrease in grain boundary voids while improving the flexoelectric response (at 0.1% strain 8.2 nC/cm) [55, 56].

6.4 *Emerging Trends in Transducer Systems*

Lead-free perovskite transducers are promising sustainable alternatives that answer IoT networks for distributed autonomous energy with a leveling capacity toward environmental resilience. The double-mode energy-harvesting characteristics of $KNbO_3$-$BaTiO_3$ composites apply 8.2 μW/cm^3 in magnetostriction (1 Hz/0.5% strain) and 15% photovoltaic efficiency under AM1.5G illumination, allowing further self-powered structural health monitoring in smart cities. Flexible $La_{1-x}Sr_xMnO_3$/PMN-PT heterostructures on PET substrates remedied mechanical strain resolution of about 0.01% for reliable gait analysis with 95% accuracy over 10^6 flexion cycles. For extreme environments, high-entropy perovskites like $(Bi,La)(Fe,Mn,Nb)O_3$ maintain thermal stability throughout 500 °C and 5% magnetostriction under pressures of Venus's atmosphere, as required for next-generation planetary probes. Nevertheless, the major bottlenecks in $KNbO_3$-$BaTiO_3$ include misalignment, above 180 °C polymerization caused by the Au/Pt electrodes and polarization screening: a 70% loss in < 10 nm films. Machine learning-guided dopant optimization (e.g., Sm^{3+}/Nb^{5+} co-doping) is shrinking R&D cycles by 65%, while ALD-enabled 8-inch wafer processing (95% uniformity vs. 30% for sol–gel) promises scalability44. These developments set the perovskite transducers in stone for resilient Industry 4.0 ecosystems [13].

These perovskite multiferroics are also enabling transformative advances in bio-integrated systems, $BiFeO_3$-based transducers are being integrated into flexible PVDF substrates for real-time monitoring of neural oscillations (0.1–500 Hz) in closed-loop neuromodulation devices15. These systems leverage the material's magnetoelectric coupling to detect weak bio magnetic fields while operating at biocompatible voltages (<100 mV) [57]. Concurrently, perovskite superlattices like $La_{0.7}Sr_{0.3}MnO_3$/PMN-PT are demonstrating magneto-optic coupling for quantum

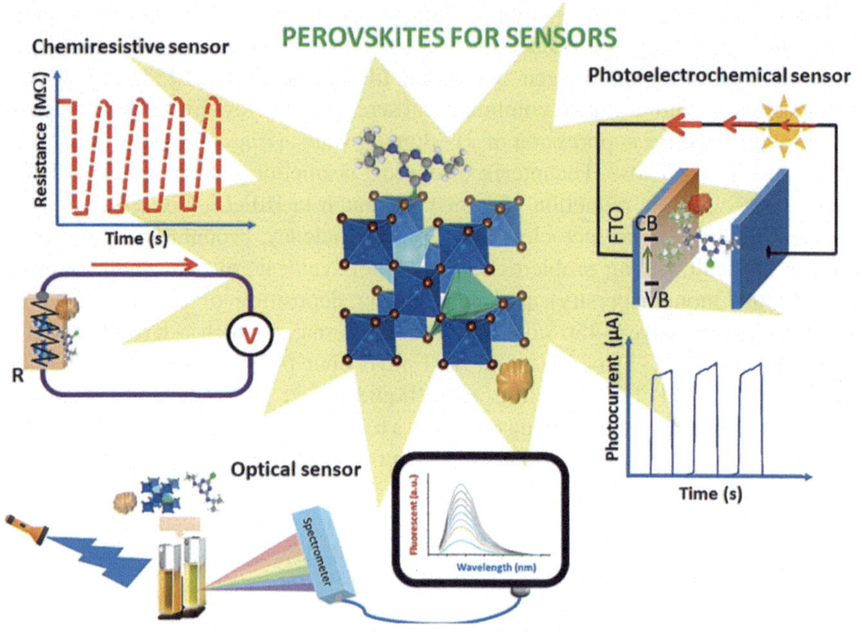

Fig. 5 Perovskite materials for sensing applications

state transduction, achieving 90% Faraday rotation efficiency at 4 K for spin-photon interfaces in quantum computing architectures21. Recent work on 2D/3D hybrid perovskites (e.g., $MoS_2/BiFeO_3$) has shown 400% magnetoresistance under 0.1 T fields, enabling cryogen-free quantum sensors with 10^{-6} rad/$\sqrt{}$Hz angular resolution for gravitational wave detection 1420 [45]. Challenges persist in minimizing interfacial losses ($<10^{-8}$ A/cm^2 leakage currents) and achieving sub-100 nm feature sizes via atomic layer deposition (ALD). Figure 5 illustrates the perovskite materials for sensing applications [58, 59].

Another transformative strategy for next-generation transducers is the integration of perovskite materials with CMOS technology. Recent advancements in the integration of CMOS and perovskite materials opened new possibilities in imaging, photonics, and energy harvesting. Monolithic X-ray imaging sensors using direct-conversion X-ray CMOS imagers with thick inorganic CsPbBr3 films (>300 μm) exhibited high spatial resolution (5.0 lp/mm) and low drift of dark current (5.2×10^{-5} nA·mm^{-1}·s^{-1}·V^{-1}) thanks to pixel-level charge-sensitive amplifiers. CsPbBr3 was integrated with Timepix4 ASICs, demonstrating compatibility with photon-counting architectures but requiring high electric fields (~1 kV/mm) to achieve efficient charge collection [60]. With the implementation of monolithic 64 × 512 CMOS-perovskite arrays, high-resolution image sensors achieved an 83.2 μm pixel pitch with excellent performance in low-light imaging. Schottky junction-based perovskite photodiodes exhibited a significant reduction of leakage currents by 78% leading to a

CMOs-compatible dark current below 10^{-9} A/cm^2. In hybrid photonic integration, $CsPbBr_3$ nanocrystals were embedded in silicon nitride nanobeam cavities to produce on-chip lasers with sub-100 nm mode volumes and low lasing thresholds (~10 μJ/cm^2) while remaining CMOS compatible. The lead-free $KNbO_3$-$BaTiO_3$ composites enabled dual-mode energy harvesting, thus generating 8.2 μW/cm^3 from mechanical sources, and giving room for a 15% photovoltaic conversion efficiency ensuring self-powered IoT nodes [61, 62]. Fabrication methods like parylene-assisted photolithography made it easy to pattern perovskite with features below 5 μm for memristors and photodetectors. In addition, HfO_2 buffer layers deliberately introduced by means of atomic layer deposition (ALD) ameliorated interfacial lattice mismatch to < 0.5% by enhancing the efficiency of strain transferring to bilayers of $BiFeO_3/CoFe_2O_4$ thus enhancing the performance of the whole device [27–29].

References

1. Aggarwal MD, Batra AK, Guggilla P, Edwards ME, Penn BG, Currie Jr JR (2010) Pyroelectric materials for uncooled infrared detectors: processing, properties, and applications. NASA, p 3
2. Aldred AT, Dunlap BD, Lam DJ, Lander GH, Mueller MH, Nowik I (1975) Magnetic properties of neptunium Laves phases: $NpMn_2$, $NpFe_2$, $NpCo_2$, and $NpNi_2$. Phys Rev B 11(1):530–544. https://doi.org/10.1103/physrevb.11.530
3. Assadi MHN, Hanaor DAH (2013) Theoretical study on copper's energetics and magnetism in TiO_2 polymorphs. J Appl Phys 113(23):233913-1–233913-5. https://arxiv.org/pdf/1304.1854
4. Ascher E et al (1966) Some properties of ferromagnetoelectric nickel-iodine boracite, $Ni_3B_7O_{13}I$. J Appl Phys 37(3):1404–1405. https://doi.org/10.1155/2010/763406
5. Attema JJ, de Wijs GA, Blake GR, de Groot RA (2005) Anionogenic ferromagnets. J Am Chem Soc 127(46):16325–16328
6. Chikazumi S (2009) Physics of ferromagnetism. English edition prepared with the assistance of C. D. Graham, Jr., 2nd edn. Oxford University Press, Oxford, p 118
7. Bichurin M, Petrov R, Sokolov O, Leontiev V, Kuts V, Kiselev D, Wang Y (2021) Magnetoelectric magnetic field sensors: a review. Sensors 21(18):6232. https://doi.org/10.3390/s21186232
8. Binek C et al (2005) Magnetoelectronics with magnetoelectrics. J Phys Condens Matter 17(2):L39–L44. https://doi.org/10.1088/0953-8984/17/2/L06
9. Astrov DN (1960) The magnetoelectric effect in antiferromagnets. Sov Phys JETP 11:708. https://doi.org/10.1088/0953-8984/23/11/113201
10. Clark NA, Lagerwall ST (1980) Submicrosecond bistable electro-optic switching in liquid crystals. Appl Phys Lett 36(11):899–901. https://doi.org/10.1063/1.98341
11. Dzyaloshinskii IE (1960) On the magneto-electrical effect in antiferromagnets. Sov Phys JETP 10:628
12. Feynman RP, Leighton RB, Sands M (1963) The Feynman lectures on physics, vol I. California Inst. of Technology, Pasadena, pp 37.5–37.6
13. Yo G-B, Lee Y-R, Choi J-H, Christensen CA, Kim TH, Thywissen JH, Pritchard DE, Ketterle W (2009) Itinerant ferromagnetism in a fermi gas of ultracold atoms. Science 325(5947):1521–1524. http://hdl.handle.net/1721.1/88507
14. Velarde G, Pandya S, Karthik J, Pesquera D, Martin LW (2021) Pyroelectric thin films—past, present, and future. APL Mater 9(1):010702
15. Goldschmidt VM (1926) The laws of crystal chemistry. Naturwissenschaften 14(21):477–485. https://doi.org/10.1007/BF01507527

16. Griffin SM, Lilienblum M, Delaney KT, Kumagai Y, Fiebig M, Spaldin NA (2012) Scaling behavior and beyond equilibrium in the hexagonal manganites. Phys Rev X 2(4):041022
17. Li H, Zhuang J, Bokov AA, Zhang N, Zhang J, Zhao J, Ren W, Ye Z-G (2021) Evolution of relaxor behavior in multiferroic $Pb(Fe_{2/3}W_{1/3})O_3$-$BiFeO_3$ solid solution of complex perovskite structure. J Eur Ceram Soc 41(1):310–318. https://doi.org/10.1016/j.jeurceramsoc.2020.07.068
18. Heron JT et al (2011) Electric-field-induced magnetization reversal in a ferromagnet-multiferroic heterostructure. Phys Rev Lett 107(21):217202. https://doi.org/10.1103/PhysRevLett.107.217202
19. Hill C, Weber MC, Lehmann J, Leinen T, Fiebig M, Kreisel J, Guennou M (2020) Role of the ferroelastic strain in the optical absorption of $BiVO_4$. APL Mater 8(8). https://doi.org/10.1063/5.0011507
20. Hill NA (2000) Why are there so few magnetic ferroelectrics? J Phys Chem B 104(29):6694–6709. https://doi.org/10.1021/accountsmr.4c00384
21. Hoffmann T et al (2011) Time-resolved imaging of magnetoelectric switching in multiferroic $MnWO_4$. Phys Rev B 84(18):184404. https://doi.org/10.1098/rsta.2012.0450
22. Khan H, Ahmad T (2024) Perspectives and scope of ABO_3 type multiferroic rare-earth perovskites. Chin J Phys 91:199–219. ISSN 0577-9073
23. Juraschek DM, Fechner M, Balatsky AV, Spaldin NA (2017) Dynamical multiferroicity. Phys Rev Mater 1(1):014401. https://doi.org/10.1103/physrevmaterials.1.014401
24. Kakekhani A, Ismail-Beigi S, Altman EI (2016) Ferroelectrics: a pathway to switchable surface chemistry and catalysis. Surf Sci 650:302–316. https://doi.org/10.1016/j.susc.2015.10.055
25. Katsura H, Balatsky AV, Nagaosa N (2007) Dynamical magnetoelectric coupling in helical magnets. Phys Rev Lett 98(2):027203
26. Kimura T et al (2003) Magnetic control of ferroelectric polarization. Nature 426(6962):55–58. https://doi.org/10.1038/nature02018
27. Lander GH, Lam DJ (1976) Neutron diffraction study of PuP: the electronic ground state. Phys Rev B 14(9):4064–4067. https://doi.org/10.1103/physrevb.14.4064
28. Laursen AB, Man IC, Trinhammer OL, Rossmeisl J, Dahl S (2011) The Sabatier principle illustrated by catalytic H_2O_2 decomposition on metal surfaces. J Chem Educ 88(12):1711–1715. https://doi.org/10.1021/acs.jchemed.0c01101
29. Dawber M, Rabe KM, Scott JF (2005) Physics of thin-film ferroelectric oxides. Rev Mod Phys 77(4):1083. https://doi.org/10.1103/RevModPhys.77.1083
30. Vopson MM (2015) Fundamentals of multiferroic materials and their possible applications. https://doi.org/10.1080/10408436.2014.992584
31. Mueller MH, Lander GH, Hoff HA, Knott HW, Reddy JF (1979) Lattice distortions measured in actinide ferromagnets PuP, $NpFe_2$, and $NpNi_2$. J Phys Colloque C4 Supplement 40(4):C4-68–C4-69
32. Zhuravlev MYe, Sabirianov RF, Jaswal SS, Tsymbal EY (2005) Giant electroresistance in ferroelectric tunnel junctions. Phys Rev Lett 94(24):246802–4. https://doi.org/10.1103/PhysRevLett.94.246802
33. Wang M et al (2017) Enhanced multiferroic properties of $YMnO_3$ ceramics fabricated by spark plasma sintering along with low-temperature solid-state reaction. Materials 10(5):474. https://doi.org/10.3390/ma10050474
34. Mostovoy M, Spaldin NA, Delaney KT, Artyukhin S (2014) Landau theory of topological defects in multiferroic hexagonal manganites. Nat Mater 13(1):42–49. https://doi.org/10.1038/nmat3786
35. Seidel J et al (2009) Conduction at domain walls in oxide multiferroics. Nat Mater 8(3):229–234. https://doi.org/10.1038/nmat2373
36. Quarterman P, Sun C, Garcia-Barriocanal J, Mahendra DC, Lv Y, Sasikanth M, Nikonov DE, Young IA, Voyles PM, Wang J-P (2018) Demonstration of Ru as the 4th ferromagnetic element at room temperature. Nat Commun 9(1):2058. https://doi.org/10.1038/s41467-018-04512-1
37. Wood IG (1984) Spontaneous birefringence of ferroelastic $BiVO_4$ and $LaNBO_4$ between 10K and T c. J Phys C Solid State Phys 17(21):L539–L543. https://doi.org/10.1088/0022-3719/17/21/003

38. Rovillain P et al (2010) Magnetoelectric excitations in multiferroic $TbMnO_3$ by Raman scattering. Phys Rev B 81(5):054428. https://doi.org/10.1103/PhysRevB.81.054428
39. Salje EKH, Hayward SA, Lee WT (2005) Ferroelastic phase transitions: structure and microstructure. Acta Crystallogr Sect A 61(1):3–18. https://doi.org/10.1107/S0108767304020318
40. Nan CW et al (2008) Multiferroic magnetoelectric composites: historical perspective, status, and future directions. J Appl Phys 103(3):031101–031101–35. https://doi.org/10.1063/1.2836410
41. Nechache R, Harnagea C, Li S, Cardenas L, Huang W, Chakrabartty J, Rosei F (2015) Bandgap tuning of multiferroic oxide solar cells. Nat Photonics 9(1):61–67. https://doi.org/10.1038/NPHOTON.2014.255
42. Parravano G (1952) Ferroelectric transitions and heterogenous catalysis. J Chem Phys 20(2):342–343. https://doi.org/10.1063/1.1700412
43. Clabel Huamán JL, Rivera VAG, Pinto AH, Marega E (2022) Multiferroic perovskite ceramics: properties and applications. https://doi.org/10.1016/B978-0-323-90586-2.00003-6
44. Salje EKH (2010) Multiferroic domain boundaries as active memory devices: trajectories towards domain boundary engineering. ChemPhysChem 11(5):940–950. https://doi.org/10.1002/cphc.200900943
45. Somasundaran P (ed) (2006) Encyclopedia of surface and colloid science, 2nd edn. Taylor & Francis, New York, p 3471
46. Känzig W (1957) Ferroelectrics and antiferroelectrics. In: Seitz F, Das TP, Turnbull D, Hahn EL (eds) Solid state physics, vol 4. Academic Press, p 5
47. Yen F, De la Cruz CR, Lorenz B, Sun YY, Wang YQ, Gospodinov MM, Chu CW (2005) Low-temperature dielectric anomalies in $HoMnO_3$: the complex phase diagram. Phys Rev B 71(18):180407
48. Spaldin NA (2017) Multiferroics: from the cosmically large to the subatomically small. Nat Rev Mater 2(5):17017. https://doi.org/10.1038/nmat2799
49. Starr MB, Shi J, Wang X (2012) Piezopotential-driven redox reactions at the surface of piezoelectric materials. Angew Chem Int Ed 51(24):5962–5966. https://doi.org/10.1002/anie.201201424
50. Sharma S, Arreguín Hernández ML, Siqueiros JM, Raymond Herrera O (2024) Recent advances in electric field-controlled magnetism using perovskite multiferroic heterostructures for spintronics applications. Mater Lett 371:136974. https://doi.org/10.1016/j.matlet.2024.136974
51. Walter J, Voigt B, Day-Roberts E, Heltemes K, Fernandes RM, Birol T, Leighton C (2020) Voltage-induced ferromagnetism in a diamagnet. Sci Adv 6(31). https://doi.org/10.1126/sciadv.abb7721
52. Gajek M et al (2007) Tunnel junctions with multiferroic barriers. Nat Mater 6(4):296–302. https://doi.org/10.1038/nmat1860
53. Cheong S-W, Kiryukhin V, Choi YJ, Lee S, Choi T (2009) Switchable ferroelectric diode and photovoltaic effect in $BiFeO_3$. Science 324(5923):63–66. https://doi.org/10.1126/science.1168636
54. Ramesh R, Spaldin NA (2007) Multiferroics: progress and prospects in thin films. Nat Mater 6(1):21–29. https://doi.org/10.1038/nmat1805
55. van den Brink J, Khomskii DI (2008) Multiferroicity due to charge ordering. J Phys Condens Matter 20(43):434217. https://doi.org/10.48550/arXiv.0803.2964
56. Chen A, Wen Y, Fang B, Zhao Y, Zhang Q, Chang Y, Li P, Wu H, Huang H, Lu Y, Zeng Z (2019) Giant nonvolatile manipulation of magnetoresistance in magnetic tunnel junctions by electric fields via magnetoelectric coupling. Nat Commun 10(1):243. https://doi.org/10.1038/s41467-018-08061-5
57. Rupp T, Truong BD, Williams S, Roundy S (2019) Magnetoelectric transducer designs for use as wireless power receivers in wearable and implantable applications. Materials (Basel, Switzerland) 12(3):512. https://doi.org/10.3390/ma12030512

58. Whatmore RW (1991) Piezoelectric and pyroelectric materials and their applications. In: Miller LS, Mullin JB (eds) Electronic materials: from silicon to organics. Springer, Boston, MA, US, pp 283–290. https://doi.org/10.1007/978-1-4615-3818-9_19
59. Yasui Y et al (2009) Studies of multiferroic system $LiCu_2O_2$: I. Sample characterization and relationship between magnetic properties and multiferroic nature. J Phys Soc Jpn 78(8):084720. https://doi.org/10.1209/0295-5075/97/37004
60. Yun Y, Altman EI (2007) Using ferroelectric poling to change adsorption on oxide surfaces. J Am Chem Soc 129(50):15684–15689. https://pubs.acs.org/doi/10.1021/acsomega.2c06828
61. Fetisov YK, Srinivasan G (2005) Electron Lett 41(19):1066–1067. https://doi.org/10.1049/el:20051653
62. Fetisov YK, Srinivasan G (2005) Appl Phys Lett 87:103502. https://doi.org/10.1063/1.2037860

Chapter 10
Applications of Perovskite Multiferroic Materials in Surface Acoustic Wave Devices

Rajveer Singh, Meenakshi Gautam, and Sukriti Atrey

1 Introduction

1.1 Overview of Surface Acoustic Wave (SAW) Devices and Their Significance in Modern Technologies

Surface acoustic waves (SAWs) are sound waves that travel along the surface of an elastic material. These are confined up to an area of around one wavelength in depth from the surface as their displacement amplitude gets diminished as they pierce through the material [1]. Mechanical deformation brought on by SAWs can cause electric fields to be produced in a piezoelectric material. A SAW can be produced by applying an appropriate oscillating signal to a specifically designed set of surface gates. These electric fields cause a change in electrostatic potential that moves with the SAW because they have no effect on the mechanical wave's propagation. Although the mechanical SAW itself is not affected, the electrostatic potential which surrounds thin metal layers or conductive areas on the surface get affected [2]. SAW devices are seen to combine electronics with the surface acoustic wave technology which operate on the principle of piezoelectricity and preparation processes [3]. These waves can be transmitted on solid surface, and their propagation speed can reach 10^3–10^5 m/s and about 85% of the energy can be transmitted along the solid surface. SAW-based sensors are basically composed of three parts—piezoelectric substrate, a pair of metal interfering transducer and a sensing area in

R. Singh (✉) · S. Atrey
Department of Physics, Atma Ram Sanatan Dharma College, University of Delhi, Dhaula Kuan New Delhi, Delhi, India
e-mail: rajveersingh@arsd.du.ac.in

M. Gautam
Chemistry Department, Satish Chandra College, Jananayak Chandrashekhar University (JNCU), Ballia, Uttar Pradesh, India

M. Kumar (ed.), *Sustainable Perovskite Multiferroic Materials*, Materials Horizons: From Nature to Nanomaterials, https://doi.org/10.1007/978-981-96-7419-0_10

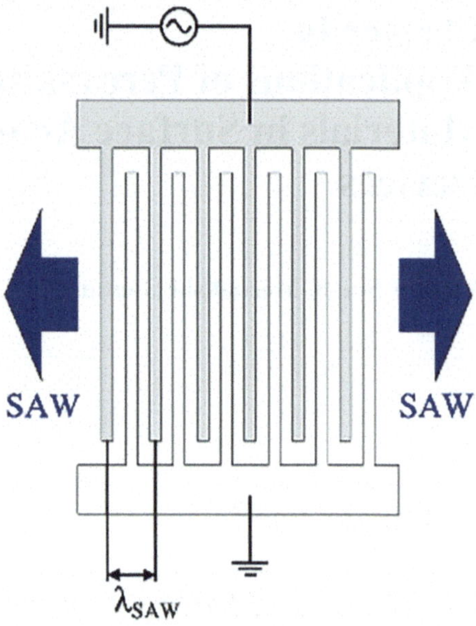

Fig. 1 Schematic representation of a SAW transducer

which the piezoelectric layer film has a crucial impact on the sensor's performance. The SAW device uses a sensor with a piezoelectric crystal vibrating at a resonance frequencies as high as 600 MHz [4–6].

The schematic diagram (Fig. 1) illustrates a typical SAW transducer, which consists of multiple pairs of interdigitated electrodes arranged in a grating-like pattern. The transducer's pitch determines the SAW wavelength. By grounding one side of the transducer and applying a signal at a frequency equal to the SAW velocity (approximately 2700 m/s for GaAs) divided by the transducer's pitch, a SAW is generated.

1.2 Role of Materials in SAW Device Performance

Materials serve an essential role in SAW devices since the performance, efficiency, and application of these devices are all directly affected by the substrate and other materials used. The materials used in SAW devices must have specific qualities such as piezoelectricity, minimal acoustic loss, and temperature stability to guarantee optimum performance. Piezoelectric materials are essential for producing and detecting SAWs as they transform electrical energy into mechanical vibrations and vice versa. As an example, Lithium Niobate ($LiNbO_3$) is preferred for high-frequency applications due to its high electromechanical coupling; Lithium Tantalate ($LiTaO_3$) is used due to its well-balanced coupling coefficient and temperature stability [7]. Energy conversion efficiency between the electrical and mechanical domains is determined by the electromechanical coupling coefficient (K^2), where larger K^2 values are preferred for better performance. Higher frequency operation is enabled by materials with higher acoustic velocity. Therefore, the device's operating frequency is influenced by the acoustic velocity. A low temperature coefficient of delay (TCD) is ideal for reliable operation since it shows how the device's performance varies with temperature [8]. For SAW devices which are used in temperature variable settings, thermal stability is crucial. Low TCD materials ensure consistent performance. Quartz, because of its near-zero TCD, is widely employed to temperature sensitive applications while Langasite ($La_3Ga_5SiO_{14}$) is preferred due to its temperature stability and high coupling coefficient [9]. Materials with low acoustic loss are essential to minimise energy dissipation and to improve the quality factor (Q-factor) of SAW devices. Quartz (SiO_2), for example, is used due to low acoustic loss while Gallium Arsenide (GaAs) can be used for high-frequency applications due to its compatibility with semiconductor fabrication and low loss [10]. Performance of SAW devices can be improved by optimising wave propagation and coupling using thin films and layered structures. For example, zinc oxide (ZnO) can be applied as a thin piezoelectric film on non-piezoelectric substrates (such as silicon) and aluminium nitride (AlN) is beneficial due to its high sonic velocity and compatibility with semiconductor processes [11]. Interdigital transducers (IDTs) produce and detect SAWs using metallic electrodes (such as Gold). Metal selection impacts the electrical conductivity and endurance of the gadgets [12]. New materials, including 2D materials (like Graphene) and sophisticated piezoelectric, are being looked into for enhancing SAW device performance and for opening up new applications. As an example, graphene can be utilised for its mechanical strength and outstanding conductivity and piezoelectric polymers such as PVDF can be used for flexible surface-acoustic waves [13]. Table 1 shows the role of multiferroics materials in SAW devices.

1.3 Introduction to Multiferroic Materials and Their Unique Properties

Materials displaying multiple primary ferroic ordering types in a single phase are known as multiferroics. Although multiferroism technically refers uniquely to materials where more than one order parameter is being coupled, researchers have recently begun to include antiferromagnetism and ferrimagnetism along with multiferroic materials. The ability of multiferroic materials to display several ferroic properties (such as ferroelectricity, ferromagnetism, and ferroelasticity) at the same time, as well as coupling effects between these qualities, is what makes them special. At an ambient temperature, bismuth ferrite ($BiFeO_3$) demonstrates both ferroelectricity and antiferromagnetism [14]. Magnetoelectric coupling, the ability of an electric field to influence magnetisation or a magnetic field to control electric polarisation, is the most unique characteristic of multiferroics. For an instance, an applied magnetic field can cause electric polarisation in rate-earth manganites (like $TbMnO_3$) [15].

Table 1 Application of various multiferroic materials in SAW devices

Properties	Role in SAW devices	Example material	References
Strong piezoelectricity	Generates and Detects SAWs efficiently	$BiFeO_3$, PZT	Jaffe et al. [24], Piezoelectric ceramics
Magnetoelectric coupling	Tunability of SAW properties generated by electric/magnetic fields	$BiFeO_3$	Wang et al. [14], Science
High electromechanical coupling	Efficiently converts high energy	$LiNbO_3$, $LiTaO_3$	Morgan [8], Surface Acoustic Wave Filters
Temperature stability	Shows stable performance under varying temperatures	$BiFeO_3$, PZT	Catalan and Scott [20], Advanced Materials
Low acoustic loss	Efficiently propagates SAWs with minimal energy loss	$LiNbO_3$, $LiTaO_3$	Ballantine et al. [9], Acoustic Wave Sensors
Tunable SAW properties	Reconfigurability of SAW devices using external fields	$BiFeO_3$	Chu et al. [17], Nature Materials
Thin-film compatibility	Incorporated into the semiconductor fabrication techniques	$BiFeO_3$, PZT	Ramesh and Spaldin [25], Nature Materials
Enhanced sensor sensitivity	Highly sensitive to strain, temperature and magnetic fields	$BiFeO_3$	Ma et al. [26], Advanced Materials
Multifunctionality	Combines filtering, sensing and signal processing in one device	$BiFeO_3$	Eerenstein et al. [27], Nature

Ferroelectricity occurs in certain multiferroics as a by-product of other order parameters, such as magnetic or structural ordering, rather than as a result of spontaneous polarisation. For example, inappropriate ferroelectricity caused by structural distortions is shown in hexagonal manganites (h-$RMnO_3$, where R = Y, Ho, etc.) [16]. Complex domain structures, where areas of various ferroic orders (such as ferroelectric and magnetic domains) coexist and interact, are frequently seen in multiferroic materials. The ferroelectric and antiferromagnetic domains of $BiFeO_3$, for instance, are connected, resulting in distinct domain patterns [17]. At particular temperatures (such as the Curie temperature for ferroelectricity or the Neel temperature for antiferromagnetism), multiferroic materials frequently experience phase transitions that alter their ferroic characteristics. As an illustration, at low temperatures, $TbMnO_3$ changes from a paraelectric to a ferroelectric phase when exposed to a magnetic field [18]. Strain mediates the interaction between ferroelectric and magnetic orders in certain multiferroics. The magnetoelectric coupling in thin-film heterostructures can be improved by strain [19]. Some materials, such as $BiFeO_3$, display multiferroic behaviour at normal temperature, making them more useful for applications than many multiferroics that only reveal their features at low temperatures [20]. Multiferroic's behaviour can be dynamically controlled by tuning their properties with external forces, such as stress, magnetic, or electric fields. An electric field can alter the magnetic characteristics of multiferroic composites [21]. Giant magnetoelectric effects, in which little variations in magnetic fields cause significant variations in electric polarisation, or vice versa, are present in certain multiferroics. As an example, giant magnetoelectric phenomena have been recorded in layered multiferroic heterostructures [22]. Because they combine mechanical, electrical, and magnetic capabilities into one material, multiferroics are naturally versatile and allow for new device applications. Spintronic devices, where spin states are controlled by electric fields, utilise multiferroics [23].

Multiferroics are very appealing for cutting-edge technological applications because of these special qualities. Rare earth manganites, ferrites, or transition metal perovskite oxides are examples of multiferroic materials. Yttrium Manganese Oxide ($YMnO_3$), Bismuth Ferrite ($BiFeO_3$), Bismuth Manganese Oxide ($BiMnO_3$), Holmium Manganese Oxide ($HoMn_2O_5$), Lutetium Iron Oxide ($LuFe_2O_4$) and Terbium Manganese Oxide ($TbMnO_3$) are a few examples. Barium Nickel Tetrafluoride ($BaNiF_4$) and spinel chalcogenides, such Zinc Chromium Selenide ($ZnCr_2Se_4$), are examples of nonoxides that are also multiferroics. There are many potential uses for multiferroic materials since they exhibit several ferroic orderings. Future memory systems with multiple control levels, sensors and actuators that respond to numerous stimuli, and spintronic devices are a few examples of these applications. The multiferroics are also divided into type I and type II multiferroics in recent research. Materials classified as type I multiferroics have a distinct source of ferroelectricity and magnetism, and their effects are largely independent of one another, but with a slight degree of interaction. The strong coupling between two states in type II materials, on the other hand, results in the occurrence of ferroelectricity due to magnetism. Nonetheless, these materials' polarisation is still quite tiny, usually less than 10^{-2} $\mu C/cm^2$ [24].

1.4 Motivation for Using Perovskite Multiferroics in SAW Applications

Surface acoustic wave (SAW) devices frequently employ perovskite multiferroics because of its special ferroelectric, magnetic, and elastic characteristics, which allow for improved functionality and performance in SAW applications. Strong piezoelectric coefficients make perovskite multiferroics, such as Lead Zirconate Titanate (PZT) and Bismuth Ferrite ($BiFeO_3$), perfect for producing and detecting SAWs [25]. SAW characteristics (such as frequency, phase, and amplitude) can be controlled with both electric and magnetic fields, thanks to the magnetoelectric interaction in perovskite multiferroics. This makes it possible for SAW devices to be versatile and adjustable. For example, the SAW propagation properties in $BiFeO_3$ can be modulated by an applied magnetic field [13]. The effectiveness of energy conversion between the electrical and mechanical domains in SAW devices is determined by the electromechanical coupling coefficient (K^2). $LiNbO_3$ and $LiTaO_3$ are examples of perovskite multiferroics that have high K^2 values, which improve their performance in SAW filters, resonators, and sensors [8]. Many perovskite multiferroics, such as $BiFeO_3$ and PZT, exhibit good temperature stability, which is crucial for SAW devices operating in varying environmental conditions [20]. Due to their low acoustic loss, perovskite multiferroics like $LiNbO_3$ and $LiTaO_3$ allow for high-quality SAW devices with little energy loss [10]. Due to the multiferroic nature of perovskite materials, external stimuli like electric or magnetic fields can be used to adjust SAW parameters (such frequency and phase). For SAW devices that can be reconfigured, this tunability is helpful. For instance, an applied electric field can change the SAW frequency and velocity in $BiFeO_3$ [17]. Perovskite multiferroics are compatible with contemporary semiconductor production procedures because they may be formed as thin films on a variety of substrates. For SAW devices to be integrated into small and portable electronics, this is essential [14]. Perovskite multiferroics' magnetoelectric coupling increases SAW sensors' sensitivity to outside stimuli including temperature, strain, and magnetic fields. They are therefore perfect for applications involving sophisticated sensing [27]. Perovskite multiferroics combine ferroelectricity, ferromagnetism, and elasticity to produce multifunctional SAW devices that can perform multiple tasks simultaneously, such as filtering, sensing, and signal processing [28].

2 Fundamentals of Perovskite Multiferroic Materials

2.1 Definition and Characteristics of Multiferroics

Multiferroic materials are a unique class of solid-state compounds where at least two ordered states, such as electric, magnetic, or piezo-elastic phases, coexist. For instance, ferromagnets exhibit piezomagnetic properties, while piezo-ferroelectrics can be considered a special category of magnetoelastic or electroelastic coupled multiferroics. Because of the coupling between ferroic phases and piezo-elastic properties, which allows for direct control of ferromagnetic and ferroelectric behaviour through applied mechanical stress, these materials are of great technological significance. Inverse effects also happen, in which mechanical strain caused by externally applied electric or magnetic fields changes the shape of the material. Figure 2 illustrates this complex interaction between electric polarisation, electric field, strain, magnetisation, magnetic field, and stress. Interestingly, piezoelectric and piezomagnetic couplings can result from the coexistence of ferroic order and piezo-elastic characteristics. Two additional multiferroic states that show cross-couplings among various order parameters are also possible: (i) elasto-magnetoelectric multiferroics, which combine all three order states—electric, magnetic, and elastic; and (ii) magnetoelectric multiferroics, in which electric and magnetic order states coexist. It should

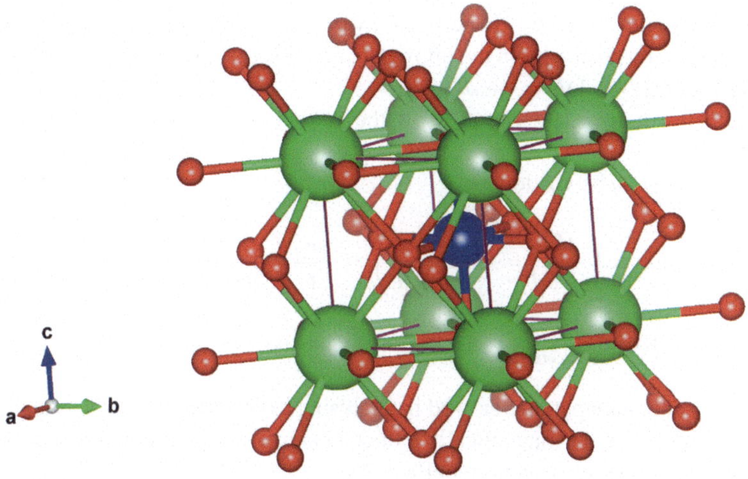

Fig. 2 Octahedral framework of ideal cubic perovskite structure

be noted that all ferroic orderings, including ferromagnetic, antiferromagnetic, ferrimagnetic, paramagnetic, superparamagnetic, and their electrical equivalents, where appropriate, are included in electric and magnetic order states.

In addition to their capacity to display multiple order states, multiferroic materials are of scientific and technological interest due to the cross-coupling effects that emerge between these states. The magnetoelectric coupling, which occurs in materials where magnetic-electric or magnetic-electric-elastic phases coexist, is another interesting and useful characteristic of some multiferroics in addition to piezo-ferro couplings. This effect enables the conversion of energy between magnetic and electric fields. As a result, magnetoelectric multiferroic materials exhibit a unique response: an applied external magnetic field modifies their electric polarisation, while an external electric field alters their net magnetisation. The magnetoelectric coupling coefficient (α) is a mathematical characterisation of the magnetoelectric effect in multiferroics, which can be induced either electrically or magnetically [29–33].

The change in the sample's magnetisation (M) brought on by the application of an electric field (E) is known as the electrically induced magneto-electric effect, and it can be expressed as follows:

$$\alpha_E = \left(\frac{\partial M}{\partial E}\right) \tag{1}$$

Similarly, the change in the sample's electric polarisation (P) brought on by the application of a magnetic field (H) is known as the magnetically induced magneto-electric effect.

$$\alpha_H = \left(\frac{\partial P}{\partial H}\right) = \epsilon_0 \epsilon_r \left(\frac{\partial E}{\partial H}\right) \tag{2}$$

where ϵ_o and ϵ_r are the dielectric permittivity of the vacuum and relative permittivity of the medium, respectively. In the above Eq. (2), it is assumed that $\epsilon_r >> 1$, so that $P = \epsilon_0 \chi E = \epsilon_0(\epsilon_r - 1)E = \epsilon_0 \epsilon_r E$.

The electric field $= \frac{V}{t}$, where 'V' is the voltage and 't' is the thickness of the dielectric structure.

Thus Eq. (2) can be written as:

$$\alpha_H = \left(\frac{\partial P}{\partial H}\right) = \epsilon_0 \epsilon_e \left(\frac{\partial E}{\partial H}\right) = \frac{\epsilon_0 \epsilon_r}{t}\left(\frac{\partial V}{\partial H}\right) = \alpha_H^V \epsilon_0 \epsilon_r. \tag{3}$$

where $\alpha_H^V = (\frac{\partial E}{\partial H}) = \frac{1}{t}(\frac{\partial V}{\partial H})$, is the magnetically induced voltage magneto-electric coefficient.

The α_H^V is the main parameter used in analysing experimental measurements. The relationship between the magnetically induced magneto-electric coupling coefficient and the voltage magneto-electric coefficient is:

$$\alpha_H = \epsilon_0 \epsilon_r \alpha_H^V \tag{4}$$

Both α_H and α_E are expressed in [s/m] in SI units [34]. Direct and indirect magneto-electric coupling are the two possible types of magneto-electric coupling. Direct magnetoelectric coupling occurs in multiferroic materials where the application of an electric field induces changes in magnetisation, and conversely, a magnetic field alters electric polarisation. This effect is a defining characteristic of a specialised class of multiferroics known as single-phase multiferroics [35]. Elasto-magnetoelectric multiferroics, sometimes referred to as composite multiferroics, are the main site of a second mechanism for triggering the magnetoelectric effect. In contrast to the direct effect, this mechanism—known as the indirect magnetoelectric effect—does not directly change electric polarisation or magnetisation when a magnetic or electric field is applied. Instead, these properties are modified indirectly through a strain-mediated magnetoelectric coupling [36].

Prussian mineralogist Gustav Rose made the initial discovery of perovskite, a mineral with the chemical formula $CaTiO_3$, in the mineral deposits of the Ural Mountains in 1839. It bears the name of Count Lev Aleksevich von Petrovski, a Russian mineralogist. Crystals of natural perovskites have a density of 4000–4300 kg/m^3 and a hardness of 5.5–6 on the mohs scale. Perovskites are a larger class of compounds with the general formula ABO_3 that bear the same name as perovskite, as many minerals do. In these structures, A and B represent positively charged cations of significantly different sizes, while O is the oxygen anion that bonds with both cations. Typically, the A cation is larger than the B cation. The smaller B^{4+} cation is surrounded by an octahedral coordination of six nearest neighbour O^{2-} ions (BO_6), while the larger A^{2+} cation can be coordinated with up to 12 nearest-neighbour O^{2-} ions. The octahedra unit of perovskite structure is displayed in Fig. 2 in which green balls indicates A^{2+} ion, blue balls represents B^{4+} ions, and oxygen ions are represented by red balls. The solid blue lines show cubic cell. The stability of this coordination depends on the relative sizes of A^{2+}, B^{4+}, and O^{2-}, which must satisfy the constraints of the well-known Goldschmidt tolerance factor [37].

$$t = \frac{r_A + r_O}{\sqrt{2}(r_A + r_O)}. \quad (5)$$

The tolerance factor decides the lattice types of perovskite and stability of the structure. It can be further influenced by temperature, pressure, electric field, etc. [38].

The origin of ferroelectricity in perovskites has been well established through soft mode theory. As the material cools, the lowest-frequency polar-phonon mode softens and eventually reaches zero frequency at the critical temperature (T_K). At this point, the material undergoes a structural transformation from an ideal cubic phase to a lower-symmetry ferroelectric phase.

The perovskite structure is currently utilised by hundreds of various compounds. The most common solid phase in the Earth's interior is bridgmanite ((Fe, Mg)SiO_3), a perovskite-structured mineral which makes up approximately 38% of the total volume. This phase occurs between 660 and 2900 km beneath the surface of the Earth, but it is only stable in extreme pressure and heat environments. When the dielectric and ferroelectric properties of barium titanate ($BaTiO_3$) were discovered in the 1940s, the importance of perovskites was widely acknowledged. Electronics quickly adopted this material, especially for transducers and capacitors. Extensive research has since been conducted to further enhance its properties.

2.2 Coupled Ferroic Properties: Ferroelectricity, Ferromagnetism, and Ferroelasticity

Multiferroics also display a characteristic hysteresis loop, featuring two equivalent response states at zero stimulus. The critical temperature is generally associated with symmetry breaking. The most commonly known ferroic orderings include: ferroelectric ordering (coupling between charge polarisation and an electric field), ferromagnetic ordering (coupling between magnetic moment and a magnetic field), and ferroelastic ordering (coupling between stress and strain). Another proposed ordering mechanism is ferrotoroidicity, where magnetic vortices arrange in a structured manner, leading to a property known as toroidisation. Figure 3 illustrates the various possible scenarios. While many materials exhibit either magnetic or electrical polarisability, only a few demonstrate both ferroelectric and ferromagnetic ordering. Materials that can be polarised both electrically and magnetically are known as magnetoelectric materials. However, materials that display both ferroelectric and ferromagnetic ordering within the same phase are specifically referred to as multiferroics.

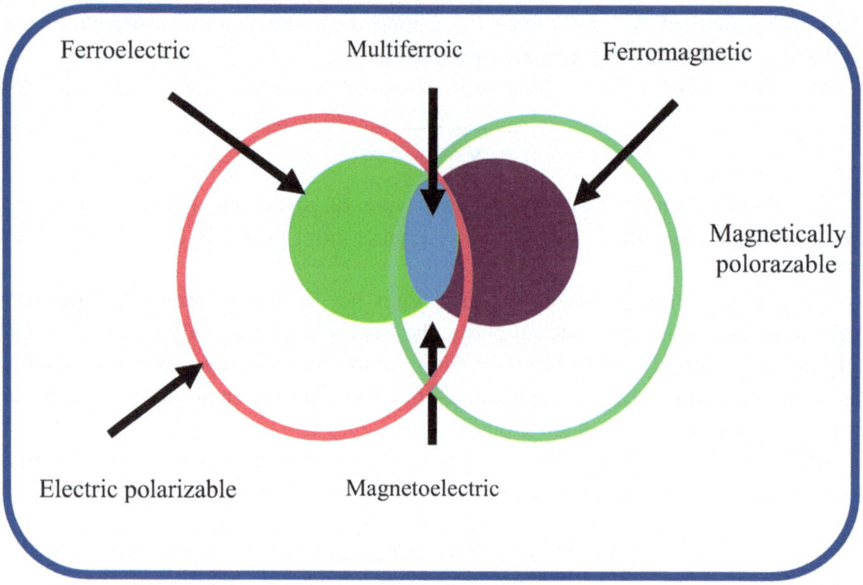

Fig. 3 Classification of multiferroic and magnetoelectric materials

3 Interaction of Acoustic Waves with Multiferroic Materials

3.1 Basics of Surface Acoustic Wave Propagation

SAW are essential in filters, sensors and signal processing systems because of their capacity to interact with external stimuli. The key components of SAWs are wave velocity, which depends on the elastic properties of the material and controls how quickly the wave propagates; attenuation, which is the loss of wave energy during propagation and affects the device's overall performance and efficiency; and wavelength, which is the distance between successive wave crests and directly affects the operating frequency of the device. Understanding these features is necessary to optimise SAW-based technologies for a variety of applications [1].

3.2 Piezoelectricity and Its Role in SAW Devices

Piezoelectricity is the capacity of a material to generate an electric charge in response to mechanical stress and vice versa. This idea is crucial for SAW devices, which produce and detect acoustic waves using piezoelectric materials. When an alternating electric field is given to a piezoelectric material, mechanical vibrations are induced and these vibrations spread as SAWs. On the other hand, the mechanical stress caused by SAWs during wave detection creates an electric field in the piezoelectric material, which can subsequently be picked up as a signal. Zinc oxide (ZnO), which is frequently utilised in thin-film SAW devices because of its compatibility with semiconductor fabrication processes; Lithium Niobate ($LiNbO_3$) which is known for its strong piezoelectric coupling and application in high-frequency devices and quartz, which is valued for its high piezoelectric coupling and low energy losses, are common piezoelectric materials used in SAW devices [39].

3.3 Magnetoelectric Coupling in Multiferroics and Its Influence on Acoustic Wave Behaviour

SAW devices are greatly impacted by magneto electric coupling, which offers improved control and tenability. The application of an electric field alters the dielectric characteristics of the material, causing changes in the wave velocity and resonant frequency. In a similar manner, magneto restriction brought on by a magnetic field changes the elastic properties and further influences wave velocity. Applying both fields at the same time enables dual tuning, which provides control over the propagation of waves. Materials that take Advantage of this effect include layered composites, which combine ferroelectric and ferromagnetic layers to improve magneto electric coupling for better performance in acoustic wave applications. For instance, $BiFeO_3$ is a perovskite multiferroic with a strong magnetoelectric coupling which makes it suited for tunable SAW devices [40].

3.4 Tunability of Acoustic Wave Propagation Using External Fields (Electric/Magnetic)

The ability of multiferroic materials to adjust the propagation of acoustic waves using external fields is a key benefit of SAW devices. This allows for real-time control over the frequency and phase of the device. There are several ways in which this tunability works. Magnetic field tuning stimulates magnetorestriction which alters the elastic properties and wave velocity. Conversely, electric field tuning shifts the resonant frequency by altering the material's dielectric constant. Dual tuning, which can be accomplished by applying both electric and magnetic fields at the same time, allows precise control over wave propagation. Multiferroic SAW devices are highly versatile due to these properties. They can be used in adaptive resonators to improve the performance of sensors and oscillators, tunable filters to dynamically adjust frequency response in communication systems and phase shifters to alter signal phases for beam forming in sophisticated 5G networks [41, 42].

4 Applications of Perovskite Multiferroics in SAW Devices

4.1 *Reconfigurable Filters: Use of Magnetoelectric Coupling for Frequency Tuning*

Reconfigurable filters, which enable real-time frequency adjusting to improve communication systems are one application where magnetoelectric coupling is used for frequency tuning. For dynamic wireless environments to adjust to shifting signals, these filters are a necessity [27].

4.2 *Design and Performance Enhancements*

Reconfigurable filters use tunable materials and complex design techniques to achieve dynamic frequency tuning, low insertion loss, high selectivity, and wide tuning ranges. Designing of reconfigurable filters involves a lot of factors, including tunable materials, filter topologies, tuning mechanisms, and integration techniques. The foundation of reconfigurable filter designs is tunable materials. $BiFeO_3$ and other multiferroics have magnetoelectric coupling [27].

4.3 *Sensors: Applications in Mass Sensing, Temperature Sensing, and Chemical Detection*

Perovskite multiferroics are especially well-suited for mass sensing, temperature sensing and chemical detection due to their high sensitivity, tunability, and ability to operate in harsh environments. When mass is deposited to the sensor surface, SAW devices use variations in the resonance frequency or wave velocity to detect mass [44]. Although there are many uses for these instruments, biosensing—the identification of biomolecules such as proteins or DNA—benefits greatly from their use. By aiding in the detection of aerosols or other airborne particulate matter, they also make a substantial contribution to environmental monitoring [43]. SAW devices use the temperature dependency of the acoustic wave velocity and resonant frequency for temperature sensing. The resonant frequency shifts as a result of the acoustic waves' altered propagation due to temperature variations. Because of their potent temperature-dependent characteristics, which improve the precision and sensitivity of temperature measurements, perovskite multiferroics are especially well-suited for this function. This technology is used in the automotive sector to monitor engine temperatures and in industrial monitoring, where temperature sensing is essential in tough settings [14]. Using SAW devices, chemical detection relies on the interaction of the target compounds with the sensor surface, which modifies surface characteristics or mass loading and causes measurable shifts in the resonant frequency. Because of their high sensitivity and capability for surface functionalisation, perovskite multiferroics greatly improve this procedure and enable more accurate chemical detection.

These devices are extremely useful in gas detection since they can detect toxic, explosive, and greenhouse gases. Therefore, they are suitable for industrial protection, environmental surveillance, and uses that include indoor air quality surveys, industrial plant leak detection, and urban air quality monitoring. SAW devices are also excellent liquid chemical sensors, which are crucial when tracking the quality of water and food. Due to their potential to sense hazardous compounds such as heavy metals, organic contaminants, and pathogens, they are ideal for monitoring industrial wastewater, detecting contaminants in drinking water, and food safety. Perovskite multiferroic-based SAW devices are suitable for application in biological and biomedical situations because they can sense biomolecules. They can identify a wide range of targets. They are capable of detecting a broad spectrum of target biomolecules such as pathogens, proteins, DNA, and RNA. Through enabling early disease detection, biomarker identification in body fluids, and pathogen detection in foods, this feature aids in enhancing health monitoring and safety across various medical and food companies [44].

4.4 Sensitivity Improvements with Multiferroic Integration

Multiferroic materials enhance the interaction between the sensor and target analytes, improve signal-to-noise ratios, and allow dual tuning via electric and magnetic fields. Perovskite multiferroic-based mechanisms for increasing sensitivity in SAW devices depend on a number of important variables. Since the surface of multiferroic materials can be functionalised with particular receptors (such as polymers or antibodies) to improve interactions with target analytes like gases, chemicals, or biomolecules, enhanced surface contacts also contribute to increased sensitivity. This improves the sensitivity and selectivity of identifying particular compounds [27]. For the detection of small changes in the environment of the sensor, multiferroic materials tend to possess good signal-to-noise ratios and low intrinsic noise. Finally, multiferroics can be subjected to magnetostriction (strain caused by a magnetic field) and electrostriction (strain caused by an electric field), which can be used to make the sensor highly sensitive to external stimuli. Strain-induced modifications in the material properties can greatly increase sensitivity.

4.5 Wireless Communication: Role in Frequency-Selective Components and Signal Processing

For contemporary wireless communication technologies such as 5G, Wi-Fi, and satellite communications, these materials make it possible to create tunable filters, resonators, delay lines, and phase shifters. In wireless communication, tunable filters are essential for choosing particular frequency bands. Multiferroic-based tunable filters are perfect for use in cognitive radio, where frequency band switching helps prevent interference, and 5G networks, where frequency bands must be dynamically altered for optimal signal transmission [27]. Resonators generate stable frequencies in communication systems. Perovskite multiferroics enable tunable resonators with minimal losses and high Q-factor. A SAW device's resonance frequency can be adjusted by adding electric or magnetic fields which in turn also modifies the material's dielectric and elastic characteristics. This feature is useful in frequency synthesisers where accurate frequency control is crucial for communication systems and oscillators which produce steady frequencies for wireless transceivers [42].

In signal processing applications like radar and telecommunications, delay lines are essential because they give the signal regulated temporal delay. By precisely adjusting the delay period by applying electric or magnetic fields, perovskite multiferroics improve these delay lines. In radar systems, where precise target detection depends on modifying the signal time, this tunability is very advantageous. Similar to this, tunable delay lines in telecommunications aid in compensating for signal delays, guaranteeing efficient network connection. Perovskite-based delay lines are beneficial in improving performance in radar and telecommunications due to such properties.

Phase shifters, which adjust signal phases for various applications, are important components of communication systems. Perovskite multiferroics are ideal for improving signal management since they offer reconfigurable phase shifters with low losses and high accuracy. These tunable phase shifters make it possible to precisely modify the signal phases in 5G beamforming, which improves coverage and enables directional transmission. Similar to this, they aid in compensating for phase distortions that may arise during long-distance transmission in satellite communication, guaranteeing dependable and clear communication. The efficiency of these vital applications is increased by the high performance of perovskite-based phase shifters. Changes in the properties of the material caused by strain can significantly increase sensitivity.

4.6 Potential for Low-Power, Multifunctional Devices

Perovskite multiferroics shows great promise for low-power multifunctional devices, facilitating energy efficient tuning, compact designs and low-power signal processing, all of which are crucial modern wireless communication systems. First, they enable energy-efficient tuning of SAW devices using electric and magnetic fields, lowering energy usage. This includes programmable filters for dynamically altering frequency bands in 5G networks and tunable resonators for creating stable frequencies in Internet of Things (IoT) devices [27]. Additionally, perovskite multiferroics enable low-power signal processing through effective delay lines and phase shifters, which are required for 5G beamforming and satellite communication. In addition to that, their ability to integrate several features, such as filtering, sensing and signal processing, enables compact and versatile SAW devices, making them perfect for low-power IoT and wearable applications [42]. Furthermore, they have an application in energy harvesting. It converts ambient energy from vibrations and magnetic fields into electric energy to power SAW devices, allowing for self-powered sensors and low-cost wireless sensor networks. In conclusion, perovskite multiferroics have great potential for producing next-generation, energy-efficient wireless communication devices.

5 Potential for Low-Advantages and Challenges

5.1 Advantages

Enhanced sensitivity and tunability

Perovskite multiferroics have strong magnetoelectric connection which increases the sensor's sensitivity to external stimuli. This improves the device's ability to detect small changes in its environment. For example, SAW gas sensors using $BiFeO_3$ can identify dangerous chemicals at ppm-level concentration [42]. Moreover, perovskite multiferroics allow for dual tuning of SAW devices utilising both electric and magnetic fields, giving fine control over frequency and phase. Electric fields influence the dielectric properties of a material, whereas magnetic fields influence its elastic properties. This tunability is especially useful for applications like tunable filters, which can dynamically modify frequencies for various communication bands in 5G networks [45].

Multifunctionality enabling diverse applications

Perovskite multiferroics provide multifunctional SAW devices that can perform various tasks, such as sensing, filtering, and signal processing for a small, single unit. A wide range of applications including gas, chemical, and biomolecule detection, tunable filters for wireless communication systems and signal processing components are made possible due to this adaptability. The main mechanism behind this is the strong magnetoelectric coupling in multiferroics which is what makes the integration of various into a single device possible. For example, a SAW device can operate as a gas sensor and a tunable filter at the same time, eliminating the need for numerous separate components and increasing efficiency in a variety of applications [27].

Miniaturisation and integration with microelectronics

SAW devices are naturally compact and using multiferroic materials decreases their size while maintaining great performance. For example, a miniaturised SAW sensor can be included in wearable devices to provide real-time health monitoring. Moreover, multiferroic materials can be manufactured using procedures compatible with ordinary microelectronics processes, allowing smooth integration with other RF components. This property enables the creation of small communication systems, such as a SAW filter combined with a wireless transmitter on a single chip, hence improving efficiency and performance in next-generation electronics [46].

5.2 *Challenges*

Material compatibility with existing device architectures

Integrating perovskite multiferroics into existing SAW device topologies is difficult due to variances in material characteristics, fabrication processes, and thermal expansion coefficients. One major issue is the mismatch in thermal expansion as perovskite multiferroics usually expand at different rates than traditional substrates such as quartz. This discrepancy can cause mechanical stress which can lead to delamination and decreased device dependability. Moreover, fabrication compatibility presents a hurdle, as normal microelectronics manufacturing procedures may not be immediately applicable to perovskite multiferroics. This involves the creation of new fabrication techniques or modifications to existing procedures to ensure seamless integration with SAW devices [14].

Stability under varying environmental conditions

Stability of perovskite multiferroics in different environments poses a major problem for their incorporation into SAW devices. Temperature sensitivity is a critical issue, since the properties of perovskite multiferroics can drastically vary with temperature, affecting the stability and reliability of SAW devices. To that, humidity sensitivity can be a problem as well, since certain perovskite materials are hygroscopic and could deteriorate in humid environments. Furthermore, mechanical stress can produce structural concerns like cracking or delamination of thin films which affects device durability. Addressing these problems involves improved material engineering and preventative measures to improve the environmental stability of perovskite multiferroic SAW devices [42].

Scalability of fabrication processes

Scalability of the fabrication processes of perovskite multiferroic-based SAW devices with high performance and reliability is a significant challenge. One of the key challenges is the complicated fabrication process since the synthesis of high-quality perovskite multiferroic thin films often relies on sophisticated methods like pulsed laser deposition (PLD) or molecular beam epitaxy (MBE), which are costly and hard to scale. And, uniformity and reproducibility across broad areas remain huge challenges as consistent thin-film quality is required for mass production. Integration with microelectronics necessitates the development of novel fabrication techniques and specialised equipment to ensure compatibility with standard manufacturing procedures. Overcoming these problems is important to the expansive use of perovskite multiferroic-based SAW devices in commercial applications.

6 Emerging Trends and Future Directions

6.1 Exploration of Novel Perovskite Multiferroic Materials with Superior Properties

New perovskite multiferroic materials with enhanced environmental stability, higher curie temperatures, and stronger magnetoelectric coupling are being investigated by researchers. As an example, $BiFeO_3$-based composites have enhanced magnetoelectric coupling and stability and layered perovskites provide improved tunability and low losses [14, 27].

6.2 Integration with MEMS/NEMS Technologies for Advanced Device Miniaturisation

Advanced miniaturisation of devices and performance improvement is being propelled by perovskite multiferroic integration with micro-electro-mechanical systems (MEMS) and nano-electro-mechanical systems (NEMS). Miniature sensor development for wearable technology and the internet of things (IoT) is enabled by this integration, providing enhanced functionality in very small form factors. It also makes it easier to create tunable high-Q resonators, which are essential for wireless communication as they offer greater reliability and efficiency. Also, by enhancing energy-harvesting ability, this combination enables one to be able to effectively tap ambient vibrations in order to power small electrical devices [44–46].

6.3 Development of Hybrid Devices Combining SAW Functionalities with Other Transduction Mechanisms

The development of hybrid devices that integrate SAW characteristics with other transduction mechanisms such as optical, thermal, or electrochemical is making way for multifunctional applications. SAW-optical hybrids, for example, combine SAW devices with photonic elements to provide sophisticated communication and sensing systems. On the contrary, SAW-thermal hybrids enhance temperature monitoring and energy-harvesting capabilities by combining SAW devices with thermoelectric materials. Plus, by combining SAW devices with electrochemical sensors, SAW-electrochemical hybrids improve their capacity to carry out chemical and biosensing activities, expanding the range of applications across several domains [44].

6.4 Prospects for Quantum and Spintronic Applications

Perovskite multiferroics are being thoroughly studied for quantum and spintronic applications by utilising their unique magnetoelectric properties. These materials are being used for quantum memory and qubit manipulation in quantum computing, which could lead to significant increases in processing power. Perovskite multiferroics are being developed in the field of spintronics for spin-based devices that provide effective data storage and low-power electronics. Moreover, because of their extreme sensitivity, they are perfect for quantum sensors, which can identify even the smallest changes in magnetic and electric fields, creating new opportunities for accurate measurements and technological improvements [28].

7 Conclusion

Summary of key insights into the use of perovskite multiferroics in SAW devices

Perovskite multiferroics have emerged as highly desirable materials for SAW devices because of their unique combination of magnetoelectric, ferromagnetic, and ferroelectric properties. These materials' strong magnetoelectric coupling, which enables great responsiveness to external stimuli like mass, temperature, and magnetic or electric fields gives them improved sensitivity and tunability. They are perfect for uses such as tunable filters and resonators in wireless communication systems because of this connection, which also permits dual tuning via both electric and magnetic fields. Perovskite multiferroics also make it possible to combine sensing, filtering, and signal processing into a single device, enabling multifunctionality. As a result, SAW devices are more portable and adaptable, and they can be used for everything from healthcare to environmental monitoring. Their interoperability with microelectronics and MEMS/NEMS technologies facilitates miniaturisation even more, resulting in the creation of portable, low-power devices appropriate for wearable electronics and the Internet of Things. Significant progress in SAW device technologies is being driven by emerging trends in this sector, including the investigation of novel materials, integration with MEMS/NEMS, the creation of hybrid devices, and their potential for spintronic and quantum applications [27].

Potential impact on future technology landscapes

If incorporated into SAW devices, perovskite multiferroics have the potential to significantly advance a number of technological horizons. In wireless communication, they can be used to develop tunable filters, resonators, and phase shifters that will support satellite, 5G, and Internet of Things networks. For example, 5G networks can potentially utilise adaptive frequency bands due to programmable filters. Perovskite multiferroics enable the development of highly sensitive sensors for applications in industrial safety, healthcare, and environmental monitoring. Some applications of these are biosensors for the early diagnosis of illness and gas sensors for monitoring air quality. The materials also hold the promise to drive the development of quantum and spintronics technology, enabling the development of spin-based devices, quantum sensors, and low-power electronics and quantum memory devices for quantum computing. Perovskite multiferroics may also be applied in energy harvesting to efficiently harvest energy from ambient vibrations, facilitating self-sustaining systems like wireless sensor networks and energy-harvesting sensors for the Internet of Things [28].

Final remarks on research opportunities and interdisciplinary collaboration

There are numerous research opportunities in perovskite multiferroics for surface acoustic wave (SAW) devices, and interdisciplinary collaboration is required to overcome major challenges and release new functionality. Material innovation is one of the primary subjects of research, where finding new perovskite multiferroics with improved qualities, such as higher Curie temperatures and greater magnetoelectric coupling, is important. Furthermore, it is essential to develop scalable and economical methods for large-scale production via improving fabrication procedures. Also offered are exciting opportunities in the exploration of hybrid devices that combine SAW properties with other forms of transduction, such as optical, thermal, and electrochemical. A lot of potential exists for quantum and spintronic applications, but new mechanisms for the control of qubits and spin-based devices must also be investigated. Multidisciplinary collaboration among chemists, electrical engineers, physicists, and materials scientists is required to further these advances. For example, while electric engineers are developing and optimising SAW devices in order to leverage such developments to full effect, material scientists may focus on the synthesis of new multiferroic materials [14, 44].

References

1. Campbell C (2012) Surface acoustic wave devices and their signal processing applications. Elsevier
2. Ruppel CC, Fjeldly TA (2001) Advances in surface acoustic wave technology, systems and applications. World scientific, vol 2
3. Peng W, He Y, Wen C, Ma K (2012) Surface acoustic wave ultraviolet detector based on zinc oxide nanowire sensing layer. Sens Actuat A 184:34–40

4. Constantinoiu I, Viespe C (2020) Zno metal oxide semiconductor in surface acoustic wave sensors: a review. Sensors 20(18):5118
5. Mujahid A, Dickert FL (2017) Surface acoustic wave (saw) for chemical sensing applications of recognition layers. Sensors 17(12):2716
6. Cui B, Ren Z, Wang W, Cheng L, Gao X, Huang L, Hu A, Hu F, Jin J (2024) Review of surface acoustic wave-based hydrogen sensor. Sens Actuat Rep 7:100197
7. Ikeda T (1996) Fundamentals of piezoelectricity. Oxford University Press
8. Morgan D (2010) Surface acoustic wave filters: With applications to electronic communications and signal processing. Academic Press
9. Ballantine Jr D, White RM, Martin SJ, Ricco AJ, Zellers E, Frye G, Wohltjen H (1996) Acoustic wave sensors: theory, design and physico-chemical applications. Elsevier
10. Hornsteiner J, Born E, Riha E (1997) Langasite for high temperature surface acoustic wave applications. Physica Status Solidi Appl Res 163(1):R3–R4
11. Tsubouchi K, Mikoshiba N (1985) Zero-temperature-coefficient saw devices on aln epitaxial films. Ph.D. thesis, Tohoku University
12. Hashimoto K-y, Hashimoto K-Y (2000) Surface acoustic wave devices in telecommunications. Springer, vol 116
13. Wang ZL, Song J (2006) Piezoelectric nanogenerators based on zinc oxide nanowire arrays. Science 312(5771):242–246
14. Wang J, Neaton J, Zheng H, Nagarajan V, Ogale S, Liu B, Viehland D, Vaithyanathan V, Schlom D, Waghmare U et al (2003) Epitaxial bifeo3 multiferroic thin film heterostructures. Science 299(5613):1719–1722
15. Kimura T, Goto T, Shintani H, Ishizaka K, Arima T-h, Tokura Y (2003) Magnetic control of ferroelectric polarization. Nature 426(6962):55–58
16. Fiebig M, Lottermoser T, Frohlich D, Goltsev AV, Pisarev RV (2002) Observation of coupled magnetic and electric domains. Nature 419(6909):818–820
17. Chu Y-H, Martin LW, Holcomb MB, Gajek M, Han S-J, He Q, Balke N, Yang C-H, Lee D, Hu W et al (2008) Electric-field control of local ferromagnetism using a magnetoelectric multiferroic. Nat Mater 7(6):478–482
18. Tokura Y (2006) Critical features of colossal magnetoresistive manganites. Reports Progr Phys 69(3):797
19. Zhao T, Salamanca-Riba L, Zheng H, Wang J, Lofland S, Ma Z, Mohaddes-Ardabili L, Shinde S, Ogale S, Bai F et al (2004) Multiferroic batio3-cofe2o4 nanostructures
20. Catalan G, Scott JF (2009) Physics and applications of bismuth ferrite. Adv Mater 21(24):2463–2485
21. Nan C-W, Bichurin M, Dong S, Viehland D, Srinivasan G (2008) Multifer- roic magnetoelectric composites: Historical perspective, status, and future directions. J Appl Phys 103(3)
22. Vaz CA, Hoffman J, Ahn CH, Ramesh R (2010) Magnetoelectric coupling effects in multiferroic complex oxide composite structures. Adv Mater 22(26–27):2900–2918
23. Bibes M, Barthelemy A (2008) Towards a magnetoelectric memory. Nat Mater 7(6):425–426
24. Jaffe B, Cook WR, Jaffe H (1971) The piezoelectric effect in ceramics. In: Piezo-electric ceramics, pp 7–21
25. Ramesh R, Spaldin NA (2007) Multiferroics: progress and prospects in thin films. Nat Mater 6(1):21–29
26. Ma J, Hu J, Li Z, Nan C-W (2011) Recent progress in multiferroic magneto- electric composites: from bulk to thin films (adv. mater. 9/2011). Adv Mater 23(9):1061–1061
27. Eerenstein W, Mathur N, Scott JF (2006) Multiferroic and magnetoelectric materials. Nature 442(7104):759–765
28. Folen V, Rado G, Stalder E (1961) Anisotropy of the magnetoelectric effect in Cr_2O_3. Phys Rev Lett 6(11):607
29. Rado G, Folen V (1961) Observation of the magnetically induced magnetoelectric effect and evidence for antiferromagnetic domains. Phys Rev Lett 7(8):310
30. Agyei A, Birman JL (1990) On the linear magnetoelectric effect. J Phys: Condens Matter 2(13):3007

31. Fiebig M (2005) Revival of the magnetoelectric effect. J Phys D: Appl Phys 38(8):R123
32. Lawes G, Srinivasan G (2011) Introduction to magnetoelectric coupling and multiferroic films. J Phys D Appl Phys 44(24):243001
33. Prellier W, Singh M, Murugavel P (2005) The single-phase multiferroic oxides: from bulk to thin film. J Phys: Condens Matter 17(30):R803
34. Vopsaroiu M, Stewart M, Hegarty T, Muniz-Piniella A, McCartney N, Cain M, Srinivasan G (2008) Experimental determination of the magnetoelectric coupling coefficient via piezoelectric measurements. Meas Sci Technol 19(4):045106
35. Goldschmidt VM (1926) Die gesetze der krystallochemie. Naturwissenschaften 14(21):477–485
36. Campbell C (1998) Surface acoustic wave devices for mobile and wireless communications, four-volume set. Academic Press
37. Royer D, Dieulesaint E (1999) Elastic waves in solids I: free and guided propagation. Springer Science & Business Media
38. Sreenivasulu G, Petrov V, Fetisov L, Fetisov Y, Srinivasan G (2012) Magnetoelectric interactions in layered composites of piezoelectric quartz and magnetostrictive alloys. Phys Rev B—Condens Matter Mater Phys 86(21):214405
39. Fetisov YK, Srinivasan G (2006) Electric field tuning characteristics of a ferrite- piezoelectric microwave resonator. Appl Phys Lett 88(14)
40. Wang Y, Hu J, Lin Y, Nan C-W (2010) Multiferroic magnetoelectric composite nanostructures. NPG Asia Mater 2(2):61–68
41. Lee J, Shin H, Kim J, Yoon H (2009) Filler effect of low loss dielectricity in printed circuit board material. J Electroceram 23:141–145
42. Jahns R, Knochel R, Greve H, Woltermann E, Lage E, Quandt E (2011) Magnetoelectric sensors for biomagnetic measurements. In: 2011 IEEE international symposium on medical measurements and applications. IEEE, pp 107–110
43. Srinivasan G, Tatarenko A, Fetisov Y, Gheevarughese V, Bichurin M (2006) Microwave magneto-electric interactions in multiferroics. MRS Online Proc Libr (OPL) 966:0966-T14
44. Luo B, Velvaluri P, Liu Y, Sun N-X (2024) Magnetoelectric BAW and SAW devices: a review. Micromachines 15(12):1471
45. Rebeiz GM (2004) RF MEMS: theory, design, and technology. Wiley
46. Bozzi M, Georgiadis A, Wu K (2011) Review of substrate-integrated waveguide circuits and antennas. IET Microwav Anten Propag 5(8):909–920

Chapter 11
Perovskite Multiferroic Nanomaterials for Biomedical Applications

Shweta Sharma, Vishnu Chauhan, Komal, Deepika, Muskaan Bansal, Deepika Dwivedi, Meenakshi Sharma, Naveen Jyoti, Aniket Bharti, Tarun Kumar, Rangoli Bhatnagar, and Rajesh Kumar

1 Introduction

Perovskites are the materials that have the structure in form of ABX_3, where site A and site B are occupied by cations, whereas site X is an anion site. $CaTiO_3$ was first time discovered Perovskite by Russian meteorologist perovskite in 1839. Perovskites structure can be modified by altering synthesis technique and doping [1, 2]. Perovskites can display ambipolar charge transport, facilitating the movement of both electrons and holes, and possess tunable bandgaps, allowing for customization in electronic and optoelectronic applications. Perovskite materials, due to their enhanced physical, chemical, electrical and magnetic properties, find applications in clean energy in place of conventional energy devices. Halide perovskites have been shown to use as X-ray photodetectors with very high sensitivity and response. Its

S. Sharma
Department of Botany, Sri Venkateswara College, University of Delhi, Dhaula Kuan, New Delhi, India

V. Chauhan · T. Kumar
Department of Physics, Ramjas College, University of Delhi, Delhi, India

Komal
Centre for Social Science and Community Health, School of Social Sciences, Jawaharlal Nehru University, New Delhi, India

Deepika · M. Bansal · N. Jyoti · A. Bharti · R. Kumar (✉)
University School of Basic and Applied Sciences, Guru Gobind Singh Indraprastha University, New Delhi, India
e-mail: kumarrrpi@gmail.com

D. Dwivedi · M. Sharma
Dr. B. R. Ambedkar Center for Biomedical Research, University of Delhi, Delhi, India

R. Bhatnagar
Department of Physics, Hansraj College, University of Delhi, Delhi, India

M. Kumar (ed.), *Sustainable Perovskite Multiferroic Materials*, Materials Horizons: From Nature to Nanomaterials, https://doi.org/10.1007/978-981-96-7419-0_11

characteristics include high material density, enhanced charge carrier mobility, high carrier diffusion lengths which are responsible for this. For instance, perovskites, such as $BaTiO_3$ has also been developed as transducers [3, 4] as well as multilayer high dielectric constant capacitors [5]. $CaTiO_3$-based perovskite has been found to assist growth of hydroxiapetite, found in the bones. It has highly corrosive resistance and good stability and bonding strength. Therefore, it is used in bone and soft tissue reconstruction. The ceramic materials-based perovskites have also been developed for bioimplantable device [6–8]. This has been possible due to the fact that, thin films of perovskite materials, due to their piezoelectric property, can transfer very minute mechanical motion in the interior of any organ into electrical signals. This piezoelectric property of perovskite makes them potential candidate in mechanical sensors, power producers and color image sensors [9–11]. With advancement in the field of research and new techniques available for synthesize and doping, piezoelectric perovskites have been developed that can be used in tissue engineering such as bone defect healing. These materials can use body's motion to generate power and work as nanogenerators, which can be used to power pacemakers and artificial retinas. The technological advancement in the field of piezoelectric perovskite can play a major role in health care. Perovskites' adaptable structure enables them to have a variety of compositions, resulting in number of kinds such as double perovskites, organic–inorganic hybrid perovskites and inorganic oxide perovskites. Significantly, among the various types of perovskites, multiferroic perovskite nanomaterials are found to have extensive and potential applications in biomedical field including cancer treatment, antimicrobial agents, X-ray detection and imaging, orthopedic implants, to name a few. Nanocrystals of multiferroic perovskites also find applications as of anti-viral, anti-fungal and antibacterial agents. Also, multiferroicperovskites lead to significant advancement in the use of self-powered devices and can help in real-time diagnosis and treatment, photocatalytic and antibacterial, drug delivery, glucose monitoring, real-time health tracking, CT imaging high resolution flexible image sensing as well as biomedical imaging [12–16]. Multiferroic perovskites have applications in the field of sensors and transducers [17] owing to theirpotential to switch electrical energy into mechanical signal and vice versa. Such materials are also used in energy harvesting [18], actuators in the field of robotics, memory devices as these materials show hysteresis, which make them to retain polarization state. Multiferroic perovskites are materials that show many ferroic characteristics, such as ferroelectricity and ferromagnetism. Notable multiferroic perovskites include bismuth ferrite ($BiFeO_3$), bismuth manganite ($BiMnO_3$) and rare-earth manganites like terbium manganite ($TbMnO_3$) and yttrium manganite (YMn_2O_5), and Lutetium Iron Oxide ($LuFe_2O_4$), $BaTiO_3$ or $Pb(ZrTi)O_3$ (PZT) are widely recognized as ferroelectric materials. There are the two types of Multiferroic Materials, i. e., Type-I and Type-II [19]. Ferroelectricity in $BaTiO_3$ or $Pb(ZrTi)O_3$ (PZT) or Type-I materials is created when transition metal ions are displaced off-center and create significant covalent bonds involving one or three oxygen atoms via the empty d shells [20, 21]. Moreover, ferroelectricity in Type-I multiferroic perovskites may arise due to lone pairs, due to charge ordering and geometric structure [22–24]. Multiple type-II multiferroics with greater coupling among magnetic and ferroelectric phases are being created and evaluated after these two $TbMnO_3$ and $TbMn_2O_5$ compounds [25, 26]. Moreover, type-II multiferroic materials fall into two groups: first, where ferroelectricity results from a magnetic spiral ($MnWO_4$, $TbMnO_3$ and $Ni_3V_2O_6$ are examples). In the example of $TbMnO_3$, and secondly when collinear magnetic structures showed ferroelectricity (common example is Ca_3CoMnO_6) [23, 27, 28]. The multiferroic perovskites can be synthesized by various physical and chemical techniques such as sol–gel auto-combustion method, solid-state reaction technique, chemical vapor deposition, physical vapor deposition, spark plasma sintering (SPS) technique, pulsed laser deposition, hydrothermal technique and co-precipitation method [29–33]. Multiferroic perovskite also shows numerous physical properties that make perovskite and its nanocrystals as potential candidate to be used in the field of biomedical applications.

2 International and National Scope of Perovskite Multiferroic Nanomaterials for Biomedical Applications

Over the recent past several centuries, there has been a steady growing interest in using the perovskite multiferroic nanomaterials (e.g., Bismuth Ferrite BFO, $BiFeO_3$), in various biomedical applications that mainly include Targeted Drug Delivery, Photo-ablation therapy, Magnetic Hyperthermia, Biosensors and Bio-imaging [34, 35]. Some of the applications comprise magnetic tweezers and magnetic separation of proteins and DNA molecules, while Hyperthermia, targeted medication and radioactive isotope administration for chemotherapy, radiation, and magnetic resonance imaging (MRI) contrast enhancement are examples of therapeutic uses. The grounds for this include their exceptional chemical stability, non-toxicity, biocompatibility, high saturation magnetization and high magnetic susceptibility [36]. Exceptionally accurate and sensitive biosensors based on smart nanomaterials have made it possible to develop new technologies for the early identification and diagnosis of biomarkers linked to illness [37]. As potential candidates for an extensive range of innovative multifunctional devices, multiferroic materials have garnered a lot of attention, leading to an increase in research activity due to their scientific curiosity and tremendous technical promise [38]. Since they can readily spread through the vascular system in the bloodstream when a static magnetic field is provided, the nanoparticles, which range in size from 100 to 1000 nm, may be targeted to certain organs. Nanoparticles' electric fields may interact with the signals acquired by nerve cells and their effects on the organelles and DNA as the signals produced by the particles have properties with the electric signals that propagate in the neural cells. The magnitude and shape of the signals produced by the nanoparticles can be controlled by the external magnetic field. This can remotely trigger the functions of neural cells for pain treatment [39]. Due to their high sensitivity and responsivity for X-ray, they are used for detection and imaging of organs for an individual. In addition to the targeted drug delivery of drugs for cancer to the precise location of the tumor while preventing harm to the nearby healthy cells, these are globally being accepted for magnetic hyperthermia treatment for tumors by heat them to more than 42 °C to eliminate the cells that are cancerous, without damaging the surrounding healthy tissues [40]. They have an immense potential for novel sensing applications, as evidenced by their remarkable hydration-dehydration, electronic transition, adsorption–desorption, phase transformation, and ion intercalation-declaration. Perovskites are ideal for optical sensors used in environmental monitoring and medical diagnostics because of their remarkable photodetection capabilities, which include great sensitivity and fast reaction times. Additionally employed in gas sensing applications, very sensitive gas detection has potential uses in industrial process control [41]. In the multiferroic nanoparticles, two or more properties out of ferroelectricity, ferromagnetism antiferromagnetism and ferro-elasticity can co-exist at room temperature [38, 42, 43]. As a result, they might serve a dual function that is determined by the frequency of the applied magnetic field. The magnetic component of the particles can be exploited to induce hyperthermia at extremely high frequencies (in the radio- or microwave range), while it can also create electric impulses to stimulate membranes or organelles at acoustic frequencies. In the future, there may be a potential application of multiferroic nanoparticles for a considerably safer cancer therapy as compared to chemotherapy or hyperthermia. The ability to manipulate electrical polarization with external magnetic fields and magnetic order with electric fields has led to a lot of interest in multiferroic materials. This property, referred to as magnetoelectric (ME) coupling, introduces an entirely novel studies of flexibility to the upcoming generation of low-dimensional devices, including solar cells, sensors and storage [44–47]. Bismuth ferrite-based multiferroic ceramic materials are preferred for synthesis of innovative devices and their implication for novel applications is significant including water treatment, energy generation, photovoltaics, piezoelectrics, high and ultrahigh frequency range devices. [8] Perovskite Metal Oxide Frameworks (MOFs) also have presented magnetic, ferroelectric, ferroelastic, multiferroic, and photovoltaic capabilities, with prospective applications in sensors, gas storage, catalysts, photoluminescence, magnets, photovoltaics, and so on [48]. Since electric fields require less energy to produce than magnetic fields, which are utilized in many magnetism-based devices and technologies, the coexistence of both electric and magnetic ordering in multiferroic materials allows for the control of magnetism by way of the use of an electric field. Presently, these materials are being used in bone tissue engineering and regenerative medicines for bone regeneration, bone growth, bone replacement, treatment of bone defects and fractures [49–52]. These can also be used as tooth filling material [53]. For the treatment of Neurotrauma and Neurodegenerative ailments also, the perovskites are widely used as computational tools in cognitive neuroscience [54]. These can monitor the overall physical health of a person. Incorporating the real-time monitoring of vital signs and detection of biomarkers, these devices work on miniaturization potential making them light, comfortable and long-term wearable. Perovskite-based drug delivery systems can deliver cancer drugs to the neurons at a very high precision [55]. These devices have versatile detection, optoelectronic properties and miniaturization potential making them light, comfortable and wearable for long term.

The use of perovskite materials in Bionic Prosthesis has revolutionized the field of material science providing individuals with absent or impaired body part, an ability to regain functionality and independence. Bionic transplants (e.g., for skin, missing body part, etc.) can be done only if a suitable donor is found which matches the recipient. Although this may be complex and immune suppressant [56]. Prior to marketing and sale of such medical devices, it is mandatory to get the regulatory approval, which may be different for different countries around the globe. Bio-implantable energy harvesting is also one of the very important applications of the perovskite nanomaterials. They work for the conversion of mechanical and thermal energies into electric power (as also shown by flexible piezoelectric and pyroelectric energy harvesters). Energy from human activities and the body may be used as inputs to power portable low-power gadgets [57]. Perovskite nanomaterials are being widely used all over the world due to their outstanding properties, for various biomedical applications. However, numerous concerns and obstacles need to be solved and thought about by the science community. Perovskite nanomaterials, used as antimicrobial, air and water purification, anti-viral, antibacterial, however, there is a growing threat of antimicrobial resistance, appropriate disposal and recycling methods need to be urgently addressed [58]. The biodegradability, biocompatibility, stability, integration with existing technology, these aspects need more study and development. Lead-based products have the potential to cause significant levels of toxicity, health risks, pollution and irreversible harm to the environment. Human behavior, learning capacity, exhaustion, joint and muscular discomfort and severe symptoms including anemia, convulsions, coma and neurological issues can all be impacted by high levels of poisoning. Newborns and children's biological systems are particularly vulnerable to heavy metal poisoning. Better would be, if lead-free alternatives can be used. More and more perovskite materials are being integrated with Biomaterial to develop new hybrid materials. The most promising prospects for a variety of cutting-edge technical applications, such as sensors, transducers, and actuators, are currently ferroelectric lead-free $NaNbO_3$-based ceramics [52]. The advancement of polycrystalline without lead solid solutions, such as (Ba, Na)(Ti, Nb)O_3 (BTNN) and (K, Na)NbO_3 (KNN) systems, offers a different approach to address these performance concerns. These solid solutions really exhibit better dielectric and piezoelectric coefficients that can satisfy the requirements needed for commercial applications [49–52]. The structure, composition, surface chemistry and stability of perovskite materials may have a potential impact on the cell behavior and response, the possible risk factors, effect of dose on the tissues, cell membrane, plasma, DNA, mitochondrial functions and toxicity effects that may follow after the perovskite exposure need to be well examined and studied [43]. Their long-term effects, degradation, effect on cells, tissues and even on the human genome, raise serious issues. Hence, their targeting, usage and dosage must also be thoroughly studied.

Lastly, there are certain stability issues also associated with these materials, as they are susceptible to deteriorate if subjected to environmental factors like oxygen, moisture, heat and light. This may lead to a shorter shelf life and decreased performance. As a result, their stability in the presence of humidity or oxygen is a substantial issue and an active topic of investigation [41]. Another important concern is the integration of biosensing platforms, which includes the need for low prices and the possibility of real-time monitoring, particularly for point-of-care applications where device simplicity is also important [40].

3 Chemical Synthesis Routes of Various Multiferroic Perovskite Nanomaterials

3.1 Combustion Method

The combustion method is a highly effective method for synthesizing multiferroic perovskite nanomaterials. In this method, the reaction starts with preparation of precursor solutions, typically comprising metal nitrates or oxalates dissolved in a suitable solvent, i.e., DI water or ethanol, to ensure homogeneity. A fuel such as glycine, citric acid or urea is then added to the solution, serving as a reducing agent to drive the combustion reaction and provides the sufficient energy necessary for the combustion process. Upon heating on higher temperatures, the fuel and precursor mixture undergoes an exo-thermic reaction where the fuel ignites, generating high localized temperatures that facilitate the formation of nanosized multiferroic perovskite nanoparticles [59]. In the similar approach, Samar Layek et al. synthesized $BiFeO_3$ multiferroic perovskite nanomaterial using the combustion method, employing bismuth nitrate ($Bi(NO_3)_3{\cdot}5H_2O$), ferric nitrate ($Fe(NO_3)_3{\cdot}9H_2O$) and citric acid ($C_6H_8O_7$) as starting materials. The mixture undergoes the combustion process at a temperature of 150 °C, resulting in highly crystalline, uniform and well-defined nanoparticles [60]. In contrast, M. Sukumar et al. utilized L-alanine as a fuel instead of citric acid and employed microwave heating as the energy source. This approach yielded highly pure multiferroic $BiFeO_3$ perovskite nanoparticles [61]. Additionally, Y.M. Abbas et al. synthesized a yellow colored $LaYFeO_3$ multiferroic perovskites using starting materials lanthanum nitrate ($La(NO_3)_30.6H_2O$), ferric nitrate ($Fe(NO_3)_30.9H_2O$), yttrium nitrate ($Y(NO_3)_30.6H_2O$) and citric acid monohydrate ($C_6H_8O_7.H_2O$) [62]. K. Naveen et al. synthesized hexagonal $YInO_3$ multiferroic perovskites using yttrium oxide (Y_2O_3), indium oxide (In_2O_3) and chromium nitrate ($Cr(NO_3)_30.6H_2O$) and studied their electrical and magnetic properties [63]. Thus, the combustion method is widely utilized for synthesizing various multiferroic perovskites. It produces uniform, well-defined nanoparticles, making it ideal for various applications in memory devices, spintronics, sensors and actuators that respond to both magnetic and electric fields [64, 65].

3.2 Co-precipitation Method

The co-precipitation approach is simple to perform and cheap for synthesizing multiferroic perovskites with controlled particle size and uniform composition. This method offers a significantly higher yield compared to other techniques, as the precipitates are formed efficiently at lower temperatures, making it an energy-efficient and scalable approach for producing high-purity multiferroic perovskite materials. In this method, metal precursors are dissolved in water, and a precipitating agent is then added to form a co-precipitate. The technique is executed under regulated conditions pH and temperature conditions to achieve the desired perovskite phase with enhanced crystallinity and morphology. Earlier studies have demonstrated the successful synthesis of multiferroic perovskites using the co-precipitation method. For instance, H. Ke et al. prepared pure-phase $BiFeO_3$ by dissolving $Bi(NO_3)_30.5H_2O$ and $Fe(NO_3)_30.9H_2O$ precursor powders in nitric acid. To ensure precise control of the chemical precipitation process, they maintained a constant pH by synchronously adding ammonia and nitrate solutions, which helped to achieve uniform and controlled precipitation of $BiFeO_3$ [66]. Similarly, L.Wang et al. synthesized multiferroic perovskite $CoFe_2O_4$–$BaTiO_3$ nanocomposites using cobalt ferrite ($CoFe_2O_4$), barium hydroxide octahydrate ($Ba(OH)_20.8H_2O$), titanium butoxide and sodium citrate as precursors [67]. In addition to $BaTiO_3$ and its composites, H. Wang et al. synthesized multiferroic $BiMnO_3$ using the same approach to enhance electron transport behavior. They used concentrated nitric acid (HNO_3), sodium hydroxide (NaOH), manganese chloride ($MnCl_20.4H_2O$), bismuth oxide (BiO_3) and polyglycol during synthesis [68]. These investigations highlight the capability of the co-precipitation method in producing high-quality multiferroic materials with tailored compositions and characteristics. This technique makes the synthesis of materials with specific properties suited for a wide range of applications, including enhanced transport mechanisms and improved dielectric properties. The precise control over chemical processes and the ability to achieve uniform precipitation make the co-precipitation method an excellent choice for engineering multiferroic materials for advanced technological uses [69].

3.3 Sol–Gel Method

The sol–gel methodology is a flexible wet-chemical approach that is commonly employed for the production of multiferroic perovskites. In this technique, the sol (or solution) progressively transforms into a network that resembles gel and is made up of both liquid and solid phases as shown in Fig. 11.1. To create a colloid, prevalent precursors include metal alkoxides and metal chlorides which undergo hydrolysis and polycondensation processes, while chelating compounds such as citric acid or ethylene glycol are often added to stabilize the solution regulate the reaction process and prevent premature hydrolysis ensuring uniform and controlled material synthesis. The material is dried to form a xerogel after gelation, and material is then processed at high temperatures(500–900 °C) to eliminate organic residues and generate a crystalline perovskite phase [70]. Using this method, M. Li et al. synthesized multiferroic $CoFe_2O_4$–$Pb(Zr,Ti)O_3$ (CFO-PZT) composite films. They prepared a mixed precursor solution using cobalt nitrate ($Co(NO_3)_2$), titanium isopropoxide ($Ti(OCH_3)_4$), lead acetate ($Pb(C_2H_3O_2)_2$), iron nitrate ($Fe(NO_3)_3$) and zirconium acetate ($Zr(C_2H_3O_2)_4$) for the CFO and PZT components. After processing the solution, both the CFO and PZT phases were homogeneously integrated, resulting in composite films with enhanced multiferroic properties [71]. Additionally, J.S. Bangruwa et al. prepared multiferroic composites of the perovskite $BiFeO_3$ and spinel ferrites $NiFe_2O_4$, $ZnFe_2O_4$, and $CoFe_2O_4$ with varying compositions. These composites were studied for their structural, magnetic and ferroelectric properties [72]. Furthermore, Jo et al. prepared multiferroic BFO/PZT multilayer films, investigating their structural and ferroelectric properties for uncooled infrared detector applications [73]. Ji et al. also employed the same technique to synthesize polycrystalline perovskite Aurivillius structure $Bi_5FeTi_3O_{15}$ (BFTO) films on Pt/Ti/SiO_2/Si substrates. The process utilized bismuth nitrate, iron acetylacetonate, titanium n-butoxide, glacial acetic acid, and n-propanol as precursors [43]. These studies highlight the efficiency of the sol–gel process in producing high-quality films with regulated composition and structure. Moreover, the method's scalability makes it a promising option for large-scale production of multiferroic perovskite materials.

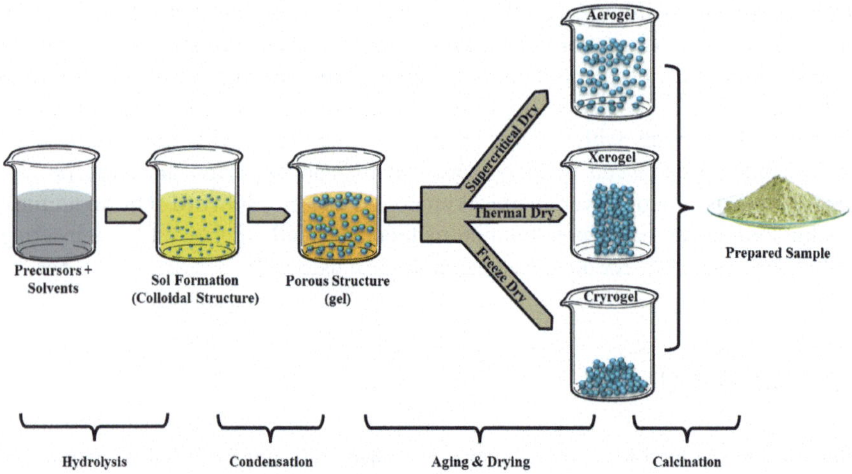

Fig. 11.1 Illustration of synthesis of nanoparticles using sol–gel method

3.4 Hydrothermal Method

The hydrothermal approach is a popular technique for creating multiferroic perovskite materials due to its ability to create high-purity, well-crystallized nanoparticles at relatively low temperatures. In this process, precursors are dissolved in a solvent, such as ethanol or deionized (DI) water as displayed in Fig. 11.2. The resulting solution mixture is then placed in an autoclave and heated to a fixed temperatures ranging in between (150–250 °C) under self-generated high pressure, which facilitates controlled growth of the perovskite phase. This method offers significant advantages in producing uniform, high-quality nanomaterials, making it a preferred approach for synthesizing multiferroic perovskites [74]. Using this technique, M. Shang et al. synthesized multiferroic $YFeO_3$ with an orthorhombic perovskite structure. They employed $Y(NO_3)_3$ and $Fe(NO_3)_30.9H_2O$ as precursors dissolved in 30 mL of distilled water with KOH as the mineralizing agent. The solution that had been created was then moved to a Teflon-lined stainless steel autoclave and kept at 240 °C for 3 days(72 h), enabling the controlled crystallization of the desired phase [75]. Similarly, S. Han et al. prepared multiferroic $BiFeO_3$ powders using bismuth nitrate ($Bi(NO_3)_3{\cdot}5H_2O$), iron nitrate ($Fe(NO_3)_3{\cdot}9H_2O$) and potassium hydroxide (KOH) [76]. Likewise, Zhang et al. synthesized the perovskite orthoferrite $SmFeO_3$ using the same approach, employing $Sm(NO_3)_3{\cdot}6H_2O$ and $Fe(NO_3)_3{\cdot}6H_2O$ with KOH as the mineralizing agent, to study its electric, magnetic and dielectric properties [77]. The method enables precisely control material properties and phase formation, which is essential for tailoring electric, magnetic, and dielectric properties. Moreover, its adaptability to various precursor combinations and reaction conditions has demonstrated by the successful synthesis of a variety of perovskite materials, including $YFeO_3$, $BiFeO_3$ $SmFeO_3$, etc., demonstrating its broad applicability and reliability in materials research.

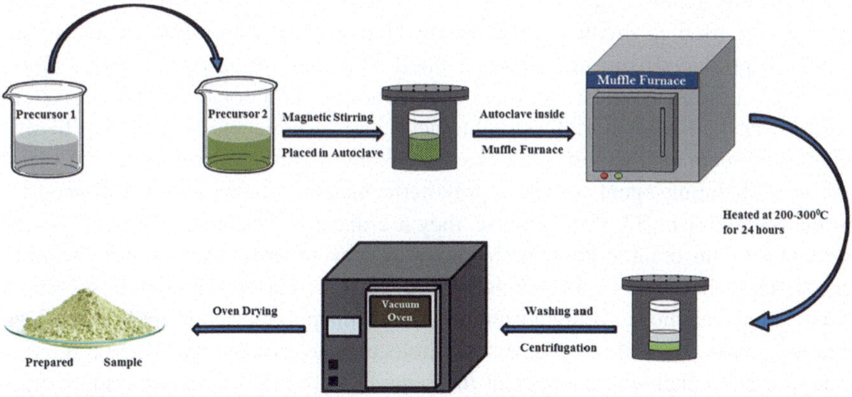

Fig. 11.2 Schematic diagram showing the preparation of nanoparticles by hydrothermal approach

3.5 Chemical Vapor Deposition

Chemical Vapor Deposition (CVD) is an emerging approach for synthesizing multiferroic perovskite materials, which provides precise control over film composition and thickness. In CVD, volatile precursors are involves introducing into a reaction chamber, where they break down and react to form a solid perovskite film on a substrate. This technique is particularly useful for producing high-quality, uniform thin films, making it perfect for applications in spintronics, memory devices and sensors [78]. Using this method, M.R. Catalano et al. synthesized perovskite multiferroic $BiFeO_3$ (BFO) films on $SrTiO_3$ (STO) substrates using Metal–Organic Chemical Vapor Deposition (MOCVD). In addition to pristine BFO films, they also synthesized Dy-doped $BaTiO_3$ films. The study revealed that both pristine and doped films exhibited ferroelectric and piezoelectric properties. Notably, the piezoelectric properties were retained after doping, highlighting the potential of these materials for applications in sensors, actuators and memory devices [79]. Furthermore, B. Meunier et al. prepared a resistive switching (RS) device using a multiferroic perovskite $LaMnO_3$ thin film with a thickness of 25 nm deposited on a SiO_2/Si substrate through the Chemical Vapor Deposition (CVD) technique. This study demonstrates the potential of $LaMnO_3$ thin films in RS applications, attributed to their unique multiferroic and functional properties [80]. In a similar approach, Nakaki et al. synthesized 2.8 μm thick $PbTiO_3$ films on Nb-doped $SrTiO_3$ substrates using CVD at 600 °C resulting in the deposition of high-quality perovskite $PbTiO_3$ films, showcasing the versatility and effectiveness of CVD for fabricating functional oxide films [81].

3.6 Pechini Method

This versatile synthesis technique, also referred to as the polymeric precursor approach, is used to create a wide range of materials. Perovskite materials are commonly produced using the Pechini method. This solution-based synthesis method specifically used in synthesis for oxides with complex compositions [82]. Excellent control over the final material's stoichiometry and homogeneity is provided by this polymeric precursor method. This process begins with the complexation of metal ions in a chelating agent within a polymeric matrix, which is then followed by thermal breakdown [83, 84]. Initially, they are dissolved in an appropriate solvent like water or an organic solvent. In order to prevent precipitation, the chelating agent is used that forms compounds with metal ions. Further, this gel is subjected to some heat treatment. At higher temperatures, this polymeric matrix breaks down forming volatile byproducts. Solid-state molecules are created when metal ions go through certain chemical changes at the same time [83–85]. Various studies illustrate that multiferroic perovskites can be synthesized using this method such as Qian et al. synthesized $0.8BiFeO_3$–$0.2SrTiO_3$ single-phase perovskite ceramics using the Pechini method and studies their multiferroic properties. They also showed that the improvement of the multiferroic characteristics and dielectric performance of $0.8BiFeO_3$–$0.2SrTiO_3$ ceramics is significantly influenced by the proper annealing temperature. The annealing temperature for the $0.8BiFeO_3$–$0.2SrTiO_3$ solid solution ceramic that produces the best multiferroic characteristics is approximately 1300 K [86]. Liu et al. synthesized Spinel-perovskite multiferroic of $NiFe_2O_4$/$BiFeO_3$ nanocomposite. They also emphasized that synthesized particle size ranges from 40 to 100 nm, and they were found to exhibit both tuning magnetic properties and a ME effect after being annealed at 700 °C for two hours. Also, the magnetic field frequency and magnetic bias had a significant impact on the ME effect of the nanocomposites [81]. Ummer et al. synthesized (1-x)$BiFeO_3$-x$NaNbO_3$ multiferroic nanocomposite using Pechini method. Due to their effective optical limiting, these samples may find use in photonic devices [87]. Huo et al. synthesized $BiFeO_3$–$Na_{0.5}K_{0.5}NbO_3$ multiferroic ceramic using Pechini method. They looked into the magneto-dielectric effect, which showed how the magnetic and electric orders are coupled, which is crucial for multiferroic materials [88].

3.7 Solid-State Reaction Method

This solid-state synthesis technique, also referred to as "shake-n-bake" or traditional high temperature synthesis, necessitates temperatures that are adequate to produce the most stable products [89]. In this technique, the raw materials are dehydrated at about 180 degrees Celsius. The dehydrated raw material is ground for few hours in mortar and pastel and then they are calcined for three hours at about 1200 degrees to create a homogenous mixture. Following a thorough mixing process using binder (polyvinyl alcohol (PVA)) solution, the mixture is compressed into disk samples. Lastly, pellets are sintered at about 1300 degree Celsius for three hours after PVA is burned out. To create conductive electrodes for electrical measurements and dielectric measurements, sintered pellets are covered on both sides with silver paste and fired for one hour at 250 degrees Celsius [90]. Various studies have been conducted that reveals solid-state reaction method can be employed for synthesizing high-purity multiferroic perovskites. $BiFeO_3K0._5Na0._5NbO_3$ and Mn-doped $BiFeO_3K0._5Na0.5$-NbO_3 ceramic samples were synthesized by Furuhashi et al., by solid-state reaction techniques [91]. An altered solid-state reaction technique was used by Yamazoe et al., to create perovskite-type $KNbO_3$ at low temperatures (<600 °C) [92]. $CaTiO_3$ – $NdAlO_3$ perovskite solid was synthesized by Jancar [93] using solid-state reaction method. LaCoO3 nanocrystals of the perovskite type with grain sizes ranging from 15 to 40 nm was synthesized by Li et al. using solid-state synthesis method [94]. Perovskite-type $LaWN_3$ was synthesized by Matsuishi, S. et al. by high-pressure solid-state metathesis reaction [95]. Mostafavi et al. created nanostructured multiferroic bismuth ferrite generated by the solid-state reaction method. Resultant perovskite BFO particles range from 40 to 62 nm. Additionally, they showed that the molar ratios of Bi_2O_3/Fe_2O_3 are a significant factor in the solid-state reaction method for the synthesis of BFO perovskite [96].

3.8 *Mechanochemical Synthesis*

This technique that unifies chemical and physical processes is mechanochemical synthesis. This novel and energy-efficient technique for making perovskite materials allows them to be formed at room temperature by mechanical activation, carrying out away the need for high-temperature processing and encouraging environmental sustainability and scalability [97]. The use of mechanical forces, like grinding or milling, which physically break bonds and start chemical reactions, is part of its physical component. On the other hand, mechanical energy causes solid-state chemical reactions between precursor materials, which results in the creation of the desired molecule like perovskites, is the chemical aspect. Bhatti et al. used this method to create lead titanate ($PbTiO_3$) nanopowders from lead oxide and titanium dioxide in which the first signs of the perovskite phase were found after five hours of synthesis [98].

4 Physical Synthesis Routes of Various Multiferroic Perovskite Nanomaterials

A range of methods are used in the physical synthesis of multiferroic perovskites to produce superior materials with desired structure and characteristics. These techniques found to provide exact control over their morphology, structure and functional characteristics so they could be incorporated into innovative spintronics, memory and sensor applications.

4.1 *High-energy Ball Milling (HEBM)*

High-energy ball milling (HEBM) is a powder treatment method that creates composite structures by mixing, dispersing, activating by using the action of the milling devices. The designs of the milling devices can vary but generally they are divided into two main groups: high-energy followed by low-energy mills [99]. The processes that occur during high-energy ball milling (HEBM) include defect accumulation, deformed welding, structural refinement, breakage and the breakdown or creation of chemical bonds [100]. Numerous studies have been conducted on high-energy ball milling, exploring its potential in synthesizing nanomaterials, developing novel alloys and enhancing material properties through mechanical activation. H. Ramírez, et al. show how $BiFeO_3$ is formed at the low temperature of 500 °C. They also emphasized that ferroelectric and magnetic characteristics are similar to those of the bismuth ferrite produced using longer reaction times or greater temperatures [101]. Trejo et al. created ferroelectric and ferromagnetic multiferroic $BaTiO_3$ by high-energy ball milling for 15–60 min. This process primarily produced a secondary $Ba_{12}Fe_{28}Ti_{15}O_{84}$ (BFTO) phase and a BTO phase with a tetragonal structure [102]. Moure, et al. synthesized perovskite structure of multiferroic ceramics based on $BiFeO_3$ and Sr-doped $BiFeO_3$ by using high energetic milling followed by thermal treatments at lower temperatures. Additionally, they stated that as the number of oxygen vacancies increases, ceramics with higher dielectric permittivity and conductivity are produced in comparison to non-doped materials [103].

4.2 Spark Plasma Sintering (SPS) Technique

Fine ceramics, composites, novel durable materials, thermoelectric semiconductors, bio-materials, homogeneous, extremely dense, nanostructured sintering compact, functionally graded materials (FGMs) and more are all made using this well-known state-of-the-art processing method [104]. The vertical SPS sintering press machine includes a single-axis pressurization procedure, a water-cooled vacuum chamber, a vacuum/air/argon-gas environment regulate system a unique DC-pulsed power generator, a cooling-water control unit, punch electrodes with a water cooler, a Z-axis location determining and control unit, the temperature determining control units, pressure display units, data analyzing units as well as multiple safety interlock devices [105]. This technique that applies a high current and ON–OFF DC pulse voltage from a special pulse generator to particles. Further, Spark plasma and spark impact pressure create a high-temperature field with sputtering phenomena when sparking, even one or two sparks, occurs. They function to get rid of impurities, oxide films, and adsorptive gases that are present on the powder particles' surface [106]. Extensive research is being conducted on the Spark Plasma Sintering (SPS) technique, focusing on its ability to densify materials rapidly, produce nanostructured composites, and enhance mechanical and functional properties through controlled processing conditions such as Song et al. synthesized $BiFeO_3$ multiferroic ceramics $BiFeO_3$ multiferroic ceramics and studied their dielectric, ferroelectric and magnetic properties. Their average grain size was found approximately 200 nm. Additionally, they suggested that the consistent microstructure produced by SPS combined with high-energy ball milling may be a viable method for producing superior $BiFeO_3$ multiferroic ceramics [107].

4.3 Sputtering

One physical method for depositing thin films is sputtering as illustrated in Fig. 11.3. It belongs to the physical vapor deposition (PVD) subset of thin film technology. Sputtering is the process by which energetic gas or plasma particles strike a solid's surface and separate tiny particles from it [108]. The phenomenon occurs naturally in outer space under high-precision conditions and may cause undesired surface wear. Plasma ions are used to blast the coating material (the sputtering target) and the particles that are eliminated move into the gas phase [109]. A very thin coating is formed when the vapor condenses on the substrate surface, where it adheres firmly. This method enables the consistent application of a wide variety of materials like metals, oxides, nitrides, etc., on a wide range of substrates from architectural glass to semiconductor wafers [110, 111]. Thus, this well-known coating technique has been used for decades in a variety of thin film applications like multiferroic $BiFeO_3$ thin films prepared by Cha, J., et al., using a traditional RF Magnetron Sputtering Method. They also stated that only for a $P(O_2)$ of 0.3 mTorr to 1.5 mTorr at a substrate temperature of 400 °C, followed by annealing at 500 °C in a N_2 ambient, were BFO thin films with a single perovskite phase produced [112]. By using RF sputtering on platinum foil at substrate temperatures between 300 and 500 °C, Okuyama et al., developed a $PbTiO_3$ thin film gate field effect transistor. They measured the dielectric constant to be 200 and the remanent polarization to be 27 $\mu C/cm^2$ [113].

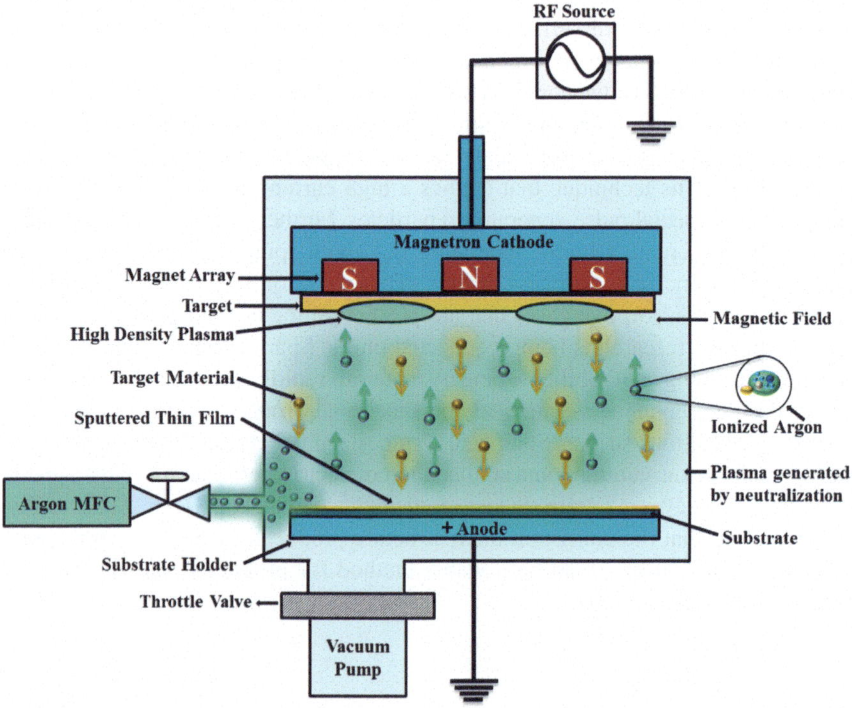

Fig. 11.3 Illustration of RF sputtering method for deposition of thin films

4.4 Pulsed Laser Deposition

Smith and Turner employed this method for the first time in 1965 to create semiconductors and dielectric thin films [114]. For a variety of reasons, pulsed laser deposition (PLD) is used as this permits the deposition of a wide variety of materials, including metals, semiconductors, oxides, nitrides, cardbides, high-temperature superconductors and even polymers or fullerenes, when paired with a stoichiometry exchange across the target as well as substrate. This method allows for the use of ambient gas and ultrahigh vacuum (UHV) because the source of energy is located outside the chamber A concentrated as well as pulsed beam of laser light assaulted elementary or alloy objects at an angle of 45 degrees in an ultrahigh vacuum (UHV) [115, 116]. The ions along with atoms that dissolve in substrates receive the target or targets. The substrates are typically connected with their surfaces parallel to the target surfaces, usually at a separation of 2–10 cm. The number of laser pulses on each target can be changed to produce multilayers with the appropriate single and bilayer thicknesses [116]. Researchers are investigating its use in producing advanced materials such as Puli et al. used pulsed laser deposition technique to grow tri-layered polycrystalline BTO/CFO/BTO nanostructures on Pt(111)/TiO_2/SiO_2/Si substrates under ideal deposition conditions, with a substrate temperature of approximately 750 °C. This tri-layered thin film has demonstrated intriguing multiferroic magnetoelectric (ME) characteristics with a polarization of 99.86 $\mu C/cm^2$ and a remnant polarization of 7.05 $\mu C/cm^2$. These films have demonstrated intriguing ferroelectric properties [117].

5 Applications

5.1 Medical Imaging

Because of their special combination of ferroelectricity, magnetism and optically tunable characteristics, perovskite multiferroic nanomaterials have enormous potential for use in medical imaging applications. They are good candidates for fluorescence and photoacoustic imaging due to their superior optical response and strong magnetoelectric coupling, which allows for increased contrast in magnetic resonance imaging (MRI). Perovskite multiferroics are especially the ones that possess both ferroelectric as well as ferromagnetic characteristics, have sparked considerable interest in the realm of medical imaging. Their distinct qualities offer novel uses that may improve imaging methods and diagnostic capacities. Magnetic sensors identify and provide with the information on the changes in the surrounding environment's magnetic fields and their properties which includes speed, angles, direction, rotation, temperature and many more, without requiring any physical contact. Magnetic sensors' output signals necessitate processed signals before being translated into the appropriate parameter. Sensors that measure magnetic fields are utilized extensively in a variety of applications with a significant influence on our society. Such applications are microscopic imaging, medical imaging, bio-medical applications, etc. Wei et al. reported the Mg^{2+}-modified $LiBaF_3$:Ni^{2+} perovskite single crystals with broadband long wavelength near IR emission for imaging human vascular structures. Conclusively, in order to create NIR single-crystal LEDs, the prepared Ni^{2+}/Mg^{2+}-doped NIR perovskite single crystals were accompanied by a commercially available, extremely effective near-UV LED chip (at 395 nm). Their promising uses with high efficiency were demonstrated in vein non-destructive imaging and NIR night-vision monitoring [118].

5.2 *Antibacterial*

Perovskite multiferroics have received great interest regarding their antibacterial activities as well as their ferroelectric along with magnetic characteristics. This section focus into the processes underlying their antibacterial activity as well as prospective uses in fighting against microbial illnesses. Ateia et al. prepared the perovskite samples using the citrate auto-combustion process. The magnetization of Co^{2+}, Cr^{3+} and Ag^{1+} ions vary based on exchanging contact strength, canting angle, along with octahedron buckling (<BO6>). The findings showed that $GdFe_{0.9}Ag_{0.1}O_3$ nanomaterials had higher antibacterial activity than $GdFe_{0.9}Co_{0.1}O_3$ and $GdFe_{0.9}Cr_{0.1}O_3$ samples. The Ag ions in $GdFeO_3$ nanoparticles may cause microbial cytotoxicity, increasing their biological impact and recommending them to be an efficient antibacterial chemical [119]. At ambient temperatures, Bismuth Ferrite (BFO), commonly referred to as $BiFeO_3$, shows ferroelectricity and antiferromagnetism. Bismuth ferrite is a key component in the multiferroics system. Sharmila et al. demonstrate the antibacterial property of BFO nanoparticles was more dynamic toward *E. coli* bacteria than *S. aureus.* Greener nanoparticles manufacturing is a more effective approach to eliminate or reduce pollutants from the environment. Honey with BFO nanomaterials can be used to increase the recycling of water productivity and develop novel medication designs [120]. Ravichandran et al. effectively synthesized magnesium-doped $BiFeO_3$ ($Bi_{1-x}Mg_xFeO_3$ ($0.00 \leq x \leq 0.15$)) nanoparticles (NPs) employing a simple solution combustion process via urea as fuel. The resulting material was sintered at 550 °C over 2 h to achieve pure-phase formation. Increased Mg levels of doping improve the antibacterial capacity of $Bi_{1-x}Mg_xFeO_3$ NPs by reducing grain size [121]. Nazeer et al. studied the influence of Ga substitutions on the characteristics of bismuth ferrite $BiGa_xFe_{1-x}O_3$ ($x = 0, 0.05, 0.10, 0.15, 0.20$, and 0.25), that was produced by a microemulsion technique. When compared to pristine BFO, the manufactured $BiGa_xFe_{1-x}O_3$ material demonstrated superior antibacterial efficacy toward S. aureus strains of bacteria [122]. R. P. Ummer et al. reported the photocatalytic and antibacterial properties of the $BiFeO_3$–$NaNbO_3$ nanocomposite system. As seen in Fig. 11.4, four compounds ($x = 0 - 0.7$) were added to each plate and incubated for eighteen hours at 37 ± 0.5 °C. As compared to other compositions, the composition $x = 0.1$ was shown to have strong antibacterial activity. Tiles,

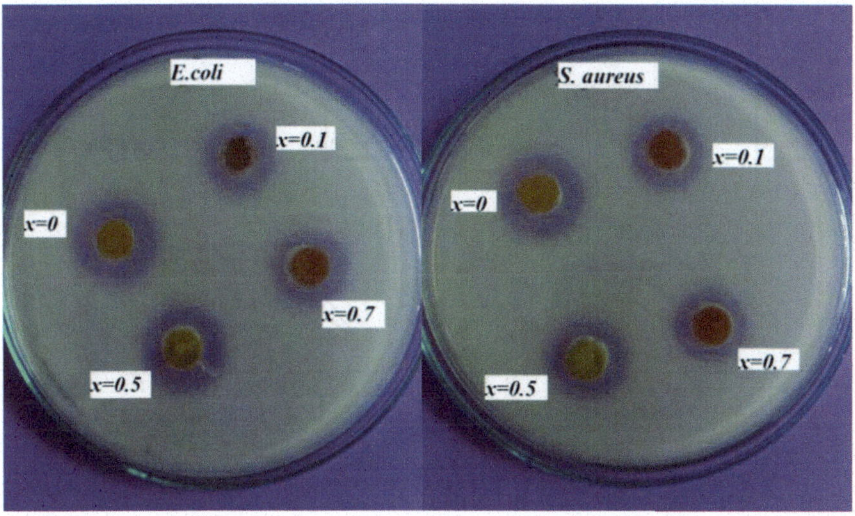

Fig. 11.4 Antibacterial activity of $(1-x)BiFeO_3$–$xNaNbO_3$ powder samples against. **a** Gram-negative *Escherichia coli (E. coli)*. **b** Gram-positive Staphylococcus aureus (*S. aureus*). *Reproduced with the permission* [13]

various building supplies, and coatings for healthcare establishments and labs can all be made from this material [13].

5.3 Drug Delivery

Perovskite multiferroics are those intended for drug delivery, have gained a lot of interest due of their distinctive characteristics, which allow for the targeted and controlled release of medicinal substances. This skill is especially useful in cancer treatment, as precise medication administration can improve effectiveness and reduce adverse effects. Shahzad et al. reveal the targeting medication release to tumor cells while protecting normal cells is a significant issue. Core–shell magnetoelectric (ME) nanoparticles have addressed this issue by exhibiting shape-dependent magnetoelectric characteristics. The colloidally stable core–shell cobalt ferrite@barium titanate (CFO@BTO) ME nanoparticles used in the in vitro investigation were synthesized using sonochemical methods. [14]. Papadopoulos et al. effectively synthesized highly crystalline pristine BFO, $BiFe_{0.97}Co_{0.03}O_3$ along with $Bi_{0.97}Sm_{0.03}FeO_3$ nanoparticles, employing a simple chemical-based co-precipitation process following the calcination at 600 °C in the air for a period of two hours. The dispersed form of $BiFe_{0.97}Co_{0.03}O_3$ nanoparticles with water at an amount of 4 mg/mL might reach the recommended therapeutic range of 41–45 °C for cancer therapy at 50 mT and 375 kHz, while the unexpected rise in temperatures of BFO-based multiferroic under the same conditions, hydrogels show promisefor an effective local treatment with

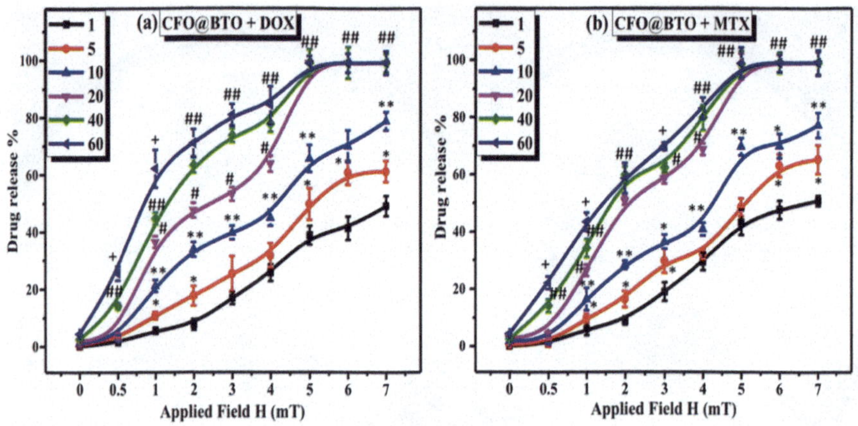

Fig. 11.5 Drug release kinetics of **a** CFO@BTO + DOX NPs and **b** CFO@BTO + MTX NPs using different magnetic field (0–7 mT) at different times 1–60 min. Reproduced with the permission [14]

low toxicity. As a result, BFO-based multiferroic hydrogels achieve controllable solid lipid particle (SLP) values, making them promising candidates for additional biological use as magnetic hyperthermia agents [123].

K. Shahzad et al. investigated the field-controlled ME core–shell $CoFe_2O_4$@$BaTiO_3$ NPs synthesized using sonochemical method proved to be effective drug carriers and drug release in vitro. The NPs were functionalized with anticancer drugs comprising doxorubicin and methotrexate up to 80% via, 1-Ethyl-3-(3-dimethylaminopropyl) carbodiimide (EDC) chemistry. According to the findings, ME CFO@BTO NPs are efficient drug transporters and have the potential to be used in cancer treatment when exposed to external magnetic fields. As illustrated in Fig. 11.5, the drug release kinetics experiment on drug-loaded CFO@BTO NPs was carried out utilizing spectrophotometric absorbance in the presence of various magnetic fields (0–7 mT) [14].

5.4 Antimicrobial Activity

Perovskite multiferroics, which show both ferroelectric as well as magnetic characteristics, have sparked interest for their possible applications in a variety of sectors, including antimicrobial technology. Abdel-Khalek et al. synthesized $SrFeO_{3-\delta}$ nanoparticle using the sol–gel technique. XRD as well as SAED patterns indicate the cubic (Pm3m) crystal form of $SrFeO_{3-\delta}$ nanoparticles, including a small tetragonal phase. Thermal along with magnetic tests confirm the existence of multiferroic properties in $SrFeO_{3-\delta}$ nanoparticle samples. The $SrFeO_{3-\delta}$ nanoparticles sample is a potential antibacterial agent and efficient wastewater treatment technique, as evidenced by microbiological tests, antimicrobial activity, MIC, and MBC investigations [124]. Hammad et al. developed multifunctional $BaTi_{0.7}Fe_{0.3}O_{3-x}NiFe_2O_4$ (BTF-xNF) nanoceramics using the sol–gel technique at 450 °C. Our research shows that multifunctional $BaTi_2Fe_4O_{11-(x)}$ $NiFe_2O_4$ nanoceramics can effectively battle microbial infections and biofilms. The $BaTi_2Fe_4O_{11-(x)}NiFe_2O_4$ nanoceramics showed remarkable antibacterial and antibiofilm effect iveness due to their distinct microstructure and spectroscopic characteristics [125]. Thankappan et al. investigated the production and study of a photoactive perovskite polymer composite film using dimethylammonium zinc formate perovskite (DMAZF), an important element of the metal–organic structure family, along with polycarbonate as a template. The produced MOF has strong antibacterial activity against both *E. coli* and *S. aureus*. However, the inhibition diameter was considerably larger for bare DMAZF than for all the composites, all samples had almost identical inhibition zones, with just a minor increase in the zone size detected with increasing Ag concentration. This work has created several potential for DMAZF and sets the way for future investigations and breakthroughs in its use throughout a variety of fields [126]. Karadirek et al. reported the Ag decorated $MnFeO_3$ catalyst and compared the investigation of visible light driven antibiotic reduction and antibacterial performance. Figure 11.6 demonstrates the clear zone, which stands for total bacterial growth suppression. The findings verified that every bacterium had a modest sensitivity to Ag–$MnFeO_3$. However, $MnFeO_3$ showed no negative effects on the bacteria, with the notable exception of *S. enteritidis* and *E. coli* ATCC 25,922. [15].

5.5 *Biosensors*

Multiferroic perovskites have arisen as interesting biosensing compounds because of their unusual mix of ferroelectric along with magnetic characteristics. These materials are capable of turning biological signals into detectable electrical or as magnetic responses, which makes them useful for a variety of biosensing systems. Caglar et al. synthesized ferromagnetic nanoparticles of $BiFeO_3$ (20–60 nm) having a rhombohedral deformed perovskite structure using the sol–gel approach. We developed a novel amperometric biosensor using cress peroxidase entrapped on an improved carbon paste electrode to detect H_2O_2. The made H_2O_2 biosensor was powered up using a chronoamperometry approach with a -0.8 V versus Ag/AgCl electrode. Under ideal circumstances, the biosensor responded linearly to H_2O_2 concentrations ranging from 2.0 10^{-7} to 1.0 10^{-5} M. The limit of both detection and quantification has been determined to be 7.0 10^{-8} M (2.7 μg L − 1) and 2.0 10 − 7 M (8.7 μg L − 1), respectively [127]. Shanthy et al. used chemical compounds to build $BoFeO_3$ (BFO) thin films over Pt/Ti/SiO_2/Si(100) structures, resulting in a low leaking current density at ambient temperature. The piezoelectric coefficient d33 was around 60 pm/V, indicating that BFO films have sensing capability and can convert mechanical vibration energy into electrical energy, which might be used in tiny biosensors [128].

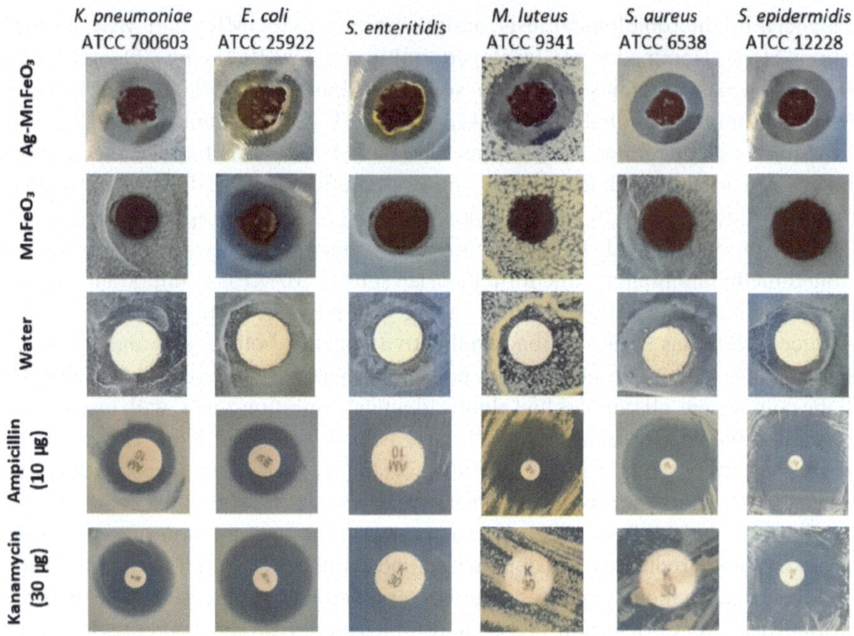

Fig. 11.6 Inhibition zone diameter of tested bacteria against $MnFeO_3$, Ag–$MnFeO_3$ water, ampicillin (10 μg/disc) and kanamycin (30 μg/disc) [15]

H. S. Magar et al. reported the nanohexagonal $Fe_{0.3}BaTi_{0.7}O_3$ nanoparticles surface-modified $NiFe_2O_4$ composite screen-printed electrode for enzymatic glucose monitoring. The sequences of GOx enzyme immobilization on the FBT/5NF@SPE surface using glutaraldehyde (2.5%) as crosslinker were examined by cyclic voltammetry and EIS spectra in order to create a glucose biosensor with high bioactivity and good performance, as illustrated in Fig. 11.7 [16].

Future Prospective and Challenges:

Perovskites find their applications in the field of optoelectronic devices such as solar cells, light-emitting diodes due to their low fabrication cost and high efficiency. But it is the use of perovskite as X-ray detection, biosensors, bioimplant devices, ultrasound imaging systems, drug delivery systems, etc., that make them more fascinating in the field of biomedical applications. Nanotechnology combined with biosynthesis or the green approach is extremely essential since biological processes are ecologically friendly and economical. Green nanoparticles for various applications are definitely not a dangerous technology because the production of metal nanoparticles is done using plant extract [129]. Progressive research and the need of humankind will unlock more opportunities in this field and use of perovskite in the near future. The potential for using multiferroic perovskite materials in biological applications is quite high because of their special blend of mechanical, magnetic, and electric properties. In radiotherapy, these materials enable magnetic-localized

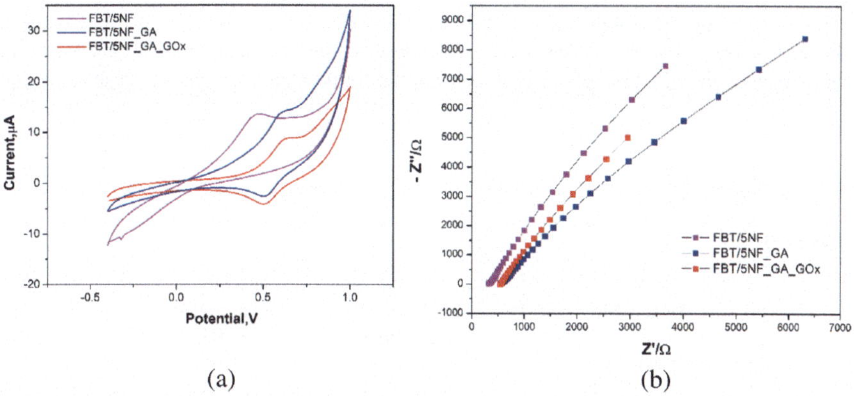

Fig. 11.7 **a** CV and **b** EIS measurements of the GOx enzyme immobilization steps on the FBT/5NF@SPE surface [16]

dose enhancement, providing a more effective and targeted therapeutic approach for cancer treatment. Applications in magnetic resonance imaging (MRI), computed tomography (CT), and magnetic hyperthermia are made possible by these materials' combination of diagnostic and therapeutic properties [130]. By serving as mediators of magnetic hyperthermia, multiferroic perovskites can generate localized heat to destroy cancer cells while minimizing damage to surrounding healthy tissues [130]. Their piezoelectric properties also make them ideal for wearable and implantable devices, enabling real time, battery-free health monitoring [131]. Furthermore, these materials are essential for tissue engineering because they use localized electric fields to promote tissue regeneration and cell differentiation, which helps with the repair of bones, nerves, and the heart [39]. Furthermore, multiferroic perovskites are developing highly sensitive biosensors for biomolecule detection and may find use in neurostimulation and brain-machine interfaces, enabling novel therapies for neurological conditions [132, 133].

However, this material faces significant challenges. Lead-based multiferroic perovskites, such as $PbTiO_3$ and $Bi_{0.5}Pb_{0.5}Fe_{0.5}Ce_{0.5}O_3$, face biocompatibility issues due to the toxic nature of lead, which limits their applications in biomedical fields. Therefore, it is essential to develop lead-free alternatives to ensure safer and more sustainable use in biomedical applications [98, 134]. Secondly, these materials possess stability issues, as they degrade rapidly and require surface functionalization or protective coatings to enhance stability, which is costly and reduces their cost-effectiveness. Furthermore, scaling difficulties for miniaturization hinder their widespread adoption [135]. Moreover, the mechanisms governing these material's interactions with biological tissues remain unclear, and further research is necessary to ensure their long-term safety and regulatory approval. Multidisciplinary research initiatives across the globe, scalable less toxic production techniques, and material design improvements will all be the necessary future aspects to meet these difficulties.

Acknowledgements One of the authors, Dr. Shweta Sharma, Assistant Professor, Sri Venkateswara College, University of Delhi, New Delhi is thankful to the Principal, Prof. Vajala Ravi, Sri Venkateswara College, University of Delhi, New Delhi for providing the facility for this research work.

References

1. Hossain KM, Hasan MZ, Ali ML (2020) Understanding the influences of Mg doping on the physical properties of $SrMoO_3$ perovskite. Results in Phys 19:103337. https://doi.org/10.1016/j.rinp.2020.103337
2. Batoo KM, Verma R, Chauhan A, Kumar R, Hadi M, Aldossary OM, Al-Douri Y (2021) Improved room temperature dielectric properties of Gd3+ and Nb5+ co-doped Barium Titanate ceramics. J Alloy Compd 883:160836. https://doi.org/10.1016/j.jallcom.2021.160836
3. Wang W, Tadé MO, Shao Z (2015) Research progress of perovskite materials in photocatalysis- and photovoltaics-related energy conversion and environmental treatment. Chem Soc Rev 44(15):5371–5408. https://doi.org/10.1039/C5CS00113G
4. Voorhoeve RJH, Johnson DW Jr, Remeika JP, Gallagher PK (1977) Perovskite oxides: materials science in catalysis: perovskite oxides, long known in solid-state chemistry and physics, find new applications in catalysis. Science 195(4281):827–833. https://doi.org/10.1126/science.195.4281.827
5. Zhu T, Yang Y, Gong X (2020) Recent advancements and challenges for low-toxicity perovskite materials. ACS Appl Mater Interfaces 12(24):26776–26811. https://doi.org/10.1021/acsami.0c02575
6. Coreno J, Coreno O (2005) Evaluation of calcium titanate as apatite growth promoter. J Biomed Mater Res Part A: An Off J Soc Biomater Jap Soc Biomater The Australian Soc Biomater the Korean Soc Biomater 75(2):478–484. https://doi.org/10.1002/jbm.a.30447
7. Song HJ, Kim MG, Moon WJ, Park YJ (2011) Formation of hydroxyapatite nanorods and anatase TiO_2 on $CaTiO_3$ powder using hydrothermal treatment. Mater Sci Eng, C 31(3):558–561. https://doi.org/10.1016/j.msec.2010.11.024
8. Wei D, Zhou Y, Jia D, Wang Y (2008) Formation of $CaTiO_3/TiO_2$ composite coating on titanium alloy for biomedical applications. J Biomed Mater Res Part B: Appl Biomat: An Official J Soc Biomater, The Jap Soc Biomater The Australian Society for Biomater Korean Soc Biomater 84(2):444–451. https://doi.org/10.1002/jbm.b.30890
9. Li L, Ye S, Qu J, Zhou F, Song J, Shen G (2021) Recent advances in perovskite photodetectors for image sensing. Small 17(18):2005606. https://doi.org/10.1002/smll.202005606
10. Hwang GT, Byun M, Jeong CK, Lee KJ (2015) Flexible piezoelectric thin-film energy harvesters and nanosensors for biomedical applications. Adv Healthcare Mater 4(5):646–658. https://doi.org/10.1002/adhm.201400642
11. Pattipaka S, Bae YM, Jeong CK, Park KI, Hwang GT (2022) Perovskite piezoelectric-based flexible energy harvesters for self-powered implantable and wearable IoT devices. Sensors 22(23):9506. https://doi.org/10.3390/s22239506
12. Yokota T, Fukuda K, Someya T (2021) Recent progress of flexible image sensors for biomedical applications. Adv Mater 33(19):2004416. https://doi.org/10.1002/adma.202004416
13. Ummer RP, Perumbilavil S, Jose J, Thomas S, Gopinath P, Kalarikkal N (2021) Exploring the optical limiting, photocatalytic and antibacterial properties of the $BiFeO_3$–$NaNbO_3$ nanocomposite system. RSC Adv 11(15):8450–8458. https://doi.org/10.1039/D0RA09776D
14. Shahzad K, Mushtaq S, Rizwan M, Khalid W, Atif M, Din FU, Ali Z (2021) Field-controlled magnetoelectric core-shell $CoFe_2O_4$@ $BaTiO_3$ nanoparticles as effective drug carriers and drug release in vitro. Mater Sci Eng, C 119:111444. https://doi.org/10.1016/j.msec.2020.111444

15. Karadirek Ş, Tuna Ö, Simsek EB, Altuntas S, Cinar AY (2024) Facile fabrication of Ag decorated $MnFeO_3$ catalyst: comparative analysis of visible light driven antibiotic reduction and antibacterial performance. J Environ Manage 358:120891. https://doi.org/10.1016/j.jenvman.2024.120891
16. Magar HS, El Nahrawy AM, Hassan RY, Abou Hammad AB (2024) Nanohexagonal iron barium titanate nanoparticles surface-modified $NiFe_2O_4$ composite screen-printed electrode for enzymatic glucose monitoring. RSC Adv 14(47):34948–34963. https://doi.org/10.1039/D4RA06689H
17. Pathania A, Thakur P, Trukhanov AV, Trukhanov SV, Panina LV, Lüders U, Thakur A (2019) Development of tungsten doped Ni-Zn nano-ferrites with fast response and recovery time for hydrogen gas sensing application. Results Phys 15:102531. https://doi.org/10.1016/j.rinp.2019.102531
18. Akmal MM, Ahmad FB, Hisham F, Hazmi AT (2021) Biopolymer-based waste for biomaterials thin film in piezoelectric application. In: Advanced technology for the conversion of waste into fuels and chemicals. Woodhead Publishing, pp 355–381. https://doi.org/10.1016/B978-0-323-90150-5.00010-8
19. Manori S, Chandra R (2021) Basics of multiferroic materials and their types. In: Ferrites and multiferroics: fundamentals to applications. Springer Singapore, Singapore, pp 129–141. https://doi.org/10.1007/978-981-16-7454-9_8
20. Khomskii D (2009) Classifying multiferroics: mechanisms and effects, Physics 2
21. Khomskii DI (2006) Multiferroics: different ways to combine magnetism and ferroelectricity. J Magn Magn Mater 306(1):1–8. https://doi.org/10.1016/j.jmmm.2006.01.238
22. Efremov DV, Van Den Brink J, Khomskii DI (2004) Bond-versus site-centred ordering and possible ferroelectricity in manganites. Nat Mater 3(12):853–856. https://doi.org/10.1038/nmat1236
23. Choi YJ, Yi HT, Lee S, Huang Q, Kiryukhin V, Cheong SW (2008) Ferroelectricity in an Ising chain magnet. Phys Rev Lett 100(4):047601. https://doi.org/10.1103/PhysRevLett.100.047601
24. Van Aken BB, Palstra TT, Filippetti A, Spaldin NA (2004) The origin of ferroelectricity in magnetoelectric $YMnO_3$. Nat Mater 3(3):164–170. https://doi.org/10.1038/nmat1080
25. Kimura T, Goto T, Shintani H, Ishizaka K, Arima TH, Tokura Y (2003) Magnetic control of ferroelectric polarization. Nature 426(6962):55–58. https://doi.org/10.1038/nature02018
26. Hur N, Park S, Sharma PA, Ahn JS, Guha S, Cheong SW (2004) Electric polarization reversal and memory in a multiferroic material induced by magnetic fields. Nature 429(6990):392–395. https://doi.org/10.1038/nature02572
27. Mostovoy M (2006) Ferroelectricity in spiral magnets. Phys Rev Lett 96:067601. https://doi.org/10.1103/PhysRevLett.96.067601
28. Mostovoy M (2006) Ferroelectricity in spiral magnets. Phys Rev Lett 96(6):067601. https://doi.org/10.1103/PhysRevLett.96.067601
29. Trukhanov AV, Kostishyn VG, Panina LV, Korovushkin VV, Turchenko VA, Thakur P, Thakur A, Yang Y, Vinnik DA, Yakovenko ES, Matzui LY (2018). Control of electromagnetic properties in substituted M-type hexagonal ferrites. J Alloys Compounds 754:247–256. https://doi.org/10.1016/j.jallcom.2018.04.150
30. Mathur P, Thakur A, Singh M (2008) Low temperature synthesis of Mn0.4Zn0.6In0. 5Fe1. 5O4 nanoferrite for high-frequency applications. J Phys Chem Solids 69(1):187–192. https://doi.org/10.1016/j.jpcs.2007.08.014
31. Thakur A, Mathur P, Singh M (2007) Study of dielectric behaviour of Mn–Zn nano ferrites. J Phys Chem Solids 68(3):378–381. https://doi.org/10.1016/j.jpcs.2006.11.028
32. Thakur P, Chahar D, Taneja S, Bhalla N, Thakur A (2020) A review on MnZn ferrites: synthesis, characterization and applications. Ceram Int 46(10):15740–15763. https://doi.org/10.1016/j.ceramint.2020.03.287
33. Xian H, Zhang X, Li X, Zou H, Meng M, Zou Z, Guo L, Tsubaki N (2010) Effect of the calcination conditions on the NOx storage behavior of the perovskite $BaFeO_{3-x}$ catalysts. Catal Today 158(3–4):215–219. https://doi.org/10.1016/j.cattod.2010.03.026

34. McNamara K, Tofail SA (2017) Nanoparticles in biomedical applications. Adv Phys: X 2(1):54–88. https://doi.org/10.1080/23746149.2016.1254570
35. Dissanayake PD, Yeom KM, Sarkar B, Alessi DS, Hou D, Rinklebe J, Noh JH, Ok YS (2023) Environmental impact of metal halide perovskite solar cells and potential mitigation strategies: a critical review. Environ Res 219:115066. https://doi.org/10.1016/j.envres.2022.115066
36. Moroz P, Jones SK, Gray BN (2002) Magnetically mediated hyperthermia: current status and future directions. Int J Hyperth 18(4):267–284. https://doi.org/10.1080/02656730110108785
37. Chen A, Chatterjee S (2013) Nanomaterials based electrochemical sensors for biomedical applications. Chem Soc Rev 42(12):5425–5438. https://doi.org/10.1039/C3CS35518G
38. Huamán JLC, Rivera VAG, Pinto AH, Marega Jr, E (2023) Multiferroic perovskite ceramics: properties and applications. Perovskite Ceramics, 339–381. https://doi.org/10.1016/B978-0-323-90586-2.00003-6
39. Kargol A, Malkinski L, Caruntu G (2012) Biomedical applications of multiferroic nanoparticles. Adv Magn Mater 89–118
40. Sharma A, Chauhan A, Verma R, Wang HMD, Thakur P, Thakur A (2024) Perovskite: a key material for the biomedical industry. In: Biomedical applications of perovskites: the era of bio-piezoelectric systems. Bentham Science Publishers, pp 82–102. https://doi.org/10.2174/97898152563831240101
41. Bui TH, Shin JH (2023) Perovskite materials for sensing applications: recent advances and challenges. Microchem J 191:108924. https://doi.org/10.1016/j.microc.2023.108924
42. Čebela M, Zagorac D, Batalović K, Radaković J, Stojadinović B, Spasojević V, Hercigonja R (2017) $BiFeO_3$ perovskites: a multidisciplinary approach to multiferroics. Ceram Int 43(1):1256–1264. https://doi.org/10.1016/j.ceramint.2016.10.074
43. Kharbanda S, Dhanda N, Sun ACA, Thakur A, Thakur P (2023) Multiferroic perovskite bismuth ferrite nanostructures: a review on synthesis and applications. J Magn Magn Mater 572:170569. https://doi.org/10.1016/j.jmmm.2023.170569
44. Carranza-Celis D, Cardona-Rodríguez A, Narváez J, Moscoso-Londono O, Muraca D, Knobel M, Ornelas-Soto N, Reiber A, Ramírez JG (2019). Control of multiferroic properties in $BiFeO_3$ nanoparticles. Sci Rep 9(1):3182. https://doi.org/10.1038/s41598-019-39517-3
45. Sarkar A, Khan GG, Chaudhuri A, Das A, Mandal K (2016) Multifunctional $BiFeO_3/TiO_2$ nano-heterostructure: photo-ferroelectricity, rectifying transport, and nonvolatile resistive switching property. Appl Phys Lett 108(3). https://doi.org/10.1063/1.4940118
46. Hong S, Choi T, Jeon JH, Kim Y, Lee H, Joo HY, Hwang I, Kim JS, Kang SO, Kalinin SV, Park BH (2013) Large resistive switching in ferroelectric $BiFeO_3$ nano-island based switchable diodes. Adv Mater 16(25):2339–2343. https://doi.org/10.1002/adma.201204839
47. Allibe J, Fusil S, Bouzehouane K, Daumont C, Sando D, Jacquet E, Deranlot C, Bibes M, Barthelemy A (2012) Room temperature electrical manipulation of giant magnetoresistance in spin valves exchange-biased with $BiFeO_3$. Nano Lett 12(3):1141–1145. https://doi.org/10.1021/nl202537y
48. Wang J, Xu XL, Li XA (2023) Recent progress of multiferroicity and magnetoelectric effects in ABX3-type perovskite metal-organic frameworks. Adv Mater Interfaces 10(25):2300123. https://doi.org/10.1002/admi.202300123
49. Pavithra AJ, Anudevi SD, Hikku GS, Ebenezar KK (2024). Bone regeneration and bone growth using perovskites. In: Biomedical applications of perovskites: the era of bio-piezoelectric systems. Bentham Science Publishers, pp 154–174. https://doi.org/10.2174/97898152563831240101
50. Bhardwaj L, Kumari S, Kumar V, Sun AC, Thakur A, Thakur P (2024) Perovskites for bone replacement. In: Biomedical applications of perovskites: the era of bio-piezoelectric systems. Bentham Science Publishers, pp 175–193. https://doi.org/10.2174/97898152563831240101
51. Kumari S, Dutta J, Bhardwaj L, Kumar V (2024) Bone tissue engineering using perovskites in regenerative medicines. In: Biomedical applications of perovskites: the era of bio-piezoelectric systems. Bentham Science Publishers, pp 194–210. https://doi.org/10.2174/97898152563831240101

52. Rosso JM, Volnistem EA, Santos IA, Bonadio TGM, Freitas VF (2022) Lead-free $NaNbO_3$-based ferroelectric perovskites and their polar polymer-ceramic composites: fundamentals and potentials for electronic and biomedical applications. Ceram Int 48(14):19527–19541. https://doi.org/10.1016/j.ceramint.2022.04.089
53. Das S (2024) Perovskite based neuromorphic devices. In: Pradhan B (eds) Perovskite optoelectronic devices. Engineering Materials. Springer, Cham. https://doi.org/10.1007/978-3-031-57663-8_12
54. Shariatinia Z, Karimzadeh Z (2024) Perovskite oxides as efficient bioactive inorganic materials in tissue engineering: a review. Mater Today Chem 35:101846. https://doi.org/10.1016/j.mtchem.2023.101846
55. Zhang L, Zhao S, Ouyang J, Deng L, Liu YN (2022) Oxygen-deficient tungsten oxide perovskite nanosheets-based photonic nanomedicine for cancer theranostics. Chem Eng J 431:133273. https://doi.org/10.1016/j.cej.2021.133273
56. Clement RGE, Bugler KE, Oliver CW (2011) Bionic prosthetic hands: a review of present technology and future aspirations. Surgeon 9(6):336–340. https://doi.org/10.1016/j.surge.2011.06.001
57. Dutta V, Radhakrishnan A, Ravelo B, Wan F, Thakur P, Thakur A (2024) Perovskite-based bio-implantable energy harvesters. In: Biomedical applications of perovskites: the era of bio-piezoelectric systems. Bentham Science Publishers, pp 304–320. https://doi.org/10.2174/978 98152563831240101
58. Pollack JR, Perou CM, Alizadeh AA, Eisen MB, Pergamenschikov A, Williams CF, Brown PO (1999) Genome-wide analysis of DNA copy-number changes using cDNA microarrays. Nat Genet 23(1):41–46
59. Layek S, Verma HC (2015) Magnetic and dielectric properties of multiferroic $BiFeO_3$ nanoparticles synthesized by a novel citrate combustion method. ArXiv preprint arXiv:1502.05797
60. Mathankumar K, Sukumar M, Dash CS, Sundararajan M, Ubaidullah M, Al-Enizi AM, Sutha A, Kausar Raza M, Arockia Dhanraj J, Kumar D (2022) Facile synthesis, characterization, catalytic and photocatalytic activity of multiferroic $BiFeO_3$ Perovskite Nanoparticles. https://doi.org/10.21203/rs.3.rs-1250457/v1
61. Abbas YM, Mansour AB, Ali SE, Ibrahim AH (2019) Investigation of structural and magnetic properties of multiferroic La1-xYxFeO$_3$ Perovskites, prepared by citrate auto-combustion technique. J Magn Magn Mater 482:66–74. https://doi.org/10.1016/j.jmmm.2019.03.056
62. Naveen K, Kumar N, Mandal TK, Babu PD, Siruguri V, Maji PK, Paul AK (2019) Multiferroic behaviour in B-site Cr-doped hexagonal $YInO_3$ perovskites: synthesis, structure and properties. J Mol Struct 1185:432–439. https://doi.org/10.1016/j.molstruc.2019.02.099
63. Zhu LF, Liu Q, Zhang BP, Cen ZY, Wang K, Li JJ, Bai Y, Wang XH, Li JF (2018). Temperature independence of piezoelectric properties for high-performance $BiFeO_3$–$BaTiO_3$ lead-free piezoelectric ceramics up to 300 °C. RSC Adv 8(62):35794–35801. https://doi.org/10.1039/C8RA07553K
64. Chahar D, Taneja S, Thakur P, Thakur A (2020) Remarkable resistivity and improved dielectric properties of Co–Zn nanoferrites for high frequency applications. J Alloy Compd 843:155681. https://doi.org/10.1016/j.jallcom.2020.155681
65. Ke H, Wang W, Wang Y, Xu J, Jia D, Lu Z, Zhou Y (2011) Factors controlling pure-phase multiferroic $BiFeO_3$ powders synthesized by chemical co-precipitation. J Alloy Compd 509(5):2192–2197. https://doi.org/10.1016/j.jallcom.2010.09.213
66. Wang L, Su Y, Cao J (2023) Room temperature multiferroic behavior of $CoFe_2O_4$–$BaTiO_3$ nanocomposites prepared by the co-precipitation method. J Mater Sci: Mater Electron 34(18):1398. https://doi.org/10.1007/s10854-023-10737-8
67. Wang HJ, Zhu YY, Zhou J (2018) Electron transport behavior of multiferroic perovskite $BiMnO_3$ prepared by co-precipitation method. Chin Phys Lett 35(2):027201. https://doi.org/10.1088/0256-307X/35/2/027201

68. Abbas Y, Kiani HF, Anis-ur-Rehman M (2024) Investigation of high-temperature transport mechanism and dielectric properties of RE-double-doped perovskite cobaltite using coprecipitation technique. J Mol Struct 1305:137710. https://doi.org/10.1016/j.molstruc.2024.137710
69. Klein LC, Garvey GJ (1980) Kinetics of the sol/gel transition. J Non-Cryst Solids 38:45–50. https://doi.org/10.1016/0022-3093(80)90392-0
70. Liu M, Li X, Lou J, Zheng S, Du K, Sun NX (2007) A modified sol-gel process for multiferroic nanocomposite films. J Appl Phys 102(8). https://doi.org/10.1063/1.2800804
71. Bangruwa JS, Vashisth BK, Singh N, Singh N, Verma V (2018) A systematic study of structural, magnetic and electric properties of perovskite-spinel composites prepared by sol-gel technique. J Alloy Compd 739:319–326. https://doi.org/10.1016/j.jallcom.2017.12.281
72. Jo SH, Lee SG, Lee SH (2012) Structural and pyroelectric properties of sol–gel derived multiferroic BFO thin films. Mater Res Bull 47(2):409–412. https://doi.org/10.1016/j.materresbull.2011.11.009
73. Ji JJ, Sun H, Mao XY, Wang W, Chen XB (2012) Structure and multiferroic properties of Bi5FeTi3O15 thin films prepared by the sol–gel method. J Sol-Gel Sci Technol 61:328–331. https://doi.org/10.1007/s10971-011-2631-4
74. Shang M, Zhang C, Zhang T, Yuan L, Ge L, Yuan H, Feng S (2013) The multiferroic perovskite $YFeO_3$. Appl Phys Lett 102(6). https://doi.org/10.1063/1.4791697
75. Han SH, Kim KS, Kim HG, Lee HG, Kang HW, Kim JS, Cheon CI (2010) Synthesis and characterization of multiferroic $BiFeO_3$ powders fabricated by hydrothermal method. Ceram Int 36(4):1365–1372. https://doi.org/10.1016/j.ceramint.2010.01.020
76. Zhang C, Shang M, Liu M, Zhang T, Ge L, Yuan H, Feng S (2016) Multiferroicity in $SmFeO_3$ synthesized by hydrothermal method. J Alloy Compd 665:152–157. https://doi.org/10.1016/j.jallcom.2016.01.042
77. Khikhlovskyi SVV, Blake G (2010). The renaissance of multiferroics: bismuth ferrite ($BiFeO_3$)–a candidate multiferroic material in nanoscience. University of Groningen 1(3)
78. Catalano MR, Spedalotto G, Condorelli GG, Malandrino G (2017) MOCVD growth of perovskite multiferroic BiFeO3 films: the effect of doping at the A and/or B sites on the structural, morphological and ferroelectric properties. Adv Mater Interfaces 4(8):1601025. https://doi.org/10.1002/admi.201601025
79. Sun B, Zhou G, Sun L, Zhao H, Chen Y, Yang F, Zhao Y, Song Q (2021) ABO_3 multiferroic perovskite materials for memristive memory and neuromorphic computing. Nanoscale Horizons 6(12):939–970. https://doi.org/10.1039/D1NH00292A
80. Nakaki H, Kim YK, Yokoyama S, Ikariyama R, Funakubo H, Streiffer SK, Nishida K, Saito K, Gruverman A (2008) Experimental evidence of strain relaxed domain structure in (100)/(001)-oriented epitaxial lead titanate thick films grown by metal organic chemical vapor deposition. J Appl Phys 104(6). https://doi.org/10.1063/1.2981193
81. Liu XM, Gao WL (2011) Synthesis and characterization of multiferroic $NiFe_2O_4/BiFeO_3$ nanocomposites by modified Pechini method. Adv Mater Res 197:456–459. https://doi.org/10.4028/www.scientific.net/AMR.197-198.456
82. Rida K, Peña MA, Sastre E, Martinez-Arias A (2012) Effect of calcination temperature on structural properties and catalytic activity in oxidation reactions of $LaNiO_3$ perovskite prepared by Pechini method. J Rare Earths 30(3):210–216. https://doi.org/10.1016/S1002-0721(12)60025-8
83. Sunde TOL, Grande T, Einarsrud MA (2016) Modified Pechini synthesis of oxide powders and thin films. Handbook of Sol-gel Sci Technol. http://hdl.handle.net/11250/2593090
84. Dimesso L (2016) Pechini processes: an alternate approach of the sol–gel method, preparation, properties, and applications. Handbook of Sol-Gel Sci Technol 2:1–22. https://doi.org/10.1007/978-3-319-19454-7_123-1
85. Qian G, Zhu C, Yin C, Wang L, Wang Y, Li C, Yuan S (2018) Multiferroic properties of single-phase perovskite structure 0.8BiFeO3–0.2SrTiO3 ceramics synthesized using the Pechini method. J Electroceram 40:190–196. https://doi.org/10.1007/s10832-018-0119-0

86. Ahmed S, Barik SK (2016) Enhanced electric and magnetic properties of (BiLi) 1/2 (Fe_2/$_3$W1/3) O_3 multiferroic as compared to $BiFeO_3$. Ceram Int 42(5):5659–5667. https://doi.org/10.1016/j.ceramint.2015.12.090
87. Ummer RP, Sreekanth P, Raneesh B, Philip R, Rouxel D, Thomas S, Kalarikkal N (2015) Electric, magnetic and optical limiting (short pulse and ultrafast) studies in phase pure (1–x)BiFeO3–x NaNbO 3 multiferroic nanocomposite synthesized by the pechini method. RSC Adv 5(82):67157–67164
88. Huo SX, Yuan SL, Qiu Y, Ma ZZ, Wang CH (2012) Crystal structure and multiferroic properties of $BiFeO_3$–$Na_{0.5}K_{0.5}NbO_3$ solid solution ceramics prepared by Pechini method. Mater Lett 68:8–10. https://doi.org/10.1016/j.matlet.2011.09.081
89. Suresh P, Srinath S (2015) Effect of synthesis route on the multiferroic properties of $BiFeO_3$: a comparative study between solid state and sol–gel methods. J Alloy Compd 649:843–850. https://doi.org/10.1016/j.jallcom.2015.07.152
90. Kumari S, Kumar A, Kumar V, Dubey SK, Goyal PK, Kumar S, Sharma AL, Arya A (2021) Structural, dielectric and ferroelectric properties of Cu2+-and Cu^{2+}/Bi^{3+}-doped BCZT lead-free ceramics: a comparative study. J Mater Sci Mater Electron 32(12):16900–16915. https://doi.org/10.1007/s10854-021-06252-3
91. Furuhashi T, Moriya M, Sakamoto W, Yogo T (2010) Synthesis and properties of perovskite $BiFeO_3$–KO. 5NaO. 5NbO3 ceramics by solid-state reaction. J Ceramic Soc Japan, 118(1380):701–705. https://doi.org/10.2109/jcersj2.118.701
92. Yamazoe S, Kawawaki T, Shibata K, Kato K, Wada T (2011) Synthetic mechanism of perovskite-type $KNbO_3$ by modified solid-state reaction process. Chem Mater 23(20):4498–4504. https://doi.org/10.1021/cm2016966
93. Jancar B, Valant M, Suvorov D (2004) Solid-state reactions occurring during the synthesis of $CaTiO_3$– $NdAlO_3$ perovskite solid solutions. Chem Mater 16(6):1075–1082. https://doi.org/10.1021/cm034972s
94. Li F, Yu X, Chen L, Pan H, Xin X (2002) Solid-state synthesis of $LaCoO_3$ perovskite nanocrystals. J Am Ceram Soc 85(9):2177–2180. https://doi.org/10.1111/j.1151-2916.2002.tb00431.x
95. Matsuishi S, Iwasaki D, Hosono H (2022) Synthesis of perovskite-type $LaWN_3$ by high-pressure solid-state metathesis reaction. J Solid State Chem 315:123508. https://doi.org/10.1016/j.jssc.2022.123508
96. Mostafavi E, Ataie A, Ahmadzadeh M (2014) Characterization of nano-structured multiferroic bismuth ferrite produced via solid state reaction route. Adv Mater Res 829:683–687. https://doi.org/10.4028/www.scientific.net/AMR.829.683
97. Kumar M, Pundir SK, Sharma S, Raymond Herrera O, Siqueiros JM, Kumar M (2024) Tuning the physical properties of perovskite multiferroic nanoparticles for green energy applications. In: Emerging applications of novel nanoparticles. Springer Nature Switzerland, Cham, pp 321–346. https://doi.org/10.1007/978-3-031-57843-4_13
98. Bhatti HS, Hussain ST, Khan FA, Hussain S (2016) Synthesis and induced multiferroicity of perovskite $PbTiO_3$; a review. Appl Surf Sci 367:291–306. https://doi.org/10.1016/j.apsusc.2016.01.164
99. Lather S, Dalal J, Gupta A, Singh S, Singh DP, Dahiya S, Maan AS, Tripathi R, Ohlan A (2019). $PbTiO_3$–NiO. 5CoO. $5Fe_2O_4$ multiferroic nanocomposites: impact of ball-milling on dielectric, magnetic and ferroelectric properties. Ceram Int 45(4):4957–4963. https://doi.org/10.1016/j.ceramint.2018.11.195
100. Elseman AM, Rashad MM, Hassan AM (2016) Easily attainable, efficient solar cell with mass yield of nanorod single-crystalline organo-metal halide perovskite based on a ball milling technique. ACS Sustain Chem Eng 4(9):4875–4886. https://doi.org/10.1021/acssuschemeng.6b01183
101. Hernández-Ramírez A, Martínez-Luévanos A, Fuentes AF, Nelson AGD, Ewing RC, Montemayor SM (2014) Molten salts activated by high-energy milling: a useful, low-temperature route for the synthesis of multiferroic compounds. J Alloy Compd 584:93–100. https://doi.org/10.1016/j.jallcom.2013.09.003

102. Reséndiz-Trejo Y, Sánchez-De Jesús F, Betancourt-Cantera LG, Reyes-Valderrama MI, Cortés-Escobedo CA, Bolarín-Miró AM (2024) Evidence of multiferroic behavior in sintered BaTiO3 obtained from high-energy ball-milled powders. Ceram Int 50(24):53533–53543. https://doi.org/10.1016/j.ceramint.2024.10.202
103. Moure A, Tartaj J, Moure C (2011) Processing and characterization of Sr doped $BiFeO_3$ multiferroic materials by high energetic milling. J Alloy Compd 509(25):7042–7046. https://doi.org/10.1016/j.jallcom.2011.03.132
104. Madhu GM (2023) Proceedings of international conference on advanced materials in engineering sciences–ICAMES
105. Munir ZA, Ohyanagi M (2021) Perspectives on the spark plasma sintering process. J Mater Sci 56(1):1–15
106. Tokita M (2021) Progress of spark plasma sintering (SPS) method, systems, ceramics applications and industrialization. Ceramics 4(2):160–198. https://doi.org/10.3390/ceramics4020014
107. Song SH, Zhu QS, Weng LQ, Mudinepalli VR (2015) A comparative study of dielectric, ferroelectric and magnetic properties of $BiFeO_3$ multiferroic ceramics synthesized by conventional and spark plasma sintering techniques. J Eur Ceram Soc 35(1):131–138. https://doi.org/10.1016/j.jeurceramsoc.2014.08.016
108. Garg R, Gonuguntla S, Sk S, Iqbal MS, Dada AO, Pal U, Ahmadipour M (2024) Sputtering thin films: materials, applications, challenges and future directions. Adv Colloid Interface Sci 103203. https://doi.org/10.1016/j.cis.2024.103203
109. Safi I (2000) Recent aspects concerning DC reactive magnetron sputtering of thin films: a review. Surf Coat Technol 127(2–3):203–218. https://doi.org/10.1016/S0257-8972(00)00566-1
110. Simon AH (2025) Sputter processing. In: Handbook of thin film deposition. Elsevier, pp 93–140. https://doi.org/10.1016/B978-0-443-13523-1.00011-9
111. Kim TC, Ojha S, Tian G, Lee SH, Jung HK, Choi JW, Kornblum L, Walker FJ, Ahn CH, Ross CA, Kim DH (2018) Self-assembled multiferroic epitaxial $BiFeO_3$–$CoFe_2O_4$ nanocomposite thin films grown by RF magnetron sputtering. J Mater Chem C 6(20):5552–5561
112. Cha J, Ahn J, Lee K, JeongOk C, JeungSun A, KwangBae L (2009) Multiferroic $BiFeO_3$ thin films prepared by using a conventional RF magnetron sputtering method. J Korean Phys Soc 54:844–848
113. Okuyama M, Matsui Y, Nakano H, Hamakawa Y (1981) $PbTiO_3$ ferroelectric thin film gate fet for infrared detection. Ferroelectrics 33(1):235–241. https://doi.org/10.1080/00150198108008091
114. Christen HM, Eres G (2008) Recent advances in pulsed-laser deposition of complex oxides. J Phys: Condens Matter 20(26):264005. https://doi.org/10.1088/0953-8984/20/26/264005
115. Krebs HU, Weisheit M, Faupel J, Süske E, Scharf T, Fuhse C, Buback M (2003) Pulsed laser deposition (PLD)--a versatile thin film technique. Adv Solid state physics, 505–518. https://doi.org/10.1007/978-3-540-44838-9_36
116. Lowndes DH, Geohegan DB, Puretzky AA, Norton DP, Rouleau CM (1996) Synthesis of novel thin-film materials by pulsed laser deposition. Science 273(5277):898–903. https://doi.org/10.1126/science.273.5277.898
117. Puli VS, Pradhan DK, Hilton A, Katiyar RS, Ghosh K, Srinivasan G, Reed AN, McConney ME, Heidger S (2023). Magnetoelectric multiferroic properties of $BaTiO_3$-$CoFe_2O_4$-$BaTiO_3$ tri-layer thin films fabricated by pulsed laser deposition. J Magnet Magnetic Mater 583:171061. https://doi.org/10.1016/j.jmmm.2023.171061
118. Wei D, Xia H, Fang L, Su L, Chen X, Song H (2024) Mg 2+-modified $LiBaF_3$: Ni2+ perovskite single crystals with broadband long wavelength near-infrared emission for imaging human vascular structures. Inorganic Chem Front 11(23):8339–8349
119. Ateia EE, Hussein B, Singh C, Arman MM (2022) Multiferroic properties of GdFeO. 9MO. 1O3 (M= Ag1+, Co2+ and Cr3+) nanoparticles and evaluation of their antibacterial activity. Eur Phys J Plus 137(4):443. https://doi.org/10.1140/epjp/s13360-022-02662-2

120. Sharmila M, Mani RJ, Parvathiraja C, Kader SA, Siddiqui MR, Wabaidur SM, Islam MA, Lai WC (2022) Visible light photocatalyst and antibacterial activity of BFO (Bismuth Ferrite) nanoparticles from honey. Water 14(10):1545. https://doi.org/10.3390/w14101545
121. Ravichandran AT, Srinivas J, Manikandan A, Baykal A (2019) Enhanced magneto-optical and antibacterial studies of Bi 1–x Mg x FeO 3 (0.0≤ x≤ 0.15)) nanoparticles. J Supercond Novel Magn 32:1663–1670. https://doi.org/10.1007/s10948-018-4842-1
122. Nazeer Z, Bibi I, Majid F, Kamal S, Alwadai N, Arshad MI, Ali A, Nouren S, Al Huwayz M, Iqbal M (2023) Optical, dielectric, magnetic, photocatalytic, and antibacterial properties of ga-Doped BiGa x Fe_{1-x} O_3 synthesized by the microemulsion approach. ACS Omega 9(1):545–558. https://doi.org/10.1021/acsomega.3c06132
123. Papadopoulos K, Myrovali E, Karfaridis D, Farle M, Wiedwald U, Angelakeris M (2023) Cation substitution and size confinement effects on structure, magnetism and magnetic hyperthermia of $BiFeO_3$-based multiferroic nanoparticles and hydrogels. J Alloy Compd 969:172337. https://doi.org/10.1016/j.jallcom.2023.172337
124. Abdel-Khalek EK, Rayan DA, Askar AA, Maksoud MA, El-Bahnasawy HH (2021) Synthesis and characterization of SrFeO 3-δ nanoparticles as antimicrobial agent. J Sol-Gel Sci Technol 97:27–38. https://doi.org/10.1007/s10971-020-05431-8
125. Abou Hammad AB, Hemdan BA, Mansour AM, El Nahrawy AM (2023) Improved antimicrobial and antibiofilm efficacy of adaptable $BaTi_2Fe_4O_{11}-(x)$ $NiFe_2O_4$ nanoceramics: investigating microstructural and spectroscopic analysis. SN Appl Sci 5(12):317. https://doi.org/10.1007/s42452-023-05547-w
126. Thankappan A, Mayeen A, Jose J, Thomas S (2023) Free standing organometallic dimethylammonium zinc formate perovskite for photonic and antimicrobial applications. Results Eng 19:101376. https://doi.org/10.1016/j.rineng.2023.101376
127. Caglar B, İçer F, Özdokur KV, Caglar S, Özdemir AO, Guner EK, Beşer BM, Altay A, Çırak Ç, Doğan B, Tabak A (2021). A novel amperometric H_2O_2 biosensor constructed by cress peroxidase entrapped on $BiFeO_3$ nanoparticles. Mater Chem Phys 262:124287. https://doi.org/10.1016/j.matchemphys.2021.124287
128. Shanthy S, Singh SK (2008) Multiferroic BiFeO3 films: candidate for room temperature micro biosensor. Integr Ferroelectr 99(1):77–85. https://doi.org/10.1080/10584580802107726
129. Bansal M, Jyoti N, Bharti A, Sharma S, Thakur M, Chauhan V, Kumar R (2025) Green synthesis of nanomaterials used as nano-fertilizer for sustainability in crop production: a overview on recent advancements and future perspectives. Plant Nano Biology, 100143.https://doi.org/10.1016/j.plana.2025.100143
130. Rajaee A, Wensheng X, Zhao L, Wang S, Liu Y, Wu Z, Wang J, Si-Shen F (2018) Multifunctional bismuth ferrite nanoparticles as magnetic localized dose enhancement in radiotherapy and imaging. J Biomed Nanotechnol 14(6):1159–1168. https://doi.org/10.1166/jbn.2018.2553
131. Hajra S, Padhan AM, Panigrahi BK, Pakawanit P, Jagličić Z, Vittayakorn N, Mishra YK, Lee S, Kim HJ (2024) Unleashing the potential of morphotropic phase boundary based hybrid triboelectric–piezoelectric nanogenerator. Journal of Materiomics 10(4):792–802. https://doi.org/10.1016/j.jmat.2023.09.011
132. Kumar V, Bhardwaj L, Saxsena A (2024) Neurotrauma and neurodegenerative treatment using perovskites. In: Biomedical applications of perovskites: the era of bio-piezoelectric systems. Bentham Science Publishers, pp 230–241. https://doi.org/10.2174/97898152563831240101
133. Sahini MG, Malima NM (2025) Multiferroic magnetoelectric-based biosensors in healthcare. In: Fundamentals of biosensors in healthcare. Academic Press, pp 337–357 https://doi.org/10.1016/B978-0-443-21658-9.00025-5
134. Parida K, Dehury SK, Choudhary RNP (2016) Structural, electrical and magneto-electric characteristics of complex multiferroic perovskite Bi0.5Pb0.5Fe0.5 Ce0.5O3. J Mater Sci: Mater Electron 27:11211–11219. https://doi.org/10.1007/s10854-016-5241-7
135. Radhakrishnan A, Srivastava R, Verma R, Thakur P, Thakur A (2024) Issues perovskites encounter in the biomedical industry. In: Biomedical applications of perovskites: The EraA of bio-piezoelectric systems. Bentham Science Publishers, pp 321–344. https://doi.org/10.2174/97898152563831240101

Chapter 12
Concluding Remarks, Future Outlook, and Emerging Ideas on the Sustainable Perovskite Multiferroic Materials

Manish Kumar and Khushboo Gupta

1 Introduction

The field of sustainable perovskite multiferroic materials has experienced remarkable growth, propelled by their unique combination of ferroelectricity, magnetism, and strong coupling effects. These materials, characterized by their environmentally friendly compositions and tunable properties, have attracted considerable attention due to their potential to revolutionize various energy and electronic technologies. As a result, perovskite multiferroics are emerging as key players in a wide range of applications, including energy harvesting, spintronics, sensors, transducers, photovoltaics, and even biomedical technologies, as depicted in Fig. 1 [1–3]. By bridging fundamental research with real-world applications, these materials offer the promise of transforming industries, leading to more efficient, sustainable, and multifunctional devices.

Recent advancements in material synthesis, structural engineering, and property optimization have expanded the functionality of perovskite multiferroics, making them even more promising candidates for next-generation sustainable technologies. This book provides a comprehensive exploration of perovskite multiferroics, covering everything from their fundamental principles to the latest breakthroughs in material synthesis, doping strategies, and device integration. The discussions begin with a historical perspective, acknowledging the rarity of multiferroic materials and delving into the mechanisms that govern their unique properties, such as lone pair effects, charge ordering, geometric constraints, and spin-driven interactions.

M. Kumar (✉)
Experimental Research Laboratory, Department of Physics, ARSD College, University of Delhi, Dhaula Kuan, New Delhi, India
e-mail: mkumar2@arsd.du.ac.in

K. Gupta
Materials Analysis and Research Laboratory, Department of Physics, NSUT, Dwarka, New Delhi, India

M. Kumar (ed.), *Sustainable Perovskite Multiferroic Materials*, Materials Horizons: From Nature to Nanomaterials, https://doi.org/10.1007/978-981-96-7419-0_12

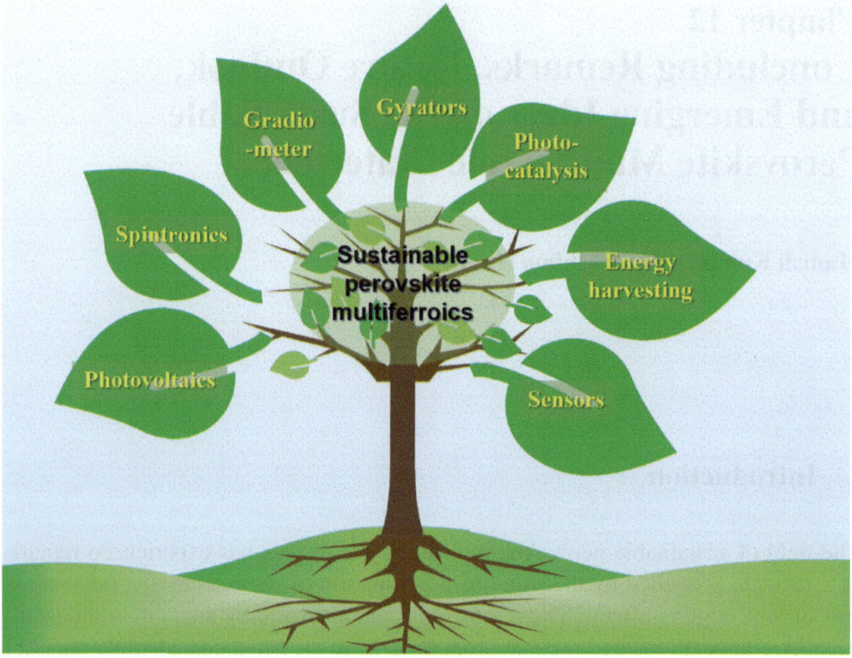

Fig. 1 Schematics to illustrate the applications of sustainable perovskite multiferroics

A thorough examination of type-I and type-II multiferroics, as well as single-phase materials and composite structures, has highlighted the diverse nature of magneto-electric (ME) coupling, along with its selection criteria and technological relevance. Multiferroic materials, which exhibit both ferroelectric and magnetic ordering, represent a class of compounds where the interplay between these two order parameters offers rich opportunities for developing novel devices with unprecedented functionalities. Type-I multiferroics, characterized by the coexistence of both ferroelectric and magnetic orders within a single phase, and type-II multiferroics, which typically involve composite structures where different materials exhibit ferroelectric and magnetic behavior, provide contrasting yet complementary avenues for designing materials with specific properties tailored for distinct applications.

The synthesis techniques discussed in this book range from conventional solid-state reactions to advanced vacuum-based deposition methods, such as molecular beam epitaxy (MBE) and pulsed laser deposition (PLD), along with emerging green synthesis techniques that emphasize eco-friendly approaches. MBE and PLD are particularly notable for their ability to produce high-quality thin films and heterostructures with atomic-scale precision, enabling the precise control of material composition and structure at the nanoscale. Green synthesis techniques, on the other hand, have gained prominence due to their environmental sustainability, utilizing nontoxic and abundant precursors to produce perovskite materials with minimal environmental impact.

The impact of interface coupling, heterojunctions, and defect engineering in modulating the structural and magnetic properties of perovskite multiferroics has also been extensively reviewed. These approaches enable the fine-tuning of material properties, leading to enhanced performance and improved device efficiency. The role of doping, solid solution strategies, and composite formation mechanisms in tailoring the bandgap and functional properties of perovskite multiferroic materials have been emphasized, as these strategies allow for the optimization of electronic, optical, and magnetic properties for specific applications.

Among perovskite multiferroics, bismuth ferrite ($BiFeO_3$, or BFO) has garnered significant attention due to its remarkable ability to exhibit both ferroelectricity and antiferromagnetism at room temperature [4]. This dual functionality makes BFO a promising candidate for a wide range of applications, from memory devices to sensors [5]. The ferroelectricity in BFO arises from its spontaneous electric polarization, while its antiferromagnetic nature is attributed to the alignment of iron (Fe) ions in opposite directions within the crystal lattice. The coexistence of these two order parameters in a single material offers the potential for developing devices that leverage both electric and magnetic properties, enabling the creation of multifunctional devices for next-generation technologies.

To understand and optimize the properties of BFO, computational techniques such as density functional theory (DFT) have played a crucial role. DFT calculations provide detailed insights into the material's behavior by simulating the interactions between electrons and the atomic lattice [6]. These theoretical advancements have helped clarify the mechanisms underlying the multiferroic behavior of BFO and have guided the design of materials with enhanced or tailored properties. By offering a deeper understanding of the structural, electronic, magnetic, optical, and thermoelectric characteristics of BFO, DFT has contributed to the growing interest in perovskite-based multiferroics and their potential for sustainable applications [7–9].

Despite the promising properties of perovskite multiferroics, their commercialization has been hindered by inherent loss mechanisms, such as high leakage currents and low remanent polarization. These challenges have been addressed in this book, which highlights recent strategies to mitigate these limitations. One such strategy involves the development of novel interface engineering techniques and the exploration of heterostructures that enhance the overall performance of multiferroic materials. Additionally, the book provides a detailed discussion on the mechanisms of magnetoelectric coupling, which governs the interaction between electric and magnetic order parameters. Understanding these mechanisms is crucial for optimizing the performance of perovskite multiferroics and expanding their potential applications.

The book also explores emerging topics such as quantum criticality, which could play a significant role in the development of perovskite multiferroics for quantum computing applications. The development of ultrathin films with robustly coupled ferroelectric–magnetic order at room temperature represents a significant step toward miniaturized, high-performance multifunctional devices. These advancements in material synthesis and processing are key to realizing the full potential of perovskite multiferroics in practical applications.

Energy harvesting has emerged as one of the most promising application areas for perovskite multiferroics. These materials serve as viable candidates for magnetoelectric-based cantilever devices, resonators, and efficient photovoltaic materials. The book discusses novel approaches to enhance solar cell efficiency, including strategies to surpass the Shockley–Queisser limit, which is the theoretical maximum efficiency for traditional single-junction solar cells. The use of nontoxic, cost-effective materials such as $BiFeO_3$ and Bi_2FeCrO_6 have been explored as a way to address both the economic and environmental challenges associated with conventional solar cell technologies.

In addition to energy applications, perovskite multiferroics have demonstrated considerable promise in spintronic devices, memristive memory, and neuromorphic computing. The integration of direct and converse magnetoelectric effects has enabled the development of magnetoelectric random access memories (MERAMs), phase shifters, voltage-tunable inductors, and magnetoelectric antennas for next-generation communication and data storage systems [10]. These devices take advantage of the unique coupling between electric and magnetic fields to offer new functionalities that are not possible with conventional materials.

Perovskite multiferroics also show great potential in sensors, actuators, and microelectromechanical systems (MEMS) and nanoelectromechanical systems (NEMS)-based devices. Their multifunctionality, which combines electric, magnetic, and mechanical properties, makes them ideal candidates for a wide range of sensor applications, including pressure, temperature, and magnetic field sensors. Furthermore, their role in biomedical applications, such as drug delivery, biomagnetic sensing, and neurological applications, represents a growing frontier in this field. The concept of surface acoustic wave (SAW) devices, which utilize strain-mediated magnetoelectric coupling, has been explored for RF signal processing and acoustic-driven magnetic resonance applications [11].

As we conclude, the discussion expands beyond conventional applications, introducing green, nontoxic, and sustainable approaches to multiferroic material development. Emerging theoretical concepts, such as quantum electromagnets, cosmic string simulations, and multiferroic-driven insights into early universe models, have been introduced, opening new avenues for interdisciplinary research. The role of defect chemistry, ion migration, and dynamic domain switching in optimizing functional properties has been examined in detail. This concluding chapter synthesizes the key findings and advancements discussed throughout the book, providing a vision for the future of sustainable perovskite multiferroic materials. It emphasizes the necessity for continued innovation, bridging experimental, theoretical, and application-driven research toward achieving commercially viable, eco-friendly, and technologically transformative solutions.

In conclusion, perovskite multiferroics represent a promising frontier in the development of sustainable, multifunctional materials for a wide range of applications. The advancements in material synthesis, computational modeling, and device integration discussed in this book provide a solid foundation for future research and innovation in this exciting field. As the world moves toward more sustainable and efficient technologies, perovskite multiferroics offer the potential to play a pivotal role in addressing global challenges and driving the next generation of electronic and energy devices.

2 Summary of the Present Work

Perovskite multiferroic materials have gained immense attention in recent years due to their ability to exhibit multiple ferroic properties simultaneously, such as ferroelectricity, ferromagnetism, and ferroelasticity. These materials hold significant promise for a wide range of applications, including spintronics, energy harvesting, sensors, and biomedical applications. Their exceptional structural adaptability allows for the integration of various functionalities, making them ideal candidates for next-generation electronic and energy devices. This document provides a comprehensive summary of the latest research and advancements in sustainable perovskite multiferroic materials, covering their synthesis techniques, quantum criticality, photovoltaic efficiency, and other emerging applications.

Chapter 1 focuses on introducing sustainable perovskite multiferroic materials, emphasizing their ability to exhibit multiple ferroic orders, including ferroelectricity, ferromagnetism (or antiferromagnetism), and ferroelasticity, within a single-phase structure. It discusses how these materials are rare, particularly in perovskite-structured multiferroics, and examines the perovskite structure characterized by the general formula ABO_3. The chapter highlights how the structural flexibility of perovskites allows for the coexistence of multiple ferroic orders, enabling magnetoelectric coupling, where electric and magnetic order parameters interact. Furthermore, it explores the potential applications of these properties in spintronics, memory storage, sensors, and energy-efficient information processing systems. Notable perovskite multiferroics such as $BiFeO_3$, $YMnO_3$, $TbMnO_3$, and $GdMnO_3$ are discussed, providing a comprehensive overview of their structural characteristics, magnetoelectric coupling mechanisms, and technological relevance. Furthermore, Chap. 2 discusses the synthesis techniques of perovskite multiferroic materials and their impact on material properties. It explains the fundamental perovskite formula ABX_3, where the A site is occupied by a larger cation, the B site by a smaller cation, and X is a negatively charged ion such as oxygen or sulfur. The chapter focuses on $BiFeO_3$ (BFO), a widely studied perovskite multiferroic, and details various synthesis methods, including solid-state reactions, sol–gel techniques, electrospinning, and hydrothermal synthesis. The discussion extends to how these synthesis techniques influence material morphology, phase purity, and functional properties. Characterization techniques such as scanning electron microscopy (SEM), transmission electron microscopy (TEM), and X-ray diffraction (XRD) are explored to assess structural, compositional, and performance attributes. Additionally, the chapter examines how optimized synthesis parameters enhance material properties, making them viable for energy harvesting, sensor applications, and spintronic devices.

Chapter 3 focuses on the impact of doping, solid solution, and composite formation mechanisms on perovskite multiferroic materials. It analyzes the $BiFeO_3$–$CoFe_2O_4$–$BaTiO_3$ composite ceramics synthesized via solid-phase methods and investigates the effects of varying molar ratios and sintering temperatures on structural and functional properties. Experimental results highlight how XRD intensity and grain size variations influence dielectric, ferroelectric, and magnetic properties. The chapter discusses how increased $CoFe_2O_4$ content enhances magnetic properties, while specific compositions optimize ferroelectricity and polarization. Moreover, it explores magnetoelectric coupling at room temperature, detailing how an applied magnetic field influences polarization behavior. This chapter provides a detailed assessment of how doping and composite formation can tailor the functionalities of perovskite multiferroics for advanced technological applications.

Chapter 4 highlights $BiFeO_3$ as a prominent room-temperature perovskite multiferroic material. It examines the unique properties of BFO, including its coexistence of antiferromagnetic and ferroelectric behavior at room temperature, high Curie and Neel temperatures, and significant residual polarization. The chapter discusses advancements in BFO-based materials for piezo-photocatalytic applications and addresses challenges such as magnetoelectric switching, domain wall engineering, leakage current issues, and phase stability concerns. Additionally, it explores structural, electronic, and magnetic characterizations of BFO nanostructures and their potential applications in energy-efficient devices and renewable energy technologies.

Chapter 5 explores quantum criticality in sustainable perovskite multiferroic materials, detailing how quantum critical points (QCPs) and quantum fluctuations significantly impact material properties. It discusses how composition, doping, and strain influence quantum phase transitions, leading to novel functionalities. The chapter further examines the implications of quantum critical behavior for practical applications, including low-power electronic devices and sustainable energy solutions. Strategies for manipulating quantum fluctuations to design next-generation multiferroic materials are also presented.

Chapter 6 focuses on the energy harvesting potential of perovskite multiferroics, examining their ability to harness energy through piezoelectric, thermoelectric, photovoltaic, and magnetoelectric mechanisms. It discusses how ambient energy can be converted into electrical energy via strain-induced polarization and the bulk photovoltaic effect. The chapter specifically analyzes $BiFeO_3$ for its energy conversion efficiency, structural modifications, and performance in different environmental conditions. Additionally, it addresses challenges related to environmental sensitivity, material stability, and scalability, along with ongoing research aimed at optimizing these materials for energy-efficient applications.

Chapter 7 discusses strategies for enhancing the photovoltaic energy conversion efficiency of sustainable perovskite multiferroic solar cells. It examines the role of multiferroic perovskites as absorber materials in photovoltaic applications and explains the bulk photovoltaic (BPV) effect resulting from their non-centrosymmetric structures. The chapter highlights the impact of doping strategies on photovoltaic conversion efficiency (PCE) and reviews theoretical models governing the BPV effect. Additionally, it provides insights into advancements in perovskite-based solar cells and strategies for improving efficiency through material engineering and device optimization. Chapter 8 explores the potential of perovskite multiferroics in spintronics applications, emphasizing their role in memory devices, sensors, and logic circuits. It discusses the principles of spin-charge coupling, magnetoelectric switching, and the sustainability of these materials for spintronic technologies. Future research directions are highlighted, focusing on how perovskite multiferroics can contribute to energy-efficient computing and next-generation electronic devices.

Chapter 9 focuses on perovskite multiferroic materials for sensor and transducer applications, highlighting their ability to sustain permanent electrical polarization. It discusses the underlying physical principles, including polarization-modulated Schottky-like barriers and rectifying behavior, and explores their applications in biomedical, environmental, and industrial sensing technologies. Chapter 10 discusses the integration of perovskite multiferroic materials in surface acoustic wave (SAW) devices used for telecommunications and signal processing. The chapter examines material compatibility, device scaling, and performance optimization, emphasizing tunability and sensitivity enhancement. It also addresses challenges in material integration and future prospects for SAW technology. Lastly, Chap. 11 provides an in-depth discussion on perovskite multiferroic nanomaterials for biomedical applications. It explores their potential roles in biosensing, imaging, drug delivery, and antibacterial applications. The chapter discusses challenges related to biocompatibility, synthesis scalability, and long-term stability while highlighting future advancements in medical nanotechnology. The key highlights of all 11 working chapters are tabulated in the Table 1.

3 Future Directions and Concluding Remarks

The exploration of sustainable perovskite multiferroic materials highlights their immense potential in various advanced technological applications. These materials, characterized by their unique coupling of ferroelectricity, magnetism, and ferroelasticity, offer promising prospects for energy-efficient devices, quantum technologies, spintronics, sensors, and biomedical applications. The synthesis techniques, doping strategies, and composite formations play a crucial role in tailoring their properties for specific applications, while quantum criticality and room-temperature stability further expand their usability. Their multifunctionality enables breakthroughs in energy harvesting, photovoltaic applications, and surface acoustic wave devices, making them key materials for next-generation sustainable technologies. However, challenges such as material stability, phase purity, and scalability need to be addressed through continued research and innovation. Advancements in synthesis, characterization, and application-specific modifications will drive the future development of these materials, paving the way for eco-friendly, high-performance technologies across multiple domains.

Table 1 High spots of the discussed 11 chapters of the current book

Chapters	Highlights
Chapter 1	• Perovskite multiferroics are scarce, exhibiting multiple ferroic properties • Structural flexibility enables electric and magnetic interactions • Useful for sensors, memory, and energy-efficient devices
Chapter 2	• Exhibits multiferroicity, influenced by synthesis methods • Used in energy harvesting, sensors, and spintronic devices
Chapter 3	• $BiFeO_3$–$CoFe_2O_4$–$BaTiO_3$ ceramics were synthesized via the solid-phase method • Increased $CoFe_2O_4$ enhances magnetism, affecting polarization under a magnetic field
Chapter 4	• BFO withstands extreme conditions due to its wide temperature range and perovskite structure • Used in piezo-photocatalysis, magnetoelectric switching, and multifunctional devices
Chapter 5	• Quantum effects influence ferroic properties and coupling near QCP • Governs magnetic and electric ordering, enabling advanced technologies
Chapter 6	• Perovskite multiferroics combine ferroelectricity, ferromagnetism, and ferroelasticity for various applications • They convert different forms of ambient energy like piezoelectric, thermoelectric, and photovoltaic effects • These materials have great potential for self-powered devices, renewable energy, and supporting sustainable development goals
Chapter 7	• Clean energy solutions like solar power are essential due to rising energy demand and environmental issues • Perovskite materials offer high efficiency (up to 25%) as absorber materials • Doping multiferroics boosts perovskite solar cell performance, with the bulk photovoltaic effect driving efficiency
Chapter 8	• This chapter explores perovskite multiferroics for sustainable spintronics devices • These materials exhibit multifunctionality, combining ferromagnetism, ferroelectricity, and ferroelasticity
Chapter 9	• Solar energy conversion requires efficient light absorption, charge separation, and fast extraction • Ferroelectric materials control electric fields through permanent polarization
Chapter 10	• Perovskite multiferroics combine ferroelectricity, ferromagnetism, and ferroelasticity for versatile uses • SAW devices benefit from these materials in telecommunications and medical diagnostics • The chapter covers perovskite multiferroics' role in tunable filters, frequency devices, and low-power systems
Chapter 11	• Perovskite multiferroic nanomaterials are ideal for biomedical applications • They are used in biosensing, imaging, drug delivery, and tissue engineering • The chapter covers synthesis, stability, safety, and future prospects in biomedicine

References

1. Kargol A, Malkinski L, Caruntu G (2012) Biomedical applications of multiferroic nanoparticles. Adv Magn Mater InTech. https://doi.org/10.5772/39100
2. Tablero C (2016) Photovoltaic application of the multiferroic Bi_2 $FeCrO_6$ double perovskite. Sol Energy 137:173–178. https://doi.org/10.1016/j.solener.2016.08.004
3. Zanolli Z (2016) Graphene-multiferroic interfaces for spintronics applications. Sci Rep 6:31346. https://doi.org/10.1038/srep31346
4. Raj A, Kumar M, Kumar A, Singh K, Sharma S, Singh RC et al (2023) Comparative analysis of 'La' modified BiFeO3-based perovskite solar cell devices for high conversion efficiency. Ceram Int 49:1317–1327. https://doi.org/10.1016/j.ceramint.2022.09.112
5. Chen S-H, Young S-J, Yang C-C, Liu Y-H (2025) Resistive switching behavior of perovskite $BiFeO_3$ thin films and their two-bit-per-cell applications. J Mater Sci: Mater Electron 36:478. https://doi.org/10.1007/s10854-025-14436-4
6. Zhou Z, Song J, Xie Y, Ma Y, Hu H, Li H et al (2025) DFT calculation for organic semiconductor-based gas sensors: Sensing mechanism, dynamic response and sensing materials. Chin Chem Lett 110906. https://doi.org/10.1016/j.cclet.2025.110906
7. Hoat DM, Amirian S, Alborznia H, Laref A, Reshak AH, Naseri M (2021) Strain effect on the electronic and optical properties of 2D tetrahexcarbon: a DFT-based study. Indian J Phys 95:2365–2373. https://doi.org/10.1007/s12648-020-01913-1
8. Piyanzina I, Mamin R (2022) Toward the ferroelectric field-effect transistor on $BaTiO_3$/$LaMnO_3$ heterostructure: DFT investigation. J Mater Sci 57:21620–21629. https://doi.org/10.1007/s10853-022-08000-2
9. Kumar S, Kumar N, Yadav K, Kumar A, Singh RP (2020) DFT investigations on optoelectronic spectra and thermoelectric properties of barium cadmium disulphide (BaCdS2). Optik (Stuttg) 207:163797. https://doi.org/10.1016/j.ijleo.2019.163797
10. Kopyl S, Surmenev R, Surmeneva M, Fetisov Y, Kholkin A (2021) Magnetoelectric effect: principles and applications in biology and medicine—a review. Mater Today Bio 12:100149. https://doi.org/10.1016/j.mtbio.2021.100149
11. Hashimoto K (2012) Surface acoustic wave (SAW) devices. In: Ultrasonic transducers. Elsevier, pp 331–73. https://doi.org/10.1533/9780857096302.3.331

The manufacturer's authorised representative in the EU is Springer Nature Customer Service Centre GmbH, Europaplatz 3, 69115 Heidelberg, Germany. If you have any concerns regarding our products, please contact ProductSafety@springernature.com

Printed and bound by CPI Group (UK) Ltd, Croydon, CR0 4YY
01/10/2026
02225915-0007